Modern Psychiatry
and Clinical Research

MODERN PSYCHIATRY AND CLINICAL RESEARCH

ESSAYS IN HONOR OF *Roy R. Grinker, Sr.*

EDITED BY DANIEL OFFER
AND DANIEL X. FREEDMAN

Basic Books, Inc., Publishers
NEW YORK   LONDON

Library of Congress Catalog Number: 71–174818
SBN 465–04644–4
Manufactured in the United States of America
DESIGNED BY THE INKWELL STUDIO

THE AUTHORS

AARON T. BECK is Professor of Psychiatry at the University of Pennsylvania and a member of the Psychiatric Research Society.

JARL E. DYRUD is Associate Professor of Psychiatry at the University of Chicago and a consultant to the National Institute of Mental Health. He was formerly Director of Research at Chestnut Lodge in Maryland.

DANIEL X. FREEDMAN is Chairman of the American Psychiatric Association Task Force on Drug Abuse in Youth and a consultant to the National Institute of Mental Health. He is also President of the Illinois Psychiatric Society and Editor-in-Chief of the *Archives of General Psychiatry*.

DAVID A. HAMBURG is Professor and Executive Head of the Department of Psychiatry, Stanford University and a member of the Carnegie Commission on Higher Education. He was formerly Chief of the Intramural Research Program at the National Institute of Mental Health.

SEYMOUR LEVINE is a consultant to the National Institute of Mental Health.

DANIEL OFFER is President-Elect of the American Society for Adolescent Psychiatry and Editor-in-Chief of the *Journal on Youth and Adolescence*.

JUDITH L. OFFER is a former graduate student in philosophy, Harvard University.

DONALD OKEN is a counselor of the American Psychosomatic Society and a member of the Editorial Board of the *Archives of General Psychiatry*. He was formerly Chief of the Clinical Research Branch of the National Institute of Mental Health.

TALCOTT PARSONS is Professor of Social Relations at Harvard University and President of the American Academy of Arts and Sciences. He was formerly President of the American Sociological Association.

FREDRICK C. REDLICH is a permanent member of the Board of Directors of the Foundations Fund for Research in Psychiatry. He was formerly Chairman of the Department of Psychiatry, Yale University.

MELVIN SABSHIN is a member of the Editorial Board of the *American Journal of Psychiatry* and Chairman of the Program Committee of the American Psychiatric Association. He was formerly President of the Illinois Psychiatric Society.

DAVID SHAKOW was formerly Chief of the Laboratory of Psychology of the National Institute of Mental Health.

M. BREWSTER SMITH was formerly President of the American Psychological Association. Before coming to Chicago, he was Director of the Institute for Human Development at the University of California at Berkeley.

JOHN P. SPIEGEL is Director of the Center for the Study of Violence and

Professor of Social Psychiatry at Brandeis University. He was formerly
a member of the Department of Social Relations, Harvard University.

MARVIN STEIN is Professor and Chairman, Department of Psychiatry, Mount
Sinai School of Medicine and was formerly Professor and Chairman,
Downstate Medical Center, State University of New York.

ROBERT S. WALLERSTEIN is Chief of the Department of Psychiatry, Mount
Zion Hospital and Medical Center and President of the American Psy-
choanalytic Association. He was formerly on the staff of the Menninger
Clinic.

FOREWORD

Roy R. Grinker, Sr., is a giant of modern psychiatry. To express their admiration and devotion a group of students and friends have attempted to bring into focus Grinker's major contributions. Throughout his professional life Grinker has helped to build clinical neurology and more lately—similar to Freud, who had a profound effect on him—clinical psychiatry and its basic sciences. More than any other he has helped erect a solid edifice of clinical scientific psychiatry through broad and significant contributions. His work ranges from theoretic endeavors to create a unified theory of behavior and of therapy to the explorations of the most important subjects of clinical psychiatry—stress, anxiety, depression—and, last but not least, to incisive contributions to normality. What he started has been taken up by devoted students whom he stimulated and led; many of these are today's leaders of psychiatry.

Roy Grinker's works are considered classics, his career serves as a model for the striving young researcher, and he continues to work and ask the questions that few have the courage or the stamina to ask. Rather than let the central ideas of a great investigator lie dormant, to be utilized as circumstances lead us to them, the editors happily decided to take the initiative and to help integrate his ideas with those of others and, in doing so, to hopefully further advance these ideas. In addition, we must ask for directions for the future. Where are scientific psychiatry and behavioral science going? What can be learned from the accomplishments of a great investigator to aid us in avoiding past errors and help build a firmer foundation for the future?

The aim of this book is to review and assess methodological and substantive research findings of Roy R. Grinker, Sr. This is being done by enlisting leading behavioral scientists who have integrated Dr. Grinker's contributions with present-day findings and theories. They cite Grinker's concepts and conclusions and relate them to the picture not only of psychiatry today but to the problems of the future in the various topical areas. The book thus gives a chance for a retrospective and prospective view—a thoughtful picture of "the state of the science and art."

One purpose of science is to provide society with useful information. Grinker has done this by his elucidation of some of the most important clinical problems. Such contributions have been sadly lacking and such work alone will assure Grinker a lasting place in psychiatry's hall of fame.

Another stated purpose of research is always to test theory. To what extent has behavioral science research in psychiatry affected prevailing theory? To what extent has theory been deaf to data? To what extent can Grinker's findings be utilized to document theory?

In selecting clinical and theoretical topics from Grinker's major contributions, research models of comprehensive importance to psychiatry are inevitably demonstrated. The works of Grinker illustrate major research approaches and indicate the extent to which psychiatric research can be fruitful. I trust that this book will not only inform the professional reader but stimulate researchers and provide them with a wise and provocative perspective.

Frederick C. Redlich, M.D.
Professor of Psychiatry
Dean, School of Medicine
Yale University

CONTENTS

Modern Psychiatry
and Clinical Research

Field Theory and System Theory: With Special Reference to the Relations between Psychobiological and Social Systems

It might seem strange to let conferences have an impact on one's thinking, but, for a sociologist rather deeply involved in the areas of interconnection-transaction between social and psychological systems, opportunity to participate in this long series of discussions proved to be a particular kind of intellectual bonanza. I refer to the conference on "A Unified Theory of Human Behavior," chaired by Roy R. Grinker, Sr., and attended by Jurgen Ruesch and John P. Spiegel. In addition, let me mention the following individuals who had a particularly outstanding influence on my own later thinking: Alfred Emerson, Charles Morris, Karl Deutsch, James Toman, and Anatole Rapoport.

There was a sense in which by the time of the beginning of the conference, during the fall of 1951, I was already well primed to a consideration of the complex intellectual problems involved in the theory of treating human behavior, or more technically, human action without excluding behavior, dealing with this as a single superordinate system that involved a plurality of interlocking restricted systems. I had very recently completed work on two books, one of which I coauthored and coedited, *Toward a General Theory of Action*, and one of which I authored, *The Social System*; these books had gone relatively far with this approach. I was also involved in a certain state of effervescence of further intellectual excitement concerned with theoretical developments beyond the stage reached in the two books, both of which, as it happened, were published almost coincidentally with the opening of the conference. But one will always remember the time lag between final submission of manuscripts and actual appearance in print, which in this case was substantially more than a year.

The organizers of the conference group all came from psychiatry, with its strong anchorage in biological thinking. Two of the three at least, were predominantly trained in the psychoanalytic tradition, but Freud himself was deeply steeped in this perspective, and I think it fair to say that the sociocultural emphasis was something in the nature of a residual category to Freud and that this shows in the initial approaches of the members of the conference including Roy Grinker's initial statement.[1]

The general idea of the involvement of distinguishable plural systems in the field of human behavior was still relatively new at that time. I think that most of the classical students of behavior had tended to assume that there was essentially one set of principles and were concerned with the relative advantages of their points of entry into its study with a view to arriving at formulation of those principles. Thus, there were important traditions that had taken social and cultural levels as their points of departure. Preeminent examples in this area were Max Weber and Émile Durkheim and the notable development of social anthropology in Great Britain during the previous generation, on the social side. Dealing more with the individual, the great tradition, stemming from Freud, of course, had a preeminent place, especially for many participants of the conference, but there were also the behaviorist and other schools of psychology, which had at least a certain resonance in the group. The cultural and personality school represented perhaps, above all, by Lawrence Frank, had made a start toward distinguishing and attempting to articulate different systems.

Inasmuch as, until the advent of Marion Levy, during the latter part of the conference's history, I was the only sociologist participating (I would count Florence Kluckhohn's point of view as more anthropological than sociological), I found myself tending to be on the weaker end of a spectrum, where the center of gravity was focused most on the biological as an individual area. In the collective area I had already developed, something that was to extend substantially further during subsequent years, a strong feeling of the importance of clearly distinguishing in analytical terms the points of view, or the system references, of sociologists and anthropologists on the assumption that the latter were primarily concerned with cultural rather than social systems. Broadly speaking, I think this held for, among participants, Jules Henry, Laura Thompson, and Florence Kluckhohn.

---

[1] The available document on the conference itself, and what was said, is in *Toward a Unified Theory of Human Behavior* (Grinker, 1967). In any detail, it covers only the discussions and presentations of the first four meetings. But Grinker's summary statement at the end gives a sketchy account of what was discussed in the subsequent ones.

For me personally, this came at a time when relative to my intellectual antecedents and to subsequent development of interests, psychological concerns were particularly salient. I had, however, been strongly ambivalent about the relation between the psychological level, which I tended to identify with the concept of personality as a system, and the biological in the organic sense. I had recently completed a medical layman's training in psychoanalysis and had been continuously bothered and puzzled by this problem in the psychoanalytic context, notably with reference to the component of Freud's theory, which was usually referred to as that of the "instincts." One of the most important benefits I gained from the conference was an immense clarification of the problems on this side of my field of interest. I had had a good deal of exposure to biology in earlier years, including being importantly influenced by W. B. Cannon's ideas, especially the concept of homeostasis. I found Roy Grinker's discussions of the organic side very helpful and illuminating, but what crystallized this aspect for me more than anything else was the contribution of Alfred Emerson, a biologist who had had no special concern with social science at the human level, though his concern with the social insects was obviously a most important background. Emerson's views, which are particularly set forth in Chapter 12 of the conference volume (Grinker, 1967), seemed to me to provide an extraordinarily fruitful basis for defining fundamental continuities between the organic level of biological systems and the humanly sociocultural levels, which I have come to conceptualize in terms of the phenomena of action.

Though I was not in a position to state it with full clarity at that time, it has become clear that the most important distinctive feature of action, as distinguished from both the organic and the behavioral, short of action levels, has been the role in action of the manipulation of symbols that were parts of symbolic systems, organized in terms of what have come to be called codes. The basic conception of a codified symbolic system, in this sense, was very much connected with the (relatively recent at that time) emergence of linguistics, as a salient discipline and its relation to what had, I think, for nearly the first time, been given the name "semiotic" by Charles Morris. In this connection, Morris's participation was exceedingly useful from my point of view, and I have been fortunate enough to have other associations with him when he was a visiting professor at Harvard.

It was, therefore, particularly striking to me when Emerson put forward the conception that, as he put it, the symbol was at the human cultural level the equivalent of the gene at the organic level. This suggestion, which was entirely new to me at the time, crystallized what seemed to be a major synthesis of the patterns of what might be called formal system structure at organic and action levels. The consequences of this synthesis have been reverberating, in my own thinking, during the nearly twenty years since it first entered my horizon of awareness. It is interesting that the new genetics was just taking shape at that time. The famous discovery of the chemical structure of DNA, the double helix, by Watson and Crick, had in fact occurred in Cambridge, England, during the very same year I went to Cambridge as a visiting professor of social theory, but I was not at all in touch with these developments in biology. (It was for that reason that I missed two of the conferences.) There were, however, hints in the statements of Emerson, Toman, and others of the importance of these new developments. Since then, however, the picture has been immensely clarified and a much more solid basis for the functional similarity, or analogy in the technical-biological sense in which Emerson used the term, between cultural symbolic systems and genetic systems. Both were for the respective living systems in which they operated—the highest order mechanisms of homeostatic control, as Emerson would put it, in not only the maintenance but the development of the systems of which they were parts.

Another example of the infusion of seminal ideas into the thinking of the conference from which I benefited very greatly was that of the cybernetic point of view. There were no engineers or physical scientists participating, and it is interesting that the member who did most to introduce this line of thought was a political scientist, Karl Deutsch, who, however, since he was at that time a member of the faculty of MIT, had been particularly exposed to the influence emanating from Norbert Wiener and other related groups, such as McCulloch and Pitts.

Though the famous Macy conference on feedback problems was then already under way, and I had had the good fortune to be a guest at two of its meetings, the significance of this line of thought for system theory was only beginning to penetrate my awareness. Karl Deutsch and the discussions he stimulated did much to enhance this. Not least among the important facts was that he and Emerson saw, on the whole, very closely eye to eye, and Emerson in particular felt that cybernetic feedback processes belonged very definitely in the category of homeostatic mechanisms.

An important controversial issue in the conference was that of the senses in which there was and was not a set of hierarchical relationships among the plural, quasiindependent systems being discussed. John P. Spiegel, in particular, wished at least to minimize, if not to deny, such hierarchies. Roy Grinker's initial presentation, however, certainly presupposed them, and I think this point of view came to be the dominant one, especially among the groups whose contributions were most important to me. For example, Grinker's five-step theories, beginning with the enzymatic system of the organism and ending with the psychological and then sociocultural system, set a certain tone. When the cybernetic idea had become salient, it seemed to me that this gave a very general and firm grounding for the ordering of systems in this general kind of way, though it did not solve a great many very particular problems. It could, however, set forth a hierarchy of organizational levels, whereby resources generated at the lower levels could then be utilized much more effectively at the higher. There were many illustrations of this principle; one that was particularly salient to me was Emerson's suggestion that sexual reproduction, though organizing the process socially, introduced an enormous facilitation of the evolutionary process, through its relation to the combination of variability and control of that variability. It has seemed to me that what we call the socialization process in the family at the human level is another instance of essentially the same principle, though what is produced here is not organic form and combinations of genes but the cultural component of personality. Here the incest taboo is in certain respects analogous to the differentiation of the sexes at the organic level in that it ensures that the culture of a family will come from two relatively independent sources, that is, the families of orientation of the two marriage partners.

The conference was especially concerned with the interrelations of these plural systems and, hence, with the nature of the boundaries between them, and, since they constitute environments to each other, more generally with the system in environment relationship. I think I should like to register what may be a disagreement with Grinker, especially in statements he has made in later papers. The point I have in mind also relates to the significance of the concept of homeostasis. Grinker's formula, especially in his recent paper on medical education (Grinker, 1971), suggests that the most essential concept of system is one that includes the environment. This has the merit of stressing the all-important point of the openness of living systems, and hence, of the continual processes of interchange that go on between their internal components and aspects of their environments. The difficulty seems to me to be that the environment of a given system is a system in a different sense from that of the system of reference itself, or one of a very different kind.

In the illuminating discussion of the concept of boundaries, there was something like agreement that a boundary was a line or zone, not necessarily physically defined, that marked a difference in the state of affairs on one side from that on the other. Classic examples can be found in Cannon's famous discussions in the *Wisdom of the Body* (Cannon, 1963), starting with the differential of body temperature and environmental temperature in the mammals and birds. Not only, however, is there on the average a differential, but body temperature is held very nearly constant by a series of mechanisms that Cannon sets forth, whereas environmental temperature may vary within a wide range and still be compatible with organic life, though there are, of course, limits. This example illustrates the very general principle that within a system, there is in general both a more stable and in certain senses organized or integrated state of affairs than between it and its environments, on the one hand, or within the environment as a whole, on the other. This was clearly stated by Emerson in talking about the relation between a species in a given physical locale and the ecosystems of which it was a part. The boundary of the system of reference, then, is that of the locus of some of the principal mechanisms by which the relevant differences of state are maintained. We must, however, be careful in defining boundary. Thus, for the human organism, clearly the most important boundary is the skin. However, both food and air, when taken into the organism, do not by merely entering the mouth or nostrils become part of the organism in the biochemical sense. Hence, the insides of the digestive and respiratory tracts are in important senses external to the organism. They also are boundary zones in a sense similar to that in which this is true of the skin. Very similar considerations apply to action systems.

Another major consideration applies, I think, to all classes of living systems, but is particularly important for action systems: Across its boundaries such a system is related not to one other system but to a plurality of others. Indeed so important is this that, on occasion, I have used the term "environments" in the plural to emphasize it. Thus, I have found it essential first, as noted above, to distinguish action from both organic systems and subaction behavioral systems, if indeed a distinction between those two is essential, which I think it is for certain purposes. Clearly, the action system is articulated with both, both in the case of the individual actor's own body and its behavior and those of others with whom he interacts or transacts. Beyond that, of course, are the components of the environment consisting of nonhuman organisms and physical objects. Internally to the action system, I have found it essential to distinguish four basic systems: the behavioral organism, the personality of the individual, the social system, and the cultural system. These, of course, are classes of systems, exemplified by particular systems in each case. The distinctions are inherently analytical rather than concrete because of a critical phenomenon, which I shall turn to in a moment.

First, however, let me note that at the time of the conference, I tended to restrict action to the latter three of the four subsystems. The inclusion of the behavioral organism was in an important degree stimulated by the conference itself. As I have said above, these discussions, and particularly what Emerson said, enormously impressed me with the continuities of the three levels: organic, behavioral, and action phenomena. I was also particularly stimulated at about the same time, and a little later, by close association with James Olds, who by that time was actively working in the field of the relations of the brain to behavior; another important influence was Karl Pribram, with whom I had certain associations.

The general phenomenon referred to in the paragraph before the last is what I call interpenetration, which was discussed, but I think in a rather elementary way, in the last of my presentations to the conference

(Grinker, 1967). From one point of view, this should be regarded as a particularly important feature of the boundary relations of living systems. In the action field, the phenomenon first became saliently important to me in the connection with Freud's discussion of introjection, of moral standards as part of the content of what he called the superego. It then became evident that there had been a fundamental convergence on this idea between Freud and Durkheim—and by certain American social psychologists, particularly W. I. Thomas and George Herbert Mead —in the notion of moral constraint, or constraint by moral authority. With reference to the personality this phenomenon has been more generally called "internalization" as a different term from Freud's introjection.

Freud's original emphasis was on the cultural aspect, whereas particularly Mead's and, in a certain sense, Durkheim's was on the social. Mead's famous concept was taking the role of the other. In Freud's later work, however, it is quite clear that he included both cultural patterns and social objects in his concept of the internalized aspects of the personality (see Parsons, 1964, chap. 4). This meant, if one looked at the problem of defining the boundaries of the personality, that it interpenetrated with both social and cultural systems. If interpenetration is a fundamental aspect of the personality in its relations to social and cultural systems, this clearly should be the case with the organism; after all, a person's relations with his own body must in certain senses be far more intimate than those with other persons in a social system. Essentially, what I have been calling the behavioral organism, which is not the total organism, concerns those aspects in which the motives internalize social objects and the culture of the personality can be said also to be internalized. By far the most important single locus of this is clearly the brain, a very notable aspect being its capacity as the storehouse of memory, which underlies the whole learned component of behavior and action. The phenomenon of motor skill, however, would seem clearly to indicate that interpenetration extends to the motor systems—not least, perhaps, to those involved in symbolic communication as through speech —and the visual reception of information from the environment— notably other persons and culture. Not least important, probably, is the organic aspect of the erotic complex with its demonstrated functional importance, especially in the socialization process, but also in the stabilization of adult personalities through erotically intimate relationships.

These considerations bring me around to a question of the balance of emphasis among the systems that concerned the conference. It seemed to me that the conference itself, and indeed Grinker's later papers, manifested a certain imbalance, namely with respect to insufficient stress on the importance of cultural systems, or perhaps more accurately, insufficient stress on the importance of the analytical distinction between social and cultural systems, manifested in a tendency to bracket them together in the expression "sociocultural." This tendency was, I think, favored by the fact that the anthropologists present had on the whole not tended to make sharp distinctions in this area. I should, however, acknowledge that in his editorial summary of the session, in which I made my last recorded presentation (Grinker, 1967, p. 339) Grinker stated my view quite accurately.

Perhaps the importance of the distinction can be stressed by saying a little about the phenomenon I have called "institutionalization," a case of interpenetration parallel to that of internalization, where personality and organism are at issue. By "institutionalization," I have meant the zone of interpenetration between cultural and social systems that is essential to the very fundamental aspect of social systems called the institution. The focus here is on normative patterns, those that, to use W. I. Thomas's (1923) term, "define the situation" for action in social relationships. They take a variety of forms, such as law, custom, convention, etiquette, and of course are manifested in many different functional contexts. Thus, the institution of property is essentially a complex of norms that regulates the relations of persons and collectivities to rights of possession, control, use, and transfer of economically valued objects, whether these be physical objects of possession or a purely cultural entity, such as money, which I have long contended is essentially a symbolic system.

One of the most important reasons for the analytical distinction is that the boundaries of cultural systems are very generally not identical with those of social systems. Various components and subsystems of a cultural system may be institutionalized, for example, not in one society, but in several societies. Science itself is a preeminent example. If we speak of the United States in a technical sociological sense as constituting a society, though I think it would be quite correct to speak of various scientific disciplines being institutionalized as constituent parts of American society, it would very definitely not be correct to speak of them as confined to American society. Even when we speak, for example, of American sociology and how it differs from French sociology, what we refer to essentially are relatively minor variants rather than distinct cultural systems. In a similar way, we could speak of French Catholicism, Irish Catholicism, and Latin American Catholicism, which certainly show important differences, but are still variants of the more general cultural system, which we call the Roman Catholic religion.

It seems to me that adequate definition of the distinctiveness of the cultural system in the first instance relative to the social, but also to the psychological and organic, is a very important condition of an adequately balanced account of the relations between social and psychological systems, to which I would like to devote the remainder of this chapter. If, as I have said, I gained from the conference an immense enrichment of my understanding of the organic systems and their relation to personality, I think I could say that, in parallel in subsequent years, I have had substantially increased preoccupation with problems of the cultural system and its relation to the social and that both experiences have contributed enormously to my competence as a theorist, both of the social system itself, which after all has been my primary professional focus, and of the psychological and the relations between the two.

One contribution I was able to make to the conference, which was not made in comparable measure by the more psychologically and biologically oriented participants, was to emphasize the fact that the historic tendency to speak of society in opposition to the individual and vice versa was a dangerous oversimplification. The unit of a social system is not the concrete individual as a whole, but involves sectors of his personality and behavioral systems. The primary sociological term for such a sector is a "role," hence a "collectivity," which term I prefer to "group," is not composed of persons, but of roles, or in fuller form, persons in roles. The crucial point at issue, is that the same person is involved, not in one role, but in many roles. By becoming a member of the Grinker conference, I assumed a new, though temporary, role. The conference was a social system in one of its aspects. I did not, however, relinquish my role as member of a family, as member of a department and a faculty of Harvard University, and so forth. Of course, where the person is involved in many roles, there is very much a problem of the integration of the personality system in terms of the allocation of personality resources among the several roles, not least important among these being the allocation of time, but also various kinds of attentions, and so on.

This is one of the most important types of cases underlying my objections to Grinker's tendency to treat system and environment as *the* system. From the point of view of a given personality, all the roles in which he is involved constitute perhaps the most important part of his social environment, though, in addition to this, there are aspects of the society in which he does not participate, but which still constitute environment to him and with which, in certain respects, he interacts, as, for example, through the communications of the mass media. There are other borderline cases of what may or may not be called roles; for example, in walking on a city street, one sees other pedestrians and in various ways reacts to their presence and, of course, is reacted to on their part, but it is surely a marginal case to speak of the role of sidewalk walker, or to speak of those who happen to be on that particular sidewalk, in that particular block, at that particular time, as a collectivity.

The main point, however, is that interaction and interchange is not in the most important cases, with the environment in general, but with highly selective sectors of the environment, such as the other members of the conference as individual persons, and the conference as a collective system. There are, hence, always relevant criteria of which environmental sectors are important in some way for some particular aspect of the action of an individual person.

The other side of the picture is that larger and more complex social systems must always be conceived as composed of many subsystems. Probably the great majority of these are wholly comprised within the one macrosocial system we call the society, but a substantial number crosscut its boundaries as a society, and here I mean not so much its physical boundaries as its participation boundaries, that is, its definition as a collectivity. The most important mode of this cross-cutting lies in the fact that many members of the society have roles that involve participation with nonmembers of the society. One may, for example, go to international conferences, which constitute social systems. Virtually all the social systems organized around cultural considerations at high levels of sophistication, such as intellectual disciplines, are institutionalized on a transsocietal basis. I will hence suggest that it is not only legitimate but essential to treat a system of reference in terms of its interrelations with environing systems, but the essential point I am driving at is the selective nature of important relations, relative to what might from the most general point of view be considered the environment as a whole.

The foregoing considerations constitute the most important reasons why it is essential to treat social systems and personality systems as analytically independent of each other, no matter how we classify our types of social role on the one hand, types of personality or trait on the other. On each side, members of the same class will stand in different interrelations with systems on the other side of the boundary. Thus, members of a university department may include married persons and bachelors, men and women, parents of young children or of children who have already left the parental home. Hence, there is no direct correspondence between professional role types and the particular individual role. All such differences, however, are linked to personality differences. There are, of course, many correlations, but the uniformity expressed in such a correlation covers a substantial diversity at another level.

A particularly interesting set of problems in this connection is raised by Grinker's (1971) recent emphasis on the importance of the life cycle. For the combination of organic and psychological systems, this clearly is of highly salient importance, an importance that, in the first instance, goes back to the nature of the individual organism and the structure of its changes and transformations through time. At the extremes, the correlations with social role are very high, notably, in the earlier phases, for the simple reason that before children can participate in complex manifolds of different social roles they must acquire certain capacities, some of which we associate with a process of differentiation in the structure of the individual personality in the course of its socialization. In the middle sectors, then, there is a period in which the level of correlation between personality characteristics and social role probably declines markedly, at least at the levels at which most sociologists analyze social roles when dealing with large-scale social systems. At the level of small-scale systems, on the other hand, the correlations, of course, are substantially higher, because the role is more nearly tailored to the characteristics of individual participants in interaction.

At a more general level of systems theory, the patterning of the life cycle would, of course, have a place, but it would be balanced more fully than in Grinker's statements by other patternings more specially concerned with social and cultural systems. I happen to think that this kind of consideration leads in the direction of a stress on a theory of social and cultural evolution. Indeed, with all the profound qualifications of it which must be made, there seems to me to be an inherent soundness in the old biological generalization that ontogeny repeats phylogeny and that the relation between a life-cycle generalization, centering on organism and personality, and evolutionary generalizations, centering on society and culture, would help very much to balance the emphases within a highly general theoretical system in the action field, as well as in the most general field of the theory of living systems.

Not unrelated to this is a point that Emerson stressed, namely, that organic processes, that is, within organisms considered as living systems, proceed between physically contiguous entities, such as the various chemical constituents of the cell or across the permeable membranes that separate the cells from each other. These, of course, are predominantly chemical processes, though such processes as the propagation of nerve impulses may have a slightly different nature.

A major criterion, then, of the evolution of the behavioral, social, and cultural aspects of living systems has to do with the fact that interaction between the entities that compose them must somehow cross intervening spaces between the interacting organisms. One basis of the significance of the erotic complex is surely that this is a locus of communication through direct bodily contact, but involving two or more distinct organisms. It therefore seems to be an intermediate type between the intraorganic and what may perhaps be called the communication mode of interaction, particularly as we move to the more distal sensory complexes, from touch through smell to the auditory and the visual modes, communication, of course, being of increasing fundamental significance.

This presumably is the locus of the development of language, and here it is significant that the earlier aspects, both phylogenetically and ontogenetically, of the development of language involved auditory communication, whereas with the development of written language, which always comes later, the media shift to the visual level. These circumstances certainly have much to do with Emerson's fundamental point about the substitution of the symbol system for the gene as the basis of coordination of the larger systems.

I may then end this rather rambling set of commentaries by noting that, in my own theoretical work subsequent to my participation in the conference, I have been particularly concerned with interchange processes at the level of symbolic communication. Here it seems to me particularly significant that, centering at the social system level, it is possible to analyze ways of relating to all three of the others. There have evolved not one but a complex network of interlocking codes of communication and the corresponding capacities to communicate particularized meanings through the manipulation of symbols. Language is surely the groundwork of this system as a whole, and it seems to me that the development of the science of linguistics has been one of the most important contributions to the understanding of human action within the last generation or so. Relative to language as a kind of matrix, however, a general purpose symbolic system, there have developed many others that are analogous, it seems to me, to some of the more specialized hormones within the organism and probably, in some ways, to the operation of central nervous systems in the higher organisms.

I first became aware of this range of problems in relation to the functioning of money in economic transactions. It gradually became clear to me that money was indeed a medium of symbolic communication, that is, of the processing of information. If we look at information and cybernetic theory, we can see that not only can money serve to communicate messages, but it can also be stored and has many other properties parallel to what is dealt with in information theory. Once, however, one links these insights with the appropriate analysis of the nature of economic production and its functions in large-scale social systems, one sees that money cannot stand alone as such a medium, but must be exchangeable for, and otherwise interdependent with, a number of others. Surprisingly, it turned out that the nature and functions of political power could be approached in this context, though this required a very substantial redefinition of the ways in which the main traditions of political theory have conceptualized political power. From the establishment of this relationship, it then proved possible to extend the family of interchanged media, which operate at the level of the social system, to two additional ones, which I have called influence and value commitments, respectively (see Parsons, 1969, chaps. 14, 15, 16).

One of the virtues of this way of looking at interaction processes within social systems is that such far-reaching dynamic conceptions as inflation and deflation, which have been worked out in economic theory, and their causes and consequences can be generalized to processes involving media other than money. Rather preliminary tentative attempts to do this have been carried out.

A still more recent development, however, is the extension of this conception of generalized symbolic media of interchange from the level of the social system to that of action in general. This has been generated in addition to the general presumption that such media must exist to references from considerations that have combined psychological with sociological interests. A very important reference point has been the concept of affect as this has entered into psychological theory, especially in the psychoanalytic tradition. I have also, for many years, been very much concerned with two of W. I. Thomas's categories of wishes at a very general level—the wish for recognition and the wish for response. Finally, one cannot have had much to do with psychological theory without having some interest in the concept of intelligence as a highly generalized capacity of the behavioral organism and personality. What has finally emerged here is a synthesis of a second family of four media, which operate at what I would call the level of the general system of action: intelligence, performance-capacity, affect, and the definition of the situation.

The use of the last term will make it clear that there is a very important relation between this scheme and the scheme put forward a generation ago by W. I. Thomas of the four wishes and the definition of the situation. What I have been calling affect is the obverse of the wish for response. That is to say, it is phrased in terms of the sanction administered by the alter, who is interacting with an acting ego. Performance capacity and recognition are linked in a similar way. Intelligence I have taken simply from general psychological tradition, but I think the great theorist of it is Piaget. The definition of the situation comes directly from Thomas and has the very striking property, which links it with symbol and gene, in Emerson's usage, in that it is the most important stable reference point for the continuity of the system as a whole. In my own technical terminology it designates the pattern maintenance functional position and is particularly associated with the cultural system; intelligence similarly is linked with the behavioral organism, performance capacity with the personality system, and affect with the social system, a conclusion that may seem surprising to psychologically minded persons.

I cite this development of the analysis of generalized symbolic media as evidence of the stimulus value of the conference on a unified theory of human behavior. Certainly, the theoretical ingredients that have gone into it came from many sources, some of them entirely outside the scope of the conference. This was particularly true of my concern in other connections with economic theory, and hence, sensitivity to the possibility of using an analysis of money in this way. Many of the ideas, however, were very prominent in the discussions in the conference, and I think that their clarifications and my further sensitization to them was to an extremely important degree a result of my participation in the conference.

I am aware that Roy Grinker has many other achievements to his credit in his long career. Having conceived the idea of such a conference, gone through the enormous work of organizing it, of chairing its discussions, and finally, of editing the volume that came out of it will, however, go down as one of the most notable of his achievements. From the point of view of this one participant, it was one of the great and timely intellectual experiences of my career.

REFERENCES

Cannon, W. 1963. *Wisdom of the Body*. New York: Norton.
Grinker, R. R., Sr., ed. 1967. *Toward a Unified Theory of Human Behavior* (1956), 2d ed. New York: Basic Books.
Grinker, R. R., Sr. 1971. "Biomedical Education as a System." *Archives of General Psychiatry* 24:291–297.
Parsons, T. 1964. *Social Structure and Personality*. New York: The Free Press.
Parsons, T. 1969. *Politics and Social Structure*. New York: The Free Press.
Thomas, W. I. 1923. *The Unadjusted Girl*. Boston: Little, Brown.

Transactional Theory
and Social Change

The direction of scientific research and the ideas of the researcher are too often described as if they occurred purely in the interests of the pursuit of knowledge or technique, quite apart from the personalities and experiences of the researcher, to say nothing of the temper of the times. The Zeitgeist of the moment, with its radical questioning of the relevance of all professions and disciplines to the problems of society, makes it especially important for the researcher to be aware of who he is, how he came to do his work, and how it fits into the world outside of his own profession. As Alvin Gouldner has recently pointed out in his significant book, *The Coming Crisis of Western Sociology* (1970), we must all take account of the direction of our work, as to whether it promotes social change, resists it, or ignores it, comfortably conforming to the norms of professional associations and to suggestions coming from the sources of research support. Accordingly, I am providing some personal as well as intellectual background explaining how we came on something so rarefied, so fancy-sounding, as the term "transactional theory."

I wanted to be a surgeon. In all of medicine, surgery seemed to be the one area in which it was surely possible to do something practical and effective for sick people; and surgical research, being based on scientific principles, appeared to have a bright future. I rapidly became disenchanted with what had seemed such a glamorous occupation. Not only did my feet hurt from the hours of standing at the operating table and making rounds, but, more important, I was dismayed at the impersonal, inconsiderate, and sometimes cruel attitudes that surgeons tended to display toward their patients.

I did not continue my surgical training for long. I decided that psychiatry was to be my profession. Shortly after I began my residency in psychiatry at Michael Reese Hospital, World War II descended on us. I was voluntarily drafted as a medical officer in the U.S. air force for an unknown destination. In due course I found myself in a combat zone in North Africa treating psychiatric problems that the air force had not expected to encounter.

Dr. Grinker also joined the air force and was scheduled to be posted to northern Ireland as director of a hospital for the treatment of psychiatric disabilities. In that infinite wisdom, which it often likes to disguise as carelessness, the armed forces somehow put Dr. Grinker on the wrong boat. Instead of ending up in northern Ireland, he found himself, on disembarking, in North Africa. Someone had misread his ticket.

It was in North Africa, then, that we began our studies of the combat neuroses (Grinker and Spiegel, 1945a, 1945b). We found it necessary to understand the dynamics of the social environment as it played on the internal, psychological, and biological factors within the personality of the combat soldier. We made the effort to integrate social, psychological, and biological processes, not because we possessed any theory calling for such an approach. On the contrary, although we talked a lot about social psychology, we were in fact completely untrained and uninformed in respect to the social sciences. Our theoretical views were based principally on psychoanalysis, which had always held the social environment at a respectable distance. We were forced to deal with the environment because we had ourselves experienced it as a source of stress and because we hoped to be able to influence military policy in the areas of prevention and therapy. In the last chapter of *Men Under Stress* (Grinker and Spiegel, 1945a), entitled "General Social Implications," we even broadened the scope of our environmental survey to include the nation as a whole, commenting on the value conflicts that make life difficult for every American, even in peacetime, and especially for combat troops returning from overseas during a war. I remember thinking at the time that this was a rather daring enterprise for a pair of psychiatrists in service during wartime and who lacked expertise in the social sciences. Still, in our introduction we had made a somewhat cautious prediction, without citing any evidence, that "it is a moot question whether, in the peace which will follow the present conflict, the degree of stress on the average individual will be much less than that imposed by the war." The concluding chapter was intended to back up this prediction. We could not have foreseen that twenty-five years later, in the midst of yet another war and mounting social disorder, the prediction would be close to realization.

After the war Dr. Grinker was to focus intensively on the relation between biological and psychological processes while I was to continue my growing concern with the interaction between psychological and social factors. I did not know enough about the theory or the methods of social science and could not gain the needed knowledge by working in isolation from colleagues sharing similar interests.

This deficit began to be repaired when I was offered the opportunity to teach in the Department of Social Relations at Harvard, working with Florence Kluckhohn, Talcott Parsons, and Samuel Stauffer in a research seminar. Under the leadership of Talcott Parsons, the faculty of the Department of Social Relations was concerned with almost every aspect of human behavior and especially with the sociology of medicine. There, I began a study with Florence Kluckhohn of the relation between the cultural values of several ethnic groups, differences in the structure, function, and interaction patterns within these ethnic families, and the relation of the overall acculturation process to the mental health of the individual family member. While Florence Kluckhohn and I were wrestling with the theoretical problem of how to combine the disciplines of cultural anthropology, sociology, and psychiatry in the study of family behavior, I received an invitation from Jurgen Ruesch and Roy Grinker to join their newly organized Committee to Frame a Unified Theory of Human Behavior.

The problems faced by the Committee to Frame a Unified Theory of Human Behavior were enormous and were never really solved. In the light of our imperfect knowledge and the disparate and mostly incompatible techniques for studying human behavior, the goal was too grandiose. As Dr. Grinker stated in his introduction to the second edition of the report of the committee, and Jurgen Ruesch reemphasized in his epilogue to the same volume (Grinker, 1967), many approaches were tried and discarded as unsuitable. Grand analogies applicable across the boards to all systems of behavior, such as communications, "homeostasis," growth and decay, differentiation and dedifferentiation, boundary maintenance, integration, and conflict were suggestive, but in the end misleading. They explained too much without specifying how the same end state could be arrived at through entirely different processes for different systems of behavior. Reductionist approaches in which one science, such as physics or chemistry, was said to be basic with all other sciences depending on it as a foundation gave rise to acrimonious discussions about the significance of the word "basic" and which science had the better claim to it. Attempts to transfer the organizational principles of one field of behavior, such as cellular biology, to that of another, such as social systems among insects or humans, were also unsatisfactory.

What seemed to me to be of the greatest interest in all the provocative and often exciting discussions of the committee was the question of how the various fields of behavior might be related to each other within a unified field. Field theory is derived from the work of Einstein. It assumes that though one cannot now, if ever, explain everything under the sun or in the universe, one can define the field in which the events requiring explanation take place, locate the position of the observer with respect to the field, and describe the structure of that field, that is, how it is organized. Suppose one were to attempt to locate all the separate fields of behavior within a unified (or organized) field. What would it look like?

I was also impressed with the concept of transaction, mentioned by various members of the committee. Derived from the writings of John Dewey and Arthur Bentley (1949) and brought to the attention of behavioral scientists by the work of Hadley Cantril and W. K. Livingston (1963), transactionalism was originally concerned with a general theory of knowledge, interesting mainly to philosophers. Nevertheless, it has direct relevance to the theory and methods of the behavioral sciences. It proposed that behavior has been described in the past as the result of self-acting entities exhibiting goal-directed (purposive) behavior, or as the result of the interaction between entities, such as the collisions between atoms, billiard balls, or nations, which take place in regular ways with predictable results characterized by an end state—the atoms combined into molecules, the billiard ball in the pocket, and the warring nations at peace. In transactions, on the other hand, one can discover no entities and no persistent end states. One is observing systems in continual process with other systems, undergoing exchanges in both directions, contributing to the stability or instability, the growth or decay, of the field of behavior in which the exchanges are taking place.

Given these starting points, I postulated a field of transacting systems, which was organized in accordance with Figure 2–1. The universe includes both the macrocosmic (space-time) and microcosmic (physicochemical particles) aspects of "the world" considered as a field. Looking, for our purposes, only at human behavior, we can include under soma those biological processes that give birth to and maintain the organism within the field. Obviously, its exchanges are conducted directly with the universe for purposes of maintenance. But, for the sake of survival, the soma transacts with psyche, which includes all those systems of ad-

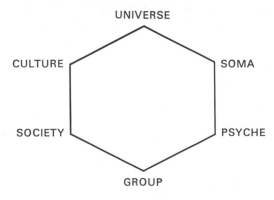

Figure 2–1.

aptation through actions that guarantee flexibility of response (learning). For human beings, however, whatever happens (or can happen) among psychological systems derives from transactions with primary groups, especially the family. The stability of such small, face-to-face groups is the product, on the one hand, of its exchanges with individual members, some of whom may have to be expelled while new members are brought in, and, on the other hand, of its transactions with the larger social system, which we have called society. The large-scale organizations of society—its political, legal, economic, religious, educational, recreational, and status systems—are in one way a vast collection of small groups transacting in accordance with the principles laid down by the institutions of the society in question. But these extended organizations create such principles through exchanges with what we have called culture, that is, the systems of values, symbols, and meanings (including language systems) that provide the society with its reasons for existing, as well as rationalizations for the kinds of institutions it maintains. On the other side of the field structure, culture also explains the existence and meaning of the universe and occasionally predicts its imminent demise through divine wrath or human carelessness, or both.

The format and assumptions of this field of transactions make it clear that processes can proceed in an orderly fashion in either direction around the circle but may not jump across from one structure to another. No structure is any more basic or important than any other. Transactions within structures, as well as between them are concerned with the maintenance of stability or change within the field as a whole. How then are we to conceive of the process of change?

This enormously complicated question can be somewhat simplified by our distinguishing three kinds of change: (1) One can consider change associated with growth, development and senescence, a process that seems to be equally characteristic of organisms, nation-states, and stars. (2) There is the change resulting from the disorganization of part-whole relationships (integration) either as a result of external damage or of internal conflict between parts of a transacting system, or a combination of both. For heuristic purposes, we often refer to this sort of change as "pathology," a term that is rather troublesome because of its tendency to evoke images of something bad, corrupt, or simply menacing. (3) We must consider change induced by man through intervening in the more or less spontaneous changes listed above. Such planned change is directed at increasing the rate of development, postponing senescence, repairing injury, or resolving conflict.

Another way to simplify the discussion of change within the total transactional field is to ask which of the six structures within the field— or the relations between them—is being looked at for purposes of inquiry. Because the act of looking at some transacting systems with a purpose in view while disregarding others is so ordinary and necessary a part of research, because to investigate means to focus down on and to keep track of one set of systems against the background of others, I have called the six labeled structures in Figure 2–1 foci within the field. The word "field" is used here in the sense defined by Kurt Lewin (1951) as "the totality of coexisting facts which are conceived of as mutually interdependent." In the interests of theory one can consider this "totality of coexisting facts" as is being attempted in this essay. In practice, however, research is either confined to one focus central to a particular discipline or combines two or three foci in an interdisciplinary study. In the usual case, change is observed through the manipulation of a few independent variables, the consequences of which are either confined to one focus or have reverberating effects on neighboring foci.

A final effort to simplify the question of change within the transaction field is concerned with linking processes. Each focus within the field is associated with neighboring foci on either side through a set of reciprocal exchanges. At the present time it is difficult to label and describe such links with precision. We can, however, discuss some bridging concepts which are plausible.

For example, soma and universe exchange processes of production and consumption, at least on earth, our localized part of the universe, and now increasingly in outer space. Man consumes what earth produces, while earth consumes, with increasing ecological discomfort, what man produces, or, at any event, wastes.

The exchanges between soma and psyche are mediated by what are loosely called instincts, drives, feelings, emotions, or motives. Whatever the label, such exchanges represent the transformation of biological information into cognitive processes and the conversion of cognitive information (perceptions) back into biological processes.

The link between psyche and group is established through group membership roles. Fathers and sons, leaders and followers, hosts and guests are names for members in formal and informal small groups, and they prescribe the kind of behavior expected from the role incumbent. Who is entitled to membership is established more or less reciprocally between the individual and the group. One person may change the nature of the group's behavior, but, more often, the group modifies the behavior of the individual.

Group and society are linked by more institutionalized social roles, specifying how the group is to behave within the differentiated legal, political, occupational, recreational, religious, and so forth institutions of the social system. Usually groups maintain fairly comfortable degrees of harmony with the rules of such large-scale systems. However, if conflict develops, a group may withdraw or rebel in the effort to change the rules, thus creating the kind of social disorder with which we are today increasingly confronted.

Reciprocal relations between society and culture are mediated by the ritualization of cultural values in the patterning of societal institutions and of styles of behavior within them. If a culture fosters warrior values ("I will not be the first President to lose a war!") while also advocating peace and democracy, the society is likely to maintain a large military establishment, called a "Department of Defense," which engages in war-like operations all over the world in the interests of peace. At the same time, certain parts of the political establishment may mount a campaign against such violent values embedded in the culture in the effort to curtail defense department operations.

Finally, culture and the universe are reciprocally related to each other through a combination of beliefs and technology, which working together generate cognitive maps of the world. For example, it was not possible to establish whether the earth was flat or round until shipbuilding and the techniques of sailing and navigation had attained sufficient development. However, the voyages of Columbus and other explorers might not have taken place without the preexisting idea circulating around Europe that the world was probably round. Reciprocally, the earth did really have to be round, that is, it had to react appropriately to this idea.

It is difficult to determine exactly how participating in the deliberations of the Committee to Frame a Unified Theory of Behavior affected all those who participated. Dr. Grinker and I had both been mildly interested in the transactional approach even before the committee was formed. I had sketched out some tentative ideas for the format of the transactional field which I presented at the first conference (Spiegel, 1967) in October 1951. Dr. Grinker (1953) discussed the transactional approach in his book on the theories and methods of psychosomatic research, but the substantive content was confined to just two foci in the field—psyche and soma.

In his subsequent study of the phenomenology of depressions (Grinker, Miller, Sabshin, Nunn, and Nunnally, 1961) the transactional field was not considered, the investigation being confined entirely to psyche. The field appeared, however, in his previous work on transactional psychotherapy (Grinker, 1959), which emphasized the importance of the reciprocal communications between patient and therapist in regard to their shared or incompatible social roles and cultural attitudes. In this work, it was assumed that the ability of the therapist to perform effectively as an agent of change depended on holding current aspects of the transactional field under continuous scrutiny. In his most recent work on the borderline syndrome (Grinker, Werbel, and Drye, 1968), the transactional field all but disappears. Some attention is paid to the family group of which the patient population are members, and the religious and racial backgrounds of the patients are listed. Nothing is said of the socioeconomic backgrounds or ethnic affiliations of the patient population. Variations in the patterning of borderline behavior correlated with social factors, differences in the understanding among patients of what role behavior is to be expected under what social conditions, and, especially, the confusion about such matters one would anticipate in borderline patients as a result of contradictory impacts from wider reaches of the transactional field—all these broader system effects remain unexamined. The focus is again almost exclusively on psyche. Evidently, it is difficult almost by definition to keep the larger field of transacting systems in view when conducting clinical studies of diagnostic entities fitted out with such labels as depression or borderline. Not only are such things as depression and borderline created by the linguistic habits that psychiatry has borrowed with a little elaboration from Western folk wisdom, not only do they isolate fixed end states that, by professional fiat, become attributes of the person, but, more importantly, the labeling process itself forces the investigator to focus on just one structure in the field—the psyche. In these respects, then, the whole enterprise of clinical studies by psychiatrists and clinical psychologists working in isolation from colleagues in the social sciences is open to question. This is said not to deprecate the clinical, diagnostic approach, which I personally regard as indispensable, but rather to point out that it so easily suffers from what the philosopher Whitehead called "the fallacy of misplaced concreteness." Even though they are concentrated mainly on psyche, the methodological precision with which the clinical studies of Dr. Grinker and his colleagues have been carried out have gone a long way toward demonstrating that the diagnostic labels do not refer to uniform sets of behaviors, but rather cover a range of behavior responsive to multiple factors in the environment of the individual.

In my own work (Spiegel, 1959; 1960; Spiegel and Bell, 1959), carried out with Dr. Florence Kluckhohn and other social scientists, the research design required that attention be paid to multiple foci in the transactional field. We wished to determine how integration or conflict in family behavior was related to the process of acculturation in ethnic families migrating to the United States from Ireland and Italy. We also wanted to investigate the effect of order or disorder in family patterns on the adaptive or maladaptive capacities of individual family members. Accordingly, we were compelled to pay attention to the relations between culture, society, group, and psyche within the transactional field. Finally, we wanted to discover how to intervene in families characterized by much conflict and psychological disturbance in individual members. Could the amount of stress experienced by the family and its members be reduced?

It was soon evident that much of the stress within disturbed families was associated with the attempt on the part of the family to engage in rapid social change; that is, to adopt American middle-class role patterns as quickly as possible. In this attempted transformation the parents tended to assume middle-class American roles as facsimiles, without real understanding of the values patterning the roles. The problem was especially evident in the expectations a wife held for her husband's behavior, and vice versa, giving rise to much role conflict, confusion, and reciprocal disappointment. Often the problem between the husband and wife was masked and displaced toward the children. By mutual if unconscious complicity, the husband and wife agreed not to raise the problem existing between them but to see it as reflected in a child's behavior and to deal with it in its displaced form, trapping the child in a problem not of his own making.

It may be helpful to be more specific about the problem of acculturation by describing the transactions between culture and society in one value dimension. According to the theory of value orientations proposed by Dr. Florence Kluckhohn (Kluckhohn and Strodtbeck, 1963), relational values pattern the relationships between group members and the forms of decision-making within the group. As with the other value dimensions proposed by the Kluckhohn theory, variation and patterning of values occurs by virtue of the order of preferences between three possible value positions. In the order of preference, the American middle-class pattern is individual over collateral over lineal. This is to say that in the styling of middle-class American roles, every individual counts and is expected to assert his wishes and opinions, to become self-reliant, and to develop self-control, in other words, not to become over dependent on the group. If difficult decisions must be made, they are decided by taking a vote. Collaterality, which is in the more restricted, second-place position, emphasizes the horizontal structure of the group. Group loyalty is stressed and the maintenance of group harmony at the expense of the individual is expected. Decisions are taken by consensus, often after prolonged discussion. The third and least preferred position, the lineal, emphasizes vertical group structure. Decisions are made by a boss and handed down the lines of authority. Superiority and inferiority assume large proportions and the struggle for status within the group hierarchy becomes intense. In official American culture, lineal orientations are reserved for very special occasions, such as situations of emergency, stress, or danger, when strict lines of authority become necessary for survival.

Native Italian culture involves a different pattern of relational values. The collateral is first, lineal second, and individual third. For the family, group loyalty and the maintenance of intimate relations throughout extended kinship lines is very important. Ordinarily, decisions are made by consensus, but if a really difficult problem emerges, the father will assume command. The individual position is rarely honored. Children, for example, are to be loved or scolded but are seldom asked their opinions.

Many Italian families migrating to this country cling to these relational orientations and experience little difficulty except for their apartness. They are so typically Italian, and not American. If they continue to live in predominantly Italian neighborhoods—in "little Italies"—even this may not become a problem. But in those families choosing the path of rapid change, a husband may expect his wife to get along in a predominantly WASP suburb, deprived of the help and warm association she has always expected from her female relatives while he works at several jobs or goes out bowling with the boys at night. Reciprocally, a wife may expect her husband to get large and rapid salary increases when he has not yet learned how to assert his individual rights and talents on the job.

What is the outside observer and would-be therapist to do about this form of rapid disorganized social change? What should be his goals: to slow down the rate of change, encourage a return to Italian values and role patterns, or teach the family members how to perform more adequately as Americans? Naturally, being therapists, we initiated none of these procedures, assuming that a therapist-knows-best attitude would be doomed to failure. Instead, we tried to help family members understand the nature of the problem so that they could decide which solution would work out best for them.

In defining our professional roles in this fashion we were, of course, manifesting our American, middle-class value pattern. And this brought us into conflict with the people we were trying to help. The Italians could not believe that we, professional people, were not going to tell them how to solve their problems. Or that we were not going to side with the wife against the husband, or vice versa. They wanted to know who was right and who was wrong. We were the experts and we would not say. For them it was quite frustrating. For the most part, however, they gradually began to adopt our point of view.

The point relevant to this essay is that we were, after all, teaching them American relational value patterns, though in an indirect and perhaps underhanded way, that is, by serving as role models. There could, I suppose, be no great objection to this mode of propagating change, except for one consideration: Are American relational values what they purport to be? How well are they actually reflected in American institutions?

This question remained in the background throughout the time I carried out research and therapy in the family area (Spiegel, 1959; 1960). When I became director of the Lemberg Center during the fall of 1966 and began to study civil disorder and riots connected with race tensions, it suddenly sprang into the foreground. It was apparent that the disorders were associated with anger and impatience over the slow rate of change for black people living in the ghetto. It was also apparent that this slow rate of change was associated with what the National Advisory Commission on Civil Disorders (Kerner Commission) called "white racism," that is, resistance among whites toward changes that would involve justice for blacks. But what is "justice"? Surveys conducted among black and white populations by the Lemberg Center on this very question—what is to be considered an appropriate and just rate of change in housing, schools, job opportunities, and so on for blacks— showed astonishing differences of opinion (Conant, Levy, and Lewis, 1969). The gap between the most liberal group of whites and the most Uncle Tomish group of blacks was staggering. Clearly, "justice" meant something different to the two racial groups. How is one to account for this difference?

Justice is obviously concerned with the ordering of cultural values. If our previous description of the links between society and culture within the transactional field are correct, one would expect to find two different and incompatible value patterns coexisting within the cultural focus, one for whites and one for blacks, each generating different and contradictory role expectations within the two communities. Anticipating that this might be the case, we included questions on value orientations in our survey of black and white populations. The analysis of this as yet unpublished data does not confirm our expectations. Though there are some interesting differences among age groups, which are too complex to be discussed here, both white and black populations generally endorse what we have called the "official," or standard, rank order of American relational orientations—individual over collateral over lineal. The agreement is clearcut, and it leaves us with no apparent explanation for the differing views of justice between the two races. Both subscribe to the value pattern that is clearly associated with the ideology of democracy and egalitarianism.

It is tempting to adopt the view now often expressed in the rhetoric of radical groups that Americans are hypocritical about their democratic values. The difficulty with this view is that it is inconsistent with the constraints previously ascribed to reverberating processes in the transactional field. We had said that reciprocating processes cannot jump across the field nor bypass foci within the field. They reverberate around the periphery. Hypocrisy, which consists of acting in ways inconsistent with one's expressed values while simultaneously denying the inconsistency, is behavior arising within psychological systems. If the values located within culture are themselves consistent, and if they pattern social role behavior, then those roles should not be taken in an inconsistent fashion simply because of operations within psyche. Either the hypocrisy theory is wrong or our descriptions of the rules governing transactions between foci in the field are wrong. Yet, there seemed to us to be something right about each.

There is a possible way of escaping from this dilemma. Perhaps "hypocrisy," a vague term at best, need not be exclusively characteristic of operations ascribed to psyche. Perhaps it can also refer to a discrepancy between the values Americans are officially taught—in school, in the family, by the media of communication—and the pattern of values actually characteristic of our major social institutions. The more we thought about this the more we were inclined to accept it as an explanation for the peculiar views of justice characteristic of our country. As we examined the institutions with which we were familiar, such as universities, hospitals, political parties, the police, and the military, and especially as we examined the foreign policies of the nation, it became apparent that the operative value pattern—as opposed to the official pattern—probably was individual over lineal over collateral. Within this arrangement, individualism functioned to mask a lineal, authoritarian, and essentially elitist definition of role relations. Thus, Americans can still say that every individual counts and all men are created equal. But after he has finished expressing his relational values in this fashion, he then goes on to rank persons and groups in vertical positions of superiority and inferiority because this is the way his institutions function (Spiegel, 1968; 1969).

If this is the way American institutions function, it is in the interests of those currently on top to maintain their status by holding others down or, at least, by preventing them from rising too rapidly. Thus, there actually are two sets of value patterns incorporated within culture, one officially sanctioned and honored verbally, the other concealed but implicitly operative within our social structure.

If this view of the actual state of the transactional field is correct, what is one to do about it, either as a citizen or as a member of a behavioral science research organization? In this view, the kind of social change for which militant black groups and radical college students are pressing is quite clear. They wish to make our operative value pattern consistent with the official pattern of values. This would require extensive transformation of our social institutions. At the Lemberg Center, as at many other research centers, we have been under intense pressure to clarify our position on these matters, and to take a stand. Where do we stand in respect to social change? What will we do about it?

Up to the present, we have said it is our job to do research, to describe the systems and processes involved in social change, and to explain to those who have the power to take action the reasons why social change must be pushed as rapidly as possible. We have been doing this with as much intensity and persuasive power as we can muster.

But is this enough? Will it be necessary to join the ranks of those who use actions rather than words. At this time, in the summer of 1971, I cannot answer the question. It is my hope that within this decade the rules for intervening in the process of social change, with a little help from transactional theory, will have become clearer.

REFERENCES

Cantril, H., and Livingston, W. K. 1963. "The Concept of Transaction in Psychology and Neurology," *Journal of Individual Psychology,* vol. 19, no. 1.

Conant, R. W., Levy, S., and Lewis, R. 1969. *Mass Polarization: Negro and White Attitudes on the Pace of Integration.* Waltham, Mass.: Lemberg Center for the Study of Violence, Brandeis University.

Dewey, J., and Bentley, A. 1949. *Knowing and the Known.* Boston: Beacon Press.

Gouldner, A. W. 1970. *The Coming Crisis of Western Sociology.* New York: Basic Books.

Grinker, R. R., Sr. 1959. "A Transactional Model of Psychotherapy," *Archives of General Psychiatry* 1:132–148.

Grinker, R. R., Sr. 1953. *Psychosomatic Research.* New York: Norton.

Grinker, R. R., Sr., ed. 1967. *Toward a Unified Theory of Human Behavior* (1956), 2d ed. New York: Basic Books.

Grinker, R. R., Sr., Miller, J., Sabshin M., Nunn, R., and Nunnally, J. C. 1961. *The Phenomena of Depression.* New York: Hoeber.

Grinker, R. R., Sr., and Spiegel, J. P. 1945a. *Men under Stress.* Philadelphia: Blakiston.

Grinker, R. R., Sr., and Spiegel, J. P. 1945b. *War Neuroses.* Philadelphia: Blakiston.

Grinker, R. R., Sr., Werble, B., and Drye, R. C. 1968. *The Borderline Syndrome.* New York: Basic Books.

Kluckhohn, F., and Strodtbeck, F. B. 1963. *Variations in Value Orientations.* Evanston, Ill.: Row, Peterson.

Lewin, K. 1951. *Field Theory in Social Science.* New York: Harper & Row.

Spiegel, J. P. 1959. "Some Cultural Aspects of Transference and Countertransference." In J. H. Masserman, ed., *Science and Psychoanalysis.* Vol. 2. *Individual and Familial Dynamics.* New York: Grune & Stratton. Pp. 160–182.

Spiegel, J. P. 1960. "The Resolution of Role Conflict within the Family." In N. W. Bell and E. F. Vogel, eds., *A Modern Introduction to the Family.* Glencoe, Ill.: The Free Press. Pp. 362–381.

Spiegel, J. P. 1967. "A Model for Relationships among Systems." In R. R. Grinker, Sr., ed., *Toward a Unified Theory of Human Behavior* (1956), 2d ed. New York: Basic Books. Pp. 16–26.

Spiegel, J. P. 1969. Theories of violence—an integrated approach. Paper presented at the annual meeting of the American Association for the Advancement of Science, Boston.

Spiegel, J. P. 1971. Toward a Theory of Collective Violence. In J. Fawcett, ed. *Dynamics of Violence.* Chicago: American Medical Association. Pp. 19–33.

Spiegel, J. P., and Bell, N. W. 1959. "The Family of the Psychiatric Patient." In S. Arieti, ed., *Handbook of American Psychiatry,* Vol. 1. New York: Basic Books. Pp. 114–149.

An Evolutionary Perspective
on Human Aggressiveness*

In 1945, I was a student in medical school, quite undecided about my field of specialization. Psychiatry was one of several possibilities. I was fortunate enough to come on the book *Men under Stress* (Grinker and Spiegel, 1945) shortly after its publication. I read it with enormous interest and decided that psychiatry was to be my field. During the late 1940's and early 1950's, I repeatedly encountered young people whose interest in psychiatry and closely related fields was strongly stimulated by *Men under Stress*. Very few publications had so much impact on the shaping of modern psychiatry as this book. One of its many contributions was an examination of man's aggressive tendencies as illuminated by combat experience. The book described aggressive and hostile reactions as one of five major patterns of response to wartime stress. Indeed, the book contains vivid descriptions and insightful comments on the psychodynamics of aggressive behavior under stress.

Growing up in the era of Adolf Hitler immediately preceding World War II, it is perhaps not surprising that I was impressed early with man's capacity to justify and rationalize depreciation, scapegoating, hatred, and violence. What was, however, personally surprising was the observation in later years that such behavior is by no means limited to certain psychotic individuals, degraded cultures, or aberrant political movements. On the contrary, I have become deeply impressed with the ubiquity of certain aggressive tendencies. Granted the enormous variability of human cultures, including many aspects of aggressive behavior, a remarkably recurrent theme emerges in many times and in many places. Let me try to state this theme briefly here.

Human societies have an extraordinary capacity to distinguish sharply

* The work described in this chapter was partly supported by National Institute of Mental Health grant, no. 1-F3-MH-36, 934-01, and partly by a grant from the Wenner-Gren Foundation for Anthropological Research.

between good and bad people, between heroes and villains, between in-groups and out-groups. This sorting tendency is so widespread, so readily learned, and so susceptible to harsh dichotomizing that we must be concerned with its consequences in the modern world.

Hostility between human groups—even subgroups of a small organization—is likely to arise when the groups perceive a conflict of interest, a difference in beliefs, or a difference in status. Such situations tend to evoke sharp in-group, out-group distinctions, with drastic depreciation of the out-group by the in-group. Perceived threat from the out-group tends to enhance in-group solidarity, ethnocentrism, tightness of group boundaries, and punishment of deviants (Campbell, 1965).

Justification for harming out-group members rests on sharp distinctions between us and them, between good men and bad men. Such justification is readily provided by assumptions regarding: (1) the damage they would do to the in-group; (2) the damage they would inadvertently do to themselves; (3) classification of the out-group as essentially nonhuman.

A great variety of political, social, economic, and pseudoscientific ideologies may be adduced in support of these hostile positions. They occur in the politics of the left as well as the right, in the black as well as the white, in professional societies as well as among blue collar workers, in Eastern as well as Western countries. Though the content of such intergroup hostility varies widely from time to time and place to place, the form of the antagonism is remarkably similar.

We may well ask whether man has a legacy of aggressive orientations, transmitted in some perplexing ways through his genes, his hormones, his brain, and his customs. It is this search for a deeper understanding of the evolution and development of human aggressiveness that has become my principal interest during recent years. My approach to this problem area is interdisciplinary, mainly psychobiological. In this orientation, I have been strongly influenced by the years of rewarding association, collaboration, and friendship with Roy Grinker.

Psychiatry, especially in its psychoanalytic tradition, has long been concerned with the ways in which man's basic nature might reflect his evolutionary heritage as transmitted through genes and customs. From time to time over the past half-century, efforts have been made to bring the then-existing anthropological data to bear on questions of this sort: For example, "What can we reconstruct of the aggressive tendencies of early man, and how might they be reflected in orientations of contemporary man?" Some of these reconstructions now look rather fanciful in the light of recent evidence, but they do reflect a serious interest in the relevance of human evolution for contemporary man.

In this context, it was my good fortune to be able to undertake a brief but informative field study with Eric Hamburg of chimpanzees and baboons in East Africa, focusing attention primarily on aggressive behavior. Much stimulation and guidance for the project came from Sherwood Washburn and Phyllis Jay Dohlinow. Grants from the National Institute of Mental Health and the Wenner-Gren Foundation made it possible; and the generosity of Jane van Lawick-Goodall and Hugo van Lawick made available a great deal of information that would otherwise have been impossible to obtain. In this essay, I wish to point out a few interesting aspects of chimpanzee aggressive behavior as observed at the Gombe Stream National Park in Tanzania and of baboon aggressive behavior as observed both at the Gombe and at the National Park near Nairobi, Kenya. Most of the summer of 1968 was spent in making these observations. They were only possible because the animals were already habituated to human observers and because of the extraordinary cooperation of the experienced field researchers with whom we were working. We were very fortunate to obtain more than 100 hours of close-range observation of chimpanzees at the Gombe; during much of this time forest-dwelling olive baboons were also present, interacting with the chimpanzees. At the Gombe, we also had the privilege of access to the rich files of chimpanzee observation accumulated by Jane van Lawick-Goodall during the years of her unique study beginning in 1960, the remarkable collection of photographs of chimpanzee and baboon behavior accumulated at the same location by Hugo van Lawick, and the perceptive observations of the several staff members working at the Gombe on both chimpanzee and baboon research. At the Nairobi Park, we also had more than 100 hours of close-range observation of savanna-dwelling olive baboons. By close-range observation I mean two or more animals within ten yards of the observer. Here we studied one troop consisting of forty-two animals that had been studied earlier by DeVore and Washburn. This habitat, characterized by vast plains covered with tall grass and occasional clumps of trees, provided a marked contrast with the forest habitat of the Gombe. Thus, the same species of baboon could be observed in two markedly different ecological settings.

In undertaking this study, we adopted the definition of aggressive behavior utilized by Carthy and Ebling (1964) in a zoological context, "An animal acts aggressively when it inflicts, attempts to inflict, or threatens to inflict damage on another animal." However, zoological definitions often rule out predation as a form of aggression. Though there is certainly a meaningful distinction between intraspecies and interspecies attacks, we did not wish to exclude the latter from our study. Being primarily interested in human aggressiveness, we wish to keep open the possible evolutionary role of carnivorous, predatory, and competitive interspecies orientations in the nature of the human organism. So we directed our attention toward any actions that clearly increased the probability of damage to other animals, in short, to threat and attack patterns, both within the species and directed toward other species.

Baboons are particularly interesting because they and their closely related species have made remarkably effective adaptations in habitats that were probably crucial in the emergence of early man (DeVore and Washburn, 1963; Washburn, 1968; Schultz, 1969). The baboon-macaque group has spread widely through Asia and Africa in numbers that are large for primates, especially those of big-body size. The baboons are the largest of all monkeys. Their ground-living capability is much greater than that of most other primates. Even in the dangerous savanna habitat, where large carnivores may lurk undetected in the tall grass, baboons cover several miles per day on the ground, often a mile or so from the nearest trees, which provide the ultimate safety for virtually all primates. Baboons sleep in trees, but must gather their food in small caloric "packages" as they make their daily rounds on the ground. Thus, their ability to cope with predators is of central adaptive significance.

The attack patterns of baboons are clear and vivid (Hall and DeVore, 1965). In general, they are similar to those of many other species of higher primates (Hall, 1964; DeVore, 1965; Washburn and Hamburg, 1968). Attack is most frequently undertaken by a high-ranking adult male, who may start the attack by leaping to his feet, raising his muzzle, and racing toward another animal. He barks loudly and often. His mane fur erects, making him look considerably larger. A chase typically ensues at high speed. If he catches the other animal, he grasps it forcefully with his hands and seizes it in his jaws, usually near the nape of the neck. Alternatively, he may beat it with his hands as it lies prone; or he may lift it up and down, slamming it against the ground. Though bites are usually not serious, deep cuts and slashes are not unusual. During the time of our study, one adult male baboon killed another by a deep bite that punctured the femoral artery (Ransom, 1970).

Threat patterns are similar to attack patterns but less extreme. They occur much more commonly. Threatening animals commonly stare at another animal, often jerking the head down and then up, flattening the ears against the head, and raising the eyebrows with a rapid blinking of the eyelids. This reaction is seen in all the macaques and baboons, and is vivid because the upper eyelids and the skin above them is white. "Yawning" is another common threat pattern, especially when two adult males come close together. This yawn is probably better described as canine display, since adult male baboons have enormous canine teeth that convey clear information to the human observer about the destructive potentiality of this organism. Presumably, similar information is conveyed to other animals.

I now wish to summarize briefly the conditions under which baboon threat and attack patterns are likely to occur. This summary is made possible by the extensive fieldwork of Hall, DeVore, Washburn, Altmann, Rowell, and Ransom, plus our own observations of baboons in two contrasting habitats (DeVore and Washburn, 1963; DeVore, 1965; DeVore and Hall, 1965; Hall and DeVore, 1965; Altmann, 1967; Rowell,

1967; Ransom, 1969). The species under consideration here is the common olive baboon of East Africa. The conditions eliciting aggressive behavior are as follows: (1) protection of the troop by adult males against predators such as lions and cheetahs; (2) protection of infants, both by their mothers and by adult males; (3) resolution of disputes within the troop by adult males; (4) formation and maintenance of consort pairs at the peak of estrus; (5) attainment of preferred sleeping sites in the trees, particularly in the presence of predators; (6) acquisition of premium foods such as figs, nuts, and bananas, especially when spatially concentrated rather than widely distributed; (7) dominance interactions, especially in the presence of premium foods, scarcity of sleeping sites, and females in full estrus; (8) exploration of strange or manifestly dangerous areas, a function largely of adult males; (9) contact between different troops, especially if such contact is infrequent. In all these circumstances, the probability of overt threat behavior and of fighting is higher than in the other circumstances of baboon life.

Evidence has accumulated from many laboratories that the chimpanzee is man's closest living relative (Sarich, 1968). During the past few years, new evidence utilizing various biological indices has reiterated this relationship. Chimpanzees have many similarities to man in chromosomes, DNA, blood proteins, immune responses, and brain structure. In respect to behavior, the newer field studies have strengthened the earlier impressions gained from laboratories and zoos that chimpanzee behavior is in some ways remarkably similar to man's own (Goodall, 1965; Reynolds and Reynolds, 1965; van Lawick, 1967, 1968a, 1968b).

These similarities do not mean that man has descended from chimpanzees. Rather, man and chimpanzees are descended from a common ancestor, and the separation of these lines occurred at least several million years ago. Even the shortest estimates of this separation leave plenty of time for behavioral divergences to evolve. Nevertheless, the general trend of evolutionary research indicates that species derived from a common ancestor tend to show important similarities, even though differences emerge as well. In directing attention to behavioral similarities, there should be no implication that man and chimpanzee inherit from this common ancestor many immutable, fixed action patterns. Chimpanzee adaptation clearly depends heavily on learning, and man's even more so (Reynolds and Reynolds, 1965; Riopelle and Rogers, 1965; Zimmerman and Torrey, 1965; van Lawick, 1967, 1968a).

Chimpanzees show a variety of threat patterns in the natural habitat. These have been exceedingly well described in Jane van Lawick-Goodall's recent monograph (van Lawick-Goodall, 1968a), which adds much new information to her previous reports. These threats are a combination of gestures, postures, and sounds. We observed them directed toward other chimpanzees, toward baboons, and toward humans: (1) Head-tipping is an upward and backward jerk of the head and is accompanied by the soft bark. (2) Arm-raising—the forearm or the entire arm is raised rapidly with the palm of the hand—is oriented toward another animal, often accompanied by a bark. (3) Hitting away is a movement with the back of the hand directed toward the threatened animal, often accompanied by a bark. (4) Flapping is a downward movement of the hand directed toward another animal. (5) Branching is the taking hold of a branch and shaking it from side to side or backward and forward. (6) Stamping and slapping is usually done on the ground, but sometimes on a tree. (7) Sticks, stones, or vegetation are thrown. (8) Bipedal arm-waving and running are directed at another individual. (9) The bipedal swagger is a movement in which the animal stands upright and sways rhythmically from foot to foot, his shoulders hunched and his arms held out and away from the body. It is typically done by males.

Attack patterns are similar to threats but more extreme. We again follow Jane van Lawick-Goodall's delineation of these patterns: (1) The attacking charge. One animal races toward another; the movement is usually quadrupedal, silent, and fast. The hair of the attacking individual is usually fully erect at this stage. Not only the attacked animal but other animals are likely to scream and run away, some into dense forest, others up nearby trees. Similar charges often occur that are not immediately directed toward another animal, though they may readily become so. Even when initially nonspecific in respect to target, they are treated with great respect by other animals, as if a direct attack were considered likely. Such aggressive displays are common, especially on the part of adult and adolescent males, and the charge is often accompanied by the brandishing of a long palm frond or tree branch, pounding on the ground or on trees, and hurling of objects. Occasionally, it is preceded by pounding on the attacker's own chest, though this is more common among gorillas (Schaller, 1963). (2) Branch-dragging. The branch may be brandished, swung, or thrown at the object of the attack. (3) Stamping on the back. The attacker seizes the victim by the hair and leaps onto its back. He then stamps on the victim with both feet. (4) Lifting, slamming, and dragging. If the victim is smaller than the attacker, it may be lifted from the ground by its hair or limbs and slammed down again, sometimes repeatedly. (5) Biting. The attacker often puts his mouth to the body of his victim, but usually does not make a deep bite. Sometimes bites are deep and serious. (6) Hair-pulling. The attacked animal may lose large handfuls of hair during a fight and be seen with bald patches for some time afterward. (7) Slapping. This is a downward movement of the arm in which the palm of the hand slaps down on the body of the victim. It occurs frequently when a female attacks another chimpanzee. Both males and females slap humans and baboons during aggressive interactions.

Aggression may be prevented, ameliorated, or terminated by a set of patterns, which, taken together, are called by J. van Lawick-Goodall submissive behavior. Here again, as with the threat and attack patterns, a rich array of behavior is apparent. Four of the prominent forms of submissive behavior are noted here: (1) presenting (that is, orienting rear toward front of other animal in lowered posture, not necessarily sexual); (2) bowing, crouching, and bobbing (much like human patterns of deference); (3) kissing (associated with crouching or bowing, often in groin); (4) hand and arm movements (especially touching at a distance; also, reaching extended arm out, palm upward, while panting loudly and slowly approaching the other animal).

Such submissive behavior is in turn likely to evoke from the aggressive animal one or more of a set of responses that van Lawick-Goodall calls reassurance behavior. Various patterns are subsumed under this heading: (1) touching; (2) patting; (3) embracing; (4) mounting; (5) kissing. In all, the typical context is one in which another animal has given clear submissive signals—as if to say "I recognize your status"— and the aggressive animal responds with an action that appears to have the effect of relieving the submissive animal's tension or apprehension. Thus, chimpanzees show a complex sequence of behavior patterns in the sphere under consideration here: aggression, submission, reassurance. In each of these three categories, a variety of behavior patterns may be utilized. That is, an aggressive orientation may be expressed in various ways; so may a submissive orientation; and so may reassuring reactions. Over all, these categories provide a framework for analyzing transactions in which one chimpanzee menaces another who signals his preference for avoiding a fight and is then signaled by the menacing animal that a fight will not be necessary after all. In this communicative repertoire of aggression-submission-reassurance there are many elements in common with human behavior in similar situations.

Since these transactions are most commonly observed in the context of dominance relations, it now becomes appropriate to consider dominance among chimpanzees. Earlier reports had little to say on this subject, and some plausible conjectures were made suggesting that chimpanzees may not relate to one another in this way. More recent observations at the Gombe greatly change the picture. Dominance relations, at least in this group of chimpanzees, are clear and vivid. This change in emphasis, as compared to early reports from the Gombe, probably mainly reflects the fact that detailed, close-range, longitudinal observations of individually recognized animals have become much richer at the Gombe during the past few years. These newer observations provide a far more adequate data base for analyzing dominance relations. In addition, the partial clearing of one small area of forest, with provision of bananas as a dietary supplement in this area, may have accentuated dominance transactions; the bananas are a highly sought-after food, present here in rather short supply and in spatially concentrated arrangement. However, the relations are basically the same elsewhere in the forest and there is no reason to believe they were created by the banana-feeding situation.

Concretely, dominance is commonly determined by observing who gives way to whom in situations where food is available, and more generally by who gives submissive signals to whom. In this way, it is possible to make the following statements about dominance relations among these chimpanzees: (1) Males are dominant over females. (2) Females are dominant over young. (3) Dominance relations are not strictly linear on an individual basis; they are also influenced by temporary associations, coalitions, and enduring preferences (Hall and DeVore, 1965; Masserman, Wechlin, and Woolf, 1968). These latter are sometimes based on sibling relationships. (4) Aggressive displays by high-ranking males seem to be used mainly as a reinforcement of status vis-à-vis other adult males, and in this respect might well be called "intimidation displays"; attacks and injuries to females and immature animals are not unusual during these displays. (5) All animals in this chimpanzee community of forty-five animals know one another well and respond differentially in terms of dominance status; indeed, van Lawick-Goodall estimates that this individual knowledge is attained by about three years of age (which is especially interesting since they do not reach full maturity until about age ten to twelve). (6) Attributes contributing to high dominance status include not only size and strength but also coordination, motivation, and ingenuity (especially in the use of objects). (7) Chimpanzees, like many other primates, show much evidence of redirected aggression. This refers to a threat situation, often occurring between two animals of similar status, in which one breaks off the transaction and attacks a lower-ranking animal—smaller, weaker, less mobile, or otherwise more vulnerable animal. (8) The phenomena of dominance are particularly vivid and turbulent among adolescent males, especially in relation to dominance changes. (9) Dominance relations may be stable for many months and serious fighting minimal during such periods. But when dominance relations are changing, fighting becomes more frequent and more intense. (10) Mothers protect their offspring in aggressive encounters with other animals. Hence, the mother's dominance status probably has a considerable bearing on the ultimate dominance status of her offspring.

Let us now briefly summarize the conditions under which chimpanzees are likely to behave aggressively (Goodall, 1964; Kawabe, 1966; van Lawick-Goodall, 1968a; Wilson and Wilson, 1968). From van Lawick-Goodall's observations, our own, and those of other workers, it appears that threat or attack patterns are likely to occur in the following contexts: (1) competition over food, especially if highly desirable foods are spatially concentrated or in short supply; (2) defense of an infant by its mother; (3) a contest over the dominance prerogatives of two individuals of similar social rank; (4) redirection of aggression (for example, when a low-ranking male has been attacked by a high-ranking male and immediately turns to attack an individual subordinate to him); (5) a failure of one animal to comply with a signal given by the aggressor (for example, when one chimpanzee does not respond to another's invitation to groom); (6) strange appearance of another chimpanzee (for example, a chimpanzee whose lower extremities became paralyzed); (7) changes in dominance status over time, especially among males; (8) the formation of consort pairs at the peak of estrus (this requires further study, since the consort pairing is quite a recent discovery among wild chimpanzees); (9) when strangers meet; (10) hunting and killing of small animals (infant baboon, infant colobus monkey). Thus, as with baboons, many of the situations eliciting aggressive behavior involve (1) protection or (2) access to valued resources.

Aggressive behavior in higher primates deserves consideration in an adaptive, evolutionary framework (Jay, 1968; Marler, 1968; Tinbergen, 1968; Tokuda and Jensen, 1968; Washburn, 1968; Washburn and Hamburg, 1968). Though we cannot be sure of the ways in which aggressive behavior may have functioned adaptively over the long course of evolution, the following possibilities merit further study: (1) increasing the effectiveness of defense; (2) providing access to valued resources such as food, water, and females in reproductive condition; (3) contributing to effective utilization of the habitat by distributing animals in relation to available resources, a spacing function of intergroup tension; (4) resolving serious disputes within the group; (5) providing a predictable social environment; (6) providing leadership for the group, particularly in dangerous circumstances; and (7) differential reproduction, it being compatible with present knowledge (though by no means proven) that relatively aggressive males are more likely to pass on their genes to subsequent generations than less aggressive males, because of the consort pairing at the peak of estrus in which male aggressiveness plays some role. Some of these factors may have given selective advantage to aggressive primates over millions of years, providing they could manage effective regulation of their aggressive behavior. Such past selective advantage for aggressiveness over long periods would make it likely that the contemporary organism inherits such tendencies. But it would certainly not ensure that this behavior is still advantageous, since environmental conditions have changed very rapidly in recent times (Hamburg, 1964; Hinde, 1970).

A remarkable finding of research on human evolution has been the discovery of distinctly man-like forms at a much earlier date than had previously been believed (Washburn, 1968). The australopithecine form must have stood upright, made simple stone tools according to an established tradition, and hunted small mammals. He lived at least 2 million years ago, probably a good deal more. For hundreds of thousands of years, his way of life seems to have remained much the same, insofar as this can be inferred from the fossil record and prehistoric archeology.

Since his canine teeth were markedly reduced in size as compared to the apes, he must have been relying on tools as weapons for a long time prior to the 2-million year level. The extent to which they used their plentiful stone tools against other australopithecines is unknown.

Bigelow (1969) has recently reviewed a great deal of zoological, archeological, and historical evidence pertaining to the aggressive behavior of early man. The synthesis that emerges from his review and analysis proceeds along the following line: Early man was probably organized in socially cooperative groups, similar to those of present baboons but on a higher level of complexity. Intragroup cooperation was probably very important in their adaptation, but intergroup cooperation was probably minimal. He believes that individuals of each small group tended to regard all other groups as potential threats to their own survival. He emphasizes the crucial nature of complex intragroup cooperation in coping with the predation of large carnivores and with the attacks of other human groups. He believes that those groups having the greatest capability for effective cooperation in attack or defense maintained themselves longest in the most fertile areas, produced more offspring and thereby additional groups. In short, he sees natural selection as favoring the groups that were best organized in attack and defense.

When population densities became extremely high in a fertile area, the necessity arose for some groups to expand outward, often driving out previous inhabitants. He documents large-scale killing that occurred under these conditions and points out that it has not been unusual in human history for displaced groups to be essentially exterminated. Bigelow says, "The essence of this thesis is that the ability to learn cooperation was actually favored by the selective force of warfare."

This suggestion of Bigelow's regarding the eliciting of intergroup hostility in "population hot centers" raises the question of whether the biological nature of early man was in some way susceptible to these environmental conditions. Research workers in the field of animal behavior have studied many environmental conditions that elicit threat and attack patterns. Among the environmental conditions that are most conducive to severe aggression across many species of birds and rodents is the crowding of strangers, especially in the presence of valued resources (such as food, sex, or nesting locations). Is this also true of primates? Are they more likely to fight when they are crowded, when they meet strangers, and when they are in the presence of valued resources? Perhaps the conjunction of these three conditions becomes an especially powerful instigation to aggression.

Our field observations of chimpanzees and baboons are quite consistent with this concept. Our observations support those of other recent workers who have conducted studies in (1) natural habitats, (2) observation and experiment in seminatural settings (for example, large compounds), and (3) laboratory experiments (van Lawick-Goodall, 1968a;

Wilson and Wilson, 1968; Southwick, 1969). One example will serve to illustrate this point. Several scholars have done extensive fieldwork of various groups of Rhesus macaque in India (Southwick, 1964; Southwick, Beg, and Siddigi, 1965; Singh, 1966; Jay, 1968; Southwick, 1969). They find marked contrast in behavior of groups living deep in forest compared with groups living in towns and cities. In general, the forest-dwelling monkeys are much less aggressive. This is probably related to the fact that they grow up in a rich environment in which neither food nor water is difficult to obtain. They have minimal contact with man and are very cautious when they contact him. In contrast, the city-dwelling monkeys live under great pressure from man and from other monkey groups. They must compete with humans for food; they are chased, stoned, and almost continually harassed. They are extremely aggressive, both in relation to other monkey groups and toward man. They show many scars of severe fighting. Some workers have analyzed crowding and stranger-contact experimentally in primates (Southwick, 1967); Singh, 1968, 1969). Elsewhere, I have reviewed this evidence and attempted to relate it to the circumstances of living in cities during historical times (Hamburg, 1970).

It may well be that millions of years of vertebrate and primate evolution have left the human species with many legacies, one of which is a readiness to react fearfully and aggressively toward strangers, especially when crowded in with them, competing for valued resources. If we have any biological tendency in this direction, it must have been accentuated within historical times by the development of living in cities. Whatever else cities may do, they crowd strangers beyond anything known in the past, on several scales: (1) the vast numbers of persons; (2) the mobility that relentlessly brings strangers into each city; (3) and the complexity of living that almost daily brings each of us into contact with many strangers, most of whom we will never see again. Moreover, cities crowd strangers on this unprecedented scale in the presence of many valued resources, often perceived as likely to be in rather short supply: valued objects, places, activities, and persons. These conditions are conducive to interindividual and intergroup conflict.

So, man very likely has a vertebrate-mammalian-primate heritage of aggressive tendencies. In particular, there is a strong tendency to dominance behavior in the line from which man is descended, and a tendency toward redirection of aggression downward. Like many higher primates, man has long been prone to respond with threat and attack in situations involving protection of or access to valued resources, including persons, objects, and activities. Moreover, man has been a predator for a long time, and easily learns to enjoy killing other animals (Lee and DeVore, 1968). In addition, man readily finds fascination in the injury and killing of other human beings, and has a remarkable capacity to justify such behavior (Freeman, 1964). In short, aggressive behavior between man and animals, between man and man, and between groups of men has been easily learned, practiced in play, encouraged by custom, and rewarded by most human societies for many thousands, and perhaps even millions of years.

The view that man carries with him a profound biological and psychosocial heritage of aggressive tendencies may seem pessimistic. I certainly do not mean to suggest the inevitability of mass destruction. Rather, it is my belief that the opportunities for practical control will be greatly enhanced by scientific understanding of man's nature, including his unique evolutionary history. Behavior that has had selective advantage for many millions of years is likely to be a part of the human heritage through both biological and social transmission, though we know very little so far about the mechanisms of such transmission. When environmental conditions change rapidly, it is difficult for organisms to change concomitantly. Genetic change is very slow if considered on the historical time scale; and even cultural change, though potentially much more rapid, is often quite sluggish, as the familiar cultural lag phenomenon attests. Rapid environmental changes have often been associated with extinction of species in the past. Many more species have become extinct than those that survive in the present; and the cardinal circumstance favoring extinction has been rapid environmental change. Yet, these changes have typically been much slower than the change we are experiencing now. I refer here primarily to the changes since the Industrial Revolution, which began a mere two centuries ago—a very short time in evolution—and accelerated during this century. I believe we can best adapt to these unprecedented conditions if we understand in detail the nature of the human organism, the main forces that shaped it through its evolution, and the ways in which recent environmental changes impinge on the very old equipment—genes, brain, and customs —that we bring with us to the new world.

REFERENCES

Altmann, S. A. 1967. *Social Communication among Primates.* Chicago: University of Chicago Press.
Bigelow, R. 1969. *The Dawn Warriors.* Boston: Little, Brown.
Campbell, D. T. 1965. "Ethnocentric and Other Altruistic Motives." In D. Levine, ed., *Nebraska Symposium on Motivation.* Lincoln: University of Nebraska Press. Pp. 283–311.
Carthy, J. D., and Ebling, F. J. eds. 1964. *The Natural History of Aggression.* New York: Academic Press.
DeVore, I. 1965. "Male Dominance and Mating Behavior in Baboons." In F. A. Beach, ed., *Sex and Behavior.* New York: Wiley. Pp. 266–289.
DeVore, I. ed. 1965. *Primate Behavior: Field Studies of Monkeys and Apes.* New York: Holt, Rinehart & Winston.

DeVore, I., and Hall, K. R. L. 1965. "Baboon Ecology." In I. DeVore, ed., *Primate Behavior: Field Studies of Monkeys and Apes.* New York: Holt, Rinehart & Winston. Pp. 20–52.

DeVore, I., and Washburn, S. L. 1963. "Baboon Ecology and Human Evolution." In F. C. Howell and F. Bourliere, eds., *African Ecology and Human Evolution.* Chicago: Aldine. Pp. 301–319.

Freeman, D. 1964. "Human Aggression in Anthropological Perspective." In J. D. Carthy and F. J. Ebling, eds., *The Natural History of Aggression.* New York: Academic Press. Pp. 109–119.

Grinker, R. R., Sr., and Spiegel, J. P. 1945. *Men under Stress.* Philadelphia: Blakiston.

Hall, K. R. L. 1964. "Aggression in Monkey and Ape Societies." In J. D. Carthy and F. J. Ebling, eds., *The Natural History of Aggression.* New York: Academic Press. Pp. 51–64.

Hall, K. R. L., and DeVore, I. 1965. "Baboon Social Behavior." In I. DeVore, ed., *Primate Behavior: Field Studies of Monkeys and Apes.* New York: Holt, Rinehart & Winston. Pp. 53–110.

Hamburg, D. A. 1962. "Relevance of Recent Evolutionary Changes to Human Stress Biology." In S. L. Washburn, ed., *Social Life of Early Man.* Chicago: Aldine. Pp. 278–288.

Hamburg, D. A. 1970. "Society, Stress, and Aggressive Behavior." In L. Levi, ed., *Society, Stress, and Disease.*

Hinde, R. A. 1971. "The Nature and Control of Aggressive Behavior," *International Social Science Journal* 23:48–52.

Jay, P. C. 1968. "Primate Field Studies and Human Evolution." In P. C. Jay, ed., *Primates: Studies in Adaptation and Variability.* New York: Holt, Rinehart & Winston. Pp. 487–503.

Jay, P. C. ed. 1968. *Primates: Studies in Adaptation and Variability.* New York: Holt, Rinehart & Winston.

Kawabe, M. 1966. "One Observed Case of Hunting Behavior among Wild Chimpanzees Living in the Savanna Woodland of Western Tanzania," *Primates* 7:393–396.

Lee, R., and DeVore, I. 1968. *Man the Hunter.* Chicago: Aldine.

Marler, P. 1968. "Aggregation and Dispersal: Two Functions in Primate Communication." In P. C. Jay, ed., *Primates: Studies in Adaptation and Variability.* New York: Holt, Rinehart & Winston. Pp. 420–438.

Masserman, J. H., Wechlin, S., and Woolf, M. 1968. "Alliances and Aggression among Rhesus Monkeys." In J. H. Masserman, ed., *Animal and Human.* New York: Grune & Stratton. Pp. 95–100.

Ransom, T. 1969. Observations of Baboons at the Gombe Stream Reserve. Presentations at Stanford University and University of California, Berkeley.

Ransom, T. 1970. Personal communication. Stanford University.

Reynolds, V., and Reynolds, F. 1965. "Chimpanzees in the Budongo Forest." In I. DeVore, ed., *Primate Behavior: Field Studies of Monkeys and Apes.* New York: Holt, Rinehart & Winston. Pp. 368–424.

Riopelle, A. J., and Rogers, C. M. 1965. "Age Changes in Chimpanzees." In A. M. Schrier, H. F. Harlow, and F. Stollnitz, eds., *Behavior of Nonhuman Primates, Modern Research Trends,* vol. 2. New York: Academic Press. Pp. 449–462.

Rowell, T. E. 1967. "A Quantitative Comparison of the Behaviour of a Wild and a Caged Baboon Group," *Animal Behavior* 15:499–509.

Sarich, V. M. 1968. "The Origin of the Hominids: An Immunological Approach." In S. L. Washburn and P. C. Jay, eds., *Perspectives on Human Evolution,* vol. 1. New York: Holt, Rinehart & Winston. Pp. 94–121.

Schaller, G. B. 1963. *The Mountain Gorilla: Ecology and Behavior.* Chicago: University of Chicago Press.

Schultz, A. H. 1969. *The Life of Primates.* New York: Universe Books.

Singh, S. D. 1966. "The Effects of Human Environment on the Social Behavior of Rhesus Monkeys," *Primates* 7(1):33–40.

Singh, S. D. 1968. "Social Interactions between the Rural and Urban Monkeys, *Macca mulatta,*" *Primates* 9:69–74.

Singh, S. D. 1969. "Urban Monkeys," *Scientific American* 221:108–116.

Southwick, C. H. 1964. "Patterns of Intergroup Social Behavior in Primates, with Special Reference to Rhesus and Howling Monkeys," *Annals of the New York Academy of Science* 102:436–454.

Southwick, C. H. 1967. "An Experimental Study of Intragroup Agonistic Behavior on Rhesus Monkeys (*Macca mulatta*)," *Behaviour* 28:182–209.

Southwick, C. H. 1969. "Aggressive Behaviour of Rhesus Monkeys in Natural and Captive Groups." In S. Garattini and E. B. Sigg, eds., *Aggressive Behaviour*. New York: Wiley. Pp. 32–43.

Southwick, C. H., Beg, M. A., and Siddigi, M. R. 1965. "Rhesus Monkeys in North India." In I. DeVore, ed., *Primate Behavior: Field Studies of Monkeys and Apes*. New York: Holt, Rinehart & Winston. Pp. 111–159.

Tinbergen, N. 1968. "On War and Peace in Animals and Man: An Ethologist's Approach to the Biology of Aggression," *Science* 160:1411–1418.

Tokuda, K., and Jensen, G. D. 1968. "The Leader's Role in Controlling Aggressive Behavior in a Monkey Group," *Primates* 9:319–322.

van Lawick-Goodall, J. 1964. "Tool-Using and Aimed Throwing in a Community of Free-Living Chimpanzees." *Nature* 201:1264–1266.

van Lawick-Goodall, J. 1965. "Chimpanzees of the Gombe Stream Reserve." In I. DeVore, ed., *Primate Behavior: Field Studies of Monkeys and Apes*. New York: Holt, Rinehart & Winston. Pp. 425–473.

van Lawick-Goodall, J. 1967. "Mother-Offspring Relationships in Free-Ranging Chimpanzees." In D. Morris, ed., *Primate Ethology*. Chicago: Aldine. Pp. 287–346.

van Lawick-Goodall, J. 1968a. "The Behaviour of Free-Living Chimpanzees in the Gombe Stream Reserve." In J. M. Cullen and C. G. Beer, eds., *Animal Behaviour Monographs*. London: Baillière, Tindall & Cassell. Pp. 165–296.

van Lawick-Goodall, J. 1968b. "A Preliminary Report on Expressive Movements and Communication in the Gombe Stream Chimpanzees." In P. C. Jay, ed., *Primates: Studies in Adaptation and Variability*. New York: Holt, Rinehart & Winston. Pp. 313–374.

Washburn, S. L. 1968. "Behavior and the Origin of Man." Huxley Memorial Lecture. In *Proceedings of the Royal Anthropological Institute of Great Britain and Ireland for 1967*. Royal Anthropological Institute of Great Britain. Pp. 21–27.

Washburn, S. L., and Hamburg, D. A. 1968. "Aggressive Behavior in Old World Monkeys and Apes." In P. C. Jay, ed., *Primates: Studies in Adaptation and Variability*. New York: Holt, Rinehart & Winston. Pp. 458–478.

Wilson, W. L., and Wilson, C. C. 1968. *Aggressive Interactions of Captive Chimpanzees Living in a Semi-Free Ranging Environment*, 6571st Aeromedical Research Laboratory Technical Report, no. ARL-TR-68-9. New Mexico: Holloman Air Force Base.

Zimmerman, R. R., and Torrey, C. C. 1965. "Ontogeny of Learning." In A. M. Schrier, H. F. Harlow, and F. Stollnitz, eds., *Behavior of Nonhuman Primates, Modern Research Trends*, vol. 2. New York: Academic Press. Pp. 405–448.

The What, How, and Why
of Psychosomatic Medicine

The concept that psychic and somatic processes are related is not a new one. Grinker (1953) has pointed out that Holland (1852) commented, "Human physiology comprises the reciprocal actions and relations of mental and bodily phenomena, as they make up the totality of life. . . . Scarcely can we have a morbid affection of body in which some feeling or function of mind is not concurrently engaged—directly or indirectly —as cause or as effect." During the past three decades many clinicians and investigators have devoted their major effort to psychosomatic medicine. Several societies have been organized with a primary interest in the psychosomatic field. Psychosomatic research has been published in journals specifically concerned with the area, such as *Psychosomatic Medicine* and the *Journal of Psychosomatic Research*, as well as in a variety of other journals and books. In spite of a considerable amount of clinical and investigative endeavors, the general field of psychosomatic medicine has not made the advances expected of it. The increased activity in psychosomatic medicine in the past twenty-five to thirty years stirred hope for greater understanding of the interrelationship between mind and body. Grinker (1969a) recently has written, "A psychosomatic breakthrough created optimism that we would find the emotional components of physical diseases and the specificity so defined would enable us to deal with pinpoint surgical skill with the causes of specific chronic degenerative diseases. *Unfortunately, this did not prove to be true.*"[1] The question must be asked why the original optimism and hopes were not fulfilled. The difficulty may well be as stated by Grinker (1969b), "the problem of all science—to ask the right questions . . . most of us have not known the significant questions and then how to get the answers."

[1] My italics—M.S.

A strategy for asking the right questions and obtaining the significant answers may lie in an approach outlined by Grinker (1969b) in "An Essay on Schizophrenia and Science." It is suggested that our understanding and knowledge may be advanced by asking what, how, and why questions. This chapter will follow Grinker's proposal and consider the what, how, and why of psychosomatic medicine.

What is psychosomatic medicine? Psychosomatic has a variety of meanings, which may *lead* to many misconceptions and misunderstandings. A definition of the term "psychosomatic medicine" is necessary before proceeding to the how and why types of question. For many psychosomatic medicine has meant a group of somatic disorders in which psychosocial processes are believed to play an important role in the etiology and course of the illness. Seven diseases have been considered to be psychosomatic: bronchial asthma, rheumatoid arthritis, ulcerative colitis, essential hypertension, neurodermatitis, thyrotoxicosis, and peptic ulcer (Alexander, French, and Pollock, 1968). Psychosomatic has also been used to refer to the relationship between emotional and biochemical or physiological responses. In the "Introductory Statement" by the editors of the first volume of *Psychosomatic Medicine* in 1939, it was stated, "Psychosomatic medicine is not a medical specialty: it designates a method of approach to problems of etiology and therapy rather than a delineation of the area." Grinker (1953) expanded on this definition and has written, " 'Psychosomatic' connotes more than a kind of illness; it is a comprehensive approach to the totality of an integrated process of transactions among many systems: somatic, psychic, social, and cultural." This definition encompasses both specific disorders as well as psychophysiologic and psychochemical phenomena in both health and illness. Grinker's definition essentially will be utilized in the present discussion of psychosomatic medicine.

In addition to defining psychosomatic medicine, it seems important and critical to pay attention to what is being studied or treated. A considerable amount of psychosomatic research has not specifically defined the disorder or response that has been investigated. In some studies, for example, there has been a lack of differentiation between lability of blood pressure and essential hypertension with a sustained elevation of the blood pressure. The need for precise definition of what is being studied is clearly illustrated in a consideration of the investigation of the psychosomatic aspects of bronchial asthma.

The primary pulmonary change in asthmatic breathing is bronchiolar obstruction, which produces the signs and symptoms (Whittenberger, 1951). Recently a variety of techniques have been developed that permit quantitative evaluation of bronchiolar function. These procedures are based on the principles of air flow in the lungs and the mechanisms of breathing. The movement of air in and out of the lungs and airways is related to the elasticity of the lung tissues and the flow-resistive properties of the lungs. The flow-resistive factor can be further divided into the resistive forces due to air flow and those due to tissue resistance. Brody and DuBois (1956) have reported that the effect of the lung-tissue resistance is small and unlikely to be altered. The major factors to be considered, therefore, in the mechanics of breathing are the elastic resistance and the resistance to air flow through the tracheobronchial tree. The elastic distensibility of the lungs can be measured by the ratio of the volume change of air to the change in intrapleural pressure at instants of no air flow. This ratio is known as compliance. As the tissues become less elastic a greater pressure is required to move the same volume, and there is a decrease in compliance. The flow-resistive component is measured by the ratio of the pressure change to the associated change in rate of flow of air, and this measure is referred to as airway resistance. As resistance to the flow of air increases, more pressure is required to maintain the same rate of flow. Bronchiolar obstruction is characterized by an increase in airway resistance and decrease in compliance (D'Silva and Lewis, 1961). These changes in airway resistance and compliance have been reported in human asthma by McIlroy and Marshall (1956) and in experimental asthma in the guinea pig by Stein, Schiavi, Ottenberg, and Hamilton (1961).

The need for a precise pathophysiological definition of asthma is evident from a review of studies of the experimental induction of respiratory changes in animals following exposure to a variety of psychological situations. Liddell (1951) has reported that respiratory dysfunction is an invariable manifestation of chronic experimental neurosis in animals and often resembles the labored breathing of bronchial asthma. Gantt (1941) produced in an experimentally neurotic dog "loud raucous breathing with quick inspiration and labored expiration accompanied by loud wheezing." Masserman and Pechtel (1953) reported that several monkeys exposed to an experimental conflictual situation exhibited "severe asthmatic attacks lasting for hours." It has been observed by Seitz (1959) that a respiratory wheezing condition develops in cats exposed to a feeding frustration test.

In each of these studies, the respiratory changes consisting of prolonged expiration and/or wheezing could be interpreted as a component of a pain-fear reaction pattern and not as bronchial asthma. Changes in respiratory rate, amplitude, and pattern have been observed in response to painful or fear-producing stimuli (Upton, 1929; Weber, 1930; Horton, 1933). It has not been demonstrated, however, that the respiratory responses following such stimuli are accompanied by bronchiolar obstruction, the definitive pulmonary change in asthma.

Schiavi, Stein, and Sethi (1961) have demonstrated that it is possible to produce in the guinea pig, in response to a pain-fear stimulus, an asthmatic-like respiratory pattern (shortened inspiration and prolonged expiration) without evidence of bronchiolar obstruction as reflected in the mechanical properties of the lungs. This was in contrast to the finding of an identical respiratory pattern which was accompanied by an increase in airway resistance and decrease in compliance in experimental allergic asthma in the guinea pig. Schiavi, Stein, and Sethi found that the asthmatic-like respiratory pattern following the pain-fear stimulus was related to the screeching of the animals. It was not observed when screeching was suppressed following tracheotomy.

Humans also may develop an asthmatic-like respiratory pattern during hyperventilation without concomitant changes in bronchiolar physiology, as observed in a study by Luparello (1971). An eighteen-year-old man presented himself as an experimental subject with a six-year history of asthma. He was given nebulized physiological saline to inhale but was told that it was an allergen to which he had previously reported sensitivity. Initially his airway resistance increased slightly and then decreased as he developed a classical hyperventilation syndrome including carpal spasm. He stated he was having an asthmatic attack. Ausculation of the chest revealed clear breathing sounds with no wheezing and the mechanical properties of the lungs as measured by body plethysmography revealed no evidence of airway obstruction. In fact, the airway resistance decreased, which is the opposite of what would have occurred had this been an attack of asthma rather than hyperventilation.

The above findings indicate that it is important to define asthma in terms of bronchiolar function as well as the usual ventilatory variables. Such respiratory criteria will permit precise evaluation of psychosocial factors that may play a role in bronchial asthma. What is being investigated must be defined before proceeding to how or why a response or disorder developed or is maintained.

Much of the literature in the general field of psychosomatic medicine has tended to consider simultaneously the how and why types of question. Theories about the etiology of specific psychosomatic disorders frequently confuse or combine mechanism and genesis. Many theoretical models have been presented and only a brief review will be outlined in this essay. More thorough and comprehensive discussions of psychosomatic theories have been made by Grinker (1953), Mendelson, Hirsch, and Webber (1956) as well as others.

An early approach to the understanding of psychosomatic medicine was to consider psychosomatic symptoms as conversion phenomena. This conceptualization has not been accepted as a meaningful model for the explanation of psychosomatic disorders or processes. Dunbar was one of the first to challenge the conversion theory and introduced a new era in the field of psychosomatic medicine. In 1935 she published *Emotions and Bodily Changes,* which reviewed and synthesized the literature available on the relationship between emotions and somatic functions. She also described specific personality profiles for a given disorder such as peptic ulcer, asthma, coronary artery disease, and so on. As pointed out by Mendelson et al. (1956), further research has led investigators in the field to question the validity of Dunbar's theory.

The next major contribution to a psychosomatic conceptual model was made by Franz Alexander (1936). He did not accept the notion that there was a relationship between personality types and specific diseases. Alexander and his colleagues, utilizing psychoanalysis as an investigative tool, proposed that a specific conflict was characteristic of each psychosomatic disorder. In addition, Alexander's conceptualization of psychosomatic processes included the hypothesis that each conflict had a specific physiological accompaniment. This differed from the conversion model of psychosomatic medicine in which the symptom was viewed as the symbolic expression of a conflict. Alexander believed that the physiological vegetative responses were associated with the specific emotional state or conflict. Just as with Dunbar's theory, there has been increasing skepticism about Alexander's conceptualization of psychosomatic disorders as more empirical data have been collected.

Wolff (1950) proposed another model to explain psychosomatic phenomena. He essentially theorized that the body reacts to stress with biological adaptive or protective responses which in part are determined by heredity. In addition to the biological reaction to stress, Wolff suggested that the bodily changes were accompanied by a specific set of feelings and attitudes. Much emphasis was placed by Wolff on the reaction to life situations without considering the idiosyncratic meaning of a given situation to an individual.

Some investigators (Szasz, 1952; Margolin, 1953), including Grinker (1953), extended the concept of regression, as defined psychoanalytically, to regressive physiological processes as a means of understanding psychosomatic phenomena. In this conceptual model of psychosomatic disorders, it was proposed that there is both a psychological and physiological regression. Szasz (1952) has suggested that the specific bodily disorder is a result of "regressive innervation." The limitations and inconsistencies of the regressive model have been throughly discussed by Mendelson et al. (1956).

In the majority of psychosomatic theories presented to date there has been little regard for the variety of transactional processes and multiple intervening variables that may be involved. The emphasis has been too heavily weighted with psychogenic considerations. This is well illustrated in a consideration of the now classic psychodynamic formulation for bronchial asthma. French and Alexander (1941) reported a study of twenty-seven asthmatic patients undergoing psychoanalytic treatment and described the central emotional problem in asthma as a fear of separation from the mother. They concluded that the asthmatic attack is equivalent to a repressed cry for the lost mother. In *Psychosomatic Specificity*, published as recently as 1968, it is stated that "The most specific feature of the patients (asthmatic), however, is their conflict about crying. . . . Asthmatic attacks can be best understood as an inhibition of the use of the expiratory act for communication, either by crying or confession." As already pointed out, expiration is not by itself a valid means of defining bronchial asthma. In the above described psychodynamic formulation of bronchial asthma, there is no mention of bronchiolar obstruction. Furthermore, there is also some question about an asthmatic attack being equivalent to a repressed cry. Experimental respiratory studies with newborns reveal that with crying there is an increase not a decrease in airway resistance (Karlberg, 1957). In the psychodynamic consideration of asthma little or no attention is paid to the transition from an abstract concept such as separation from the mother or a repressed cry to the specific pathophysiological change of asthma —bronchiolar obstruction.

It is clear from this brief discussion of the psychodynamic formulation of bronchial asthma that basically the how and why types of question have not been dealt with successfully. This may be due, as suggested by Grinker (1953) almost twenty years ago, to "a lack of critical challenge to the existing theories. . . . As a result, psychosomatic formulations have become stereotypes into which each patient's life history and situation is molded by special focusing, selective interpretation, and omission or neglect of the incongruent." In part the lack of a scientific approach appears to be related to "a tremendous jump from feelings to organ dysfunction" (Grinker, 1953). This is not to deny the importance of a thorough understanding of the psychic component or the collection of data in the therapeutic setting. It has been well demonstrated, for example, that in the consideration of an emotional situation as a precipitating factor of a physiological response it is necessary to go beyond the mere description of the situation and investigate the emotional life of the individual (Stein, 1970). At times a psychodynamic formulation offers the most appropriate and applicable conceptual model. However, as stated by Mirsky (1957), psychoanalysis and other psychological procedures provide information about the sources and effects of stress. "These methods, however, yield no data concerning the mechanisms involved in the production of physiological concomitants."

Recently, there has been more emphasis in psychosomatic medicine on a consideration of how psychic, social, cultural, and somatic processes may be involved in a complex series of integrated transactional processes. Some past and current research in relation to bronchial asthma will be cited to illustrate that experimental approaches are changing and attempts are being made to deal specifically with the how type of question.

The effect of psychological stimuli on the precipitation of asthma attacks in a laboratory setting has been evaluated sporadically over the years. MacKenzie (1886) noted bronchospasm in a patient allergic to roses when he presented her an artificial rose. Dekker and Groen (1956) exposed asthmatic subjects to "meaningful emotional stimuli" and were able to measure a decrease in vital capacity in some of the subjects. Each stimulus in that study was specific for a particular subject and represented historical events in the individual's disease process. For example, one subject reported developing asthma attacks at the sight of goldfish in a bowl. When shown an artificial representation of this by the experimenters, the subject developed bronchospasm. Another individual, who reported dust as a substance that would lead to an asthmatic attack, reacted with bronchospasm when presented with a sealed glass container filled with dust. These experiments have demonstrated that certain asthmatics are sensitive to perceptual cues that are capable of affecting bronchomotor tone. However, the heterogeneity of the stimuli employed was such that it was not possible to make meaningful comparisons within the group of subjects. In addition, the respiratory end point chosen was an indirect and relatively insensitive measure of airway obstruction.

Recently, Luparello, Lyons, Bleecker, and McFadden (1968) evaluated the effect of psychological stimuli on the bronchomotor tone of asthmatic subjects in a setting in which accurate, rapid, and reproducible measurements of airway resistance could be made. All subjects were led to believe that they were inhaling irritants or allergens that cause bronchoconstriction. The actual substance used in all instances was nebulized physiologic saline solution. Nineteen of forty asthmatics reacted to the experimental situation with a significant increase in airway resistance. Twelve of the asthmatic subjects developed full-blown attacks of bronchospasm which was reversed with a saline placebo.

It was shown further that this phenomenon could be blocked by atropine (McFadden, Luparello, Lyons, and Bleecker, 1969), indicating that the effect of the psychological stimuli on airway reactivity was mediated through efferent cholinergic (vagal) pathways. These findings are in keeping with the observations of Simonsson, Jacobs, and Nadel (1967) who have shown that stimulation of subepithelial receptors will trigger reflex airway constriction, which is eliminated by atropine blockade of the efferent limb of the reflex arc. Wan and Stein (1968) have shown that direct vagal stimulation in intact unanesthetized animals produces bronchoconstriction as evidenced by increased airway resistance and decreased compliance. The above studies not only support the notion that emotional phenomena play a role in asthma, but provide some data on how psychological processes may have a direct effect on the central nervous system control of airway reactivity independent of allergic mechanisms.

The need for a transactional approach in psychosomatic medicine is clearly shown by a further consideration of bronchial asthma as an example. The most thoroughly studied mechanism in bronchial asthma has been hypersensitivity to allergens, which has been demonstrated experimentally as well as clinically in humans and animals (Herxheimer, 1953; Stein et al., 1961; Colen, Van Arsdel, Pasnick, and Horan, 1964). Psychosocial processes may play a role in bronchial asthma by modifying the immunologic or allergic processes involved in some cases. There have been a number of experimental studies demonstrating the effect of psychosocial processes on immunologic responses, and these have been reviewed by Stein, Schiavi, and Luparello (1969). For example, Rasmussen and his coworkers have demonstrated that mice subjected to a standardized avoidance conditioning stress show an increased susceptibility to herpes virus (Rasmussen, Marsh, and Brill, 1957) as well as decreased susceptibility to passive anaphylaxis (Rasmussen, Spencer, and Marsh, 1959). It has also been reported (Vessey, 1964) that mice housed in groups have a significantly lower level of circulating antibodies than mice housed individually.

Recently, the neurophysiological mechanisms that might mediate the psychosocial influences on immunological reactions have been experimentally studied. There is a growing body of literature indicating that central nervous system processes may modify hypersensitive mechanisms. The effect of midbrain lesions on the course of anaphylaxis in the guinea pig has been investigated by Freedman and Fenichel (1958). Bilateral electrolytic lesions at the level of the superior collicolus involving the reticulothalamic tracts and deep tegmental nuclei inhibited anaphylactic death. Szentivanyi and Filipp (1958) studied the role of the hypothalamus on anaphylaxis and demonstrated that lethal anaphylaxis in the guinea pig and rabbit can be prevented by bilateral lesions in the tuberal region of the hypothalamus. Lesions in the tuberal region have also been reported to result in lower circulating antibody levels in the guinea pig (Filipp and Szentivanyi, 1958). Luparello, Stein, and Park (1964) investigated the effect of hypothalamic lesions on rat anaphylaxis and found that anterior but not posterior lesions inhibited development of lethal anaphylaxis. Recently, Macris, Schiavi, Camerino, and Stein (1971) reported protection against lethal anaphylaxis, low antibody titers, and diminished delayed cutaneous reactions in guinea pigs with lesions in the anterior hypothalamus. Lesions in the median or posterior hypothalamus did not alter these immune responses.

The degree of protection afforded by the anterior hypothalamic lesions cannot be explained solely on the basis of lower antibody titers. Guinea pigs passively sensitized with homologous as well as heterologous antibody have also been protected by hypothalamic lesions (Filipp and Szentivanyi). It is possible that hypothalamic lesions, in addition to their effect on circulating antibodies, may modify the reactivity of the organism to histamine and other substances liberated during the antigen-antibody reaction. Several studies have reported that the central nervous system modified the susceptibility of guinea pigs to exogenous histamine. Przbylski (1962) found that removal of the region of the quadrigeminal bodies or localized damage to the midbrain reticular formation (Przbylski, 1969) resulted in a decrease in histamine susceptibility when administered either intravenously or by inhalation of an aerosol. Removal of the cerebral cortex did not modify the reactivity of the animals. Szentivanyi and Szekely (1958) demonstrated a decrease in histamine toxicity following lesions in the tuberal region of the hypothalamus in guinea pigs. Schiavi, Adams, and Stein (1966) studied the effect of bilateral electrolytic lesions in the anterior and posterior hypothalamus in guinea pigs on histamine toxicity as measured by dose mortality curves and the LD_{50}. The animals with lesions in the anterior hypothalamus were afforded significant protection against histamine toxicity. These observations suggest that the antianaphylactic effects observed with midbrain and hypothalamic lesions may be related, in part, to a decrease in the susceptibility of the guinea pig to the endogenously liberated histamine during the anaphylactic reaction. Przbylski (1969) recently studied the mechanisms that mediate the influence of the midbrain reticular formation on histamine susceptibility. He found that electric stimulation of the dorsocaudal parts of the midbrain reticular formation resulted in parasympathetic activation and bronchoconstriction potentiating the constricting effects of histamine. The increased tolerance to histamine observed in guinea pigs with lesions in the dorsocaudal part of the midbrain reticular formation was explained by a decrease in the physiological tone of the airways leading to a diminished bronchoconstricting effect of histamine. Though it has been reported that anterior hypothalamic lesions are associated with decreased parasympathetic activity, there is little experimental evidence that anterior lesions modify the physiological reactivity of the bronchial tree to bronchoconstricting agents. Recently, Szentivanyi (1968) proposed that the excessive bronchial irritability observed in bronchial asthma is the result of a diminished responsiveness of the beta adrenergic receptors present in the bronchial tree. According to the beta adrenergic theory, the lesions may create an autonomic imbalance that alters the responses of the adrenergic receptor cells to the pharmacological mediators of anaphylaxis.

The influence of the central nervous system on immune mechanisms may also be due to changes in neuroendocrine function induced by the destruction of specific hypothalamic structures. It has been shown that the anterior hypothalamus is involved in the regulation of the secretion of thyroid stimulating hormone by the hypophysis (Florsheim, 1958; De-Jong and Moll, 1965). Electrolytic lesions in this area induce low plasma levels of thyroid stimulating hormone and decreased thyroid function. A number of investigators have demonstrated in the rat and guinea pig a relationship between thyroid physiology and immune processes. It has been noted that the resistance to the anaphylactic reaction is increased in thyroidectomized rats (Leger and Masson, 1947). Similar findings were observed by Nilzen (1954, 1955a, 1955b) in the guinea pig following thyroidectomy or administration of I^{131}. Suppression of thyroid activity inhibits local and systemic anaphylaxis, abolishes circulating precipitins, and decreases the susceptibility of the animals to exogenous histamine. Filipp and Mess (1969) have recently demonstrated that treatment of horse-serum sensitized guinea pigs with thyroxine after lesions in the tuberal region of the hypothalamus partially restored anaphylactic reactivity on antigen challenge.

The role of the hypothalamic pituitary adrenal axis must also be considered since exogenous adrenal steroids have been found to have an inhibitory effect on antibody formation in guinea pigs, to protect mice and rats against anaphylactic shock, and to diminish tuberculin skin sensitivity (Rose, 1959). It does not seem probable, however, that changes in adrenal steroids mediate the protective effect of anterior lesions, since lesions in the anterior hypothalamus are associated with a decrease in blood corticoid levels (Porter, 1963). Moreover, adrenal hormones do not modify significantly the susceptibility of guinea pigs to anaphylaxis and to exogenous histamine (Rose, 1959).

In reviewing the mechanisms that may be involved in bronchial asthma, it is evident that there is no single specific factor. It has been suggested that psychosocial processes may play a role in the development and course of bronchial asthma by means of two major central nervous system pathways: (1) a direct effect on the central nervous system control of airway caliber; (2) modification of immunologic or allergic processes. These two processes may occur simultaneously and probably reinforce each other. A variety of intervening variables are more than likely also involved (see Figure 4–1). In addition, consideration must be given to how psychosocial influences are transmitted to the hypothalamus. Papez (1937), Grinker (1939), and MacLean (1949) have discussed the relationship and anatomical pathways between the cortex and hypothalamus. Weiner (1964) has reviewed some of the recent studies concerned with the relationship between brain and behavior. He points out that there is currently a shift in research from gross anatomical correlations with psychological processes to the study of neuronal correlates of behavior. Cellular neurophysiological techniques may provide further understanding of the mechanisms involved in the relationship between behavior and the central nervous system. The role of psychological, social, central nervous system, respiratory, hypersensitive, and other processes must be considered in the etiology and course of asthma. Such a psychosomatic approach will permit a better understanding of asthma and lead to more meaningful research and effective treatment. The model described for asthma is only an example, and similar approaches can and are being applied to other psychosomatic disorders.

This discussion has reviewed the what and how of psychosomatic disorders and processes. Perhaps the why type of questions cannot be

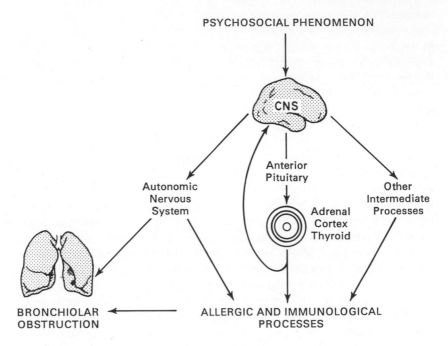

Figure 4–1. Schematic representation of the effects of psychosocial phenomena on the course and development of asthma by means of the central nervous system.

meaningfully approached until more knowledge is available about the how or mediating mechanisms. There has been, however, some attention paid to the why of psychosomatic medicine, and specifically to predisposing factors. Though for the most part speculative, particular emphasis has been placed on organ susceptibility in the conceptualization of why psychosomatic phenomena develop. For example, Adler (1924) proposed that somatic symptoms develop in neurotic patients at a point of constitutional organ inferiority. Deutsch (1922), rather than focusing on a hereditary factor, suggested that the organ neuroses occurred at the site of a previous disorder. An example of Deutsch's model would be the development of asthma in an individual who had pertussis as a child. In Alexander's view of the development of psychosomatic disorders, he stressed the constitutional or x factor (Alexander et al., 1968). Though specific psychodynamic forces were proposed to be characteristic of specific disease entities, the psychological factors do not cause the disease except when a specific organ vulnerability or x factor is present.

The research of Mirsky (1957) and his colleagues in relation to duodenal ulcer lends some support to Alexander's hypothesis. It is suggested that the organic x factor in duodenal ulcer is the hypersecretion of pepsinogen, and the psychodynamic processes center about a conflict over dependency needs. In a specific social situation an individual who is a hypersecretor and has the described psychological factor will develop an ulcer. A short term of study, reported by Weiner, Thaler, Reiser, and Mirsky (1957), was able to predict the development of duodenal ulcer in specific individuals on the basis of physiological, psychological, and social data. It is also of interest that, in addition to the hypersecretion of pepsinogen proposed as a necessary factor for the development of an ulcer, it may also play a role in the development of the required psychological processes. Replication of this important study is required, and further research is also needed to determine why the pepsinogen level is elevated, as well as why the proposed specific psychodynamic processes occur.

It has been pointed out in this chapter that a complex variety of intervening or mediating mechanisms may be involved in psychosomatic disorders or processes. Why a specific response develops may be due to predisposing factors that are genetically determined, or the result of physical trauma or early psychosocial experiences. A predisposition may occur at a variety of levels and multiple factors appear to be involved. A given predisposition to a specific psychosomatic response may be necessary but not sufficient for the response or disorder to develop unless other required psychosocial or physiological or chemical processes are present.

As cited earlier in this chapter, Grinker (1953) has stated that psychosomatic "is a comprehensive approach to the totality of an integrated process of transactions among many systems: somatic, psychic, social and cultural." In *Psychosomatic Research* Grinker (1953) comments further that psychosomatic

deals with a living process that is born, matures, and develops through differentiation and successive stages of new forms of integration of parts and other wholes. It deals with stresses, strains, and adjustments, with acute emergency mechanisms, disintegrations, and chronic defensive states or disease. In fact, "psychosomatic" refers not to physiology or pathophysiology, not to psychology or psychopathology, but to a concept of process among all living systems and their social and cultural elaborations.

Experimental studies using this transactional approach have begun to provide some answers to the what, the how, and the why of psychosomatic medicine. The research strategy currently being followed in psychosomatic medicine is changing from a primary and almost exclusive emphasis on psychogenic considerations to a more inclusive conceptual model, which considers not only the psychosocial but the vast variety of intervening and mediating mechanisms involved. Future research utilizing the what, how, and why strategy will continue to enhance our understanding of psychosomatic processes and disorders.

REFERENCES

Adler, A. 1924. *The Practice and Theory of Individual Psychology*. New York: Harcourt, Brace.

Alexander, F. 1936. *Psychosomatic Medicine*. New York: Norton.

Alexander, F., French, T. M., and Pollock, G. H. 1968. *Psychosomatic Specificity*. Vol. 1. *Experimental Study and Results*. Chicago: University of Chicago Press.

Brody, A. W., and DuBois, A. B. 1956. "Determination of Tissue, Airway and Total Resistance to Respiration in Cats," *Journal of Applied Physiology* 9:213.

Colen, J., Van Arsdel, P. P., Pasnick, L. J., and Horan, H. 1964. "Observations on the Experimental Reproduction of Asthma with Mold Antigens," *Journal of Allergy* 35:331.

DeJong, W., and Moll, J. 1965. "Differential Effects of Hypothalamic Lesions on Pituitary-Thyroid Activity in the Rat," *Acta Endocrinology* 48:522.

Dekker, E., and Groen, J. 1956. "Reproducible Psychogenic Attacks of Asthma," *Journal of Psychosomatic Research* 1:58.

Deutsch, F. 1922. "Der Gesunde und der kranke Körper in psychoanalytischer Betrachtung," *International Zeitschrift für Psychoanalyse* 8:290.

D'Silva, S. L., and Lewis, A. F. 1961. "The Measurement of Bronchoconstriction in Vivo," *Journal of Physiology* 157:611.

Dunbar, H. F. 1935. *Emotions and Bodily Changes*. New York: Columbia University Press.

Editors. 1939. "Introductory Statement," *Psychosomatic Medicine* 1:3.

Filipp, G., and Mess, B. 1969. "Role of the Thyroid Hormone System in Suppression of Anaphylaxis Due to Electrolytic Lesion of the Tuberal Region of the Hypothalamus," *Annals of Allergy* 27:500.

Filipp, G., and Szentivanyi, A. 1958. "Anaphylaxis and the Nervous System: Part III," *Annals of Allergy* 16:306.

Florsheim, N. H. 1958. "The Effect of Anterior Hypothalamic Lesions on Thyroid Function and Goiter Development in the Rat," *Endocrinology* 62:783.

Freedman, D. X., and Fenichel, G. 1958. "Effect of Midbrain Lesions in Experimental Allergy," *Archives of Neurology and Psychiatry* 79:164.

French, T. M., and Alexander, F. 1941. "Psychogenic Factors in Bronchial Asthma," *Psychosomatic Medicine*, monograph 4.

Gantt, W. H. 1941. "Experimental Basis for Neurotic Behaviour," *Psychosomatic Medicine*, monograph 3, nos. 3 and 4.

Grinker, R. R., Sr. 1939. "Hypothalamic Functions in Psychosomatic Interrelations," *Psychosomatic Medicine* 1:19.

Grinker, R. R., Sr. 1953. *Psychosomatic Research*. New York: Norton.

Grinker, R. R., Sr. 1969a. "An Editor's Farewell," *Archives of General Psychiatry* 21:641–646.

Grinker, R. R., Sr. 1969b. "An Essay on Schizophrenia and Science," *Archives of General Psychiatry* 20:1–25.

Herxheimer, H. 1953. "Induced Asthma in Humans," *International Archives of Allergy* 3:192.

Holland, H. 1852. *Mental Physiology*. London: Longmans, Brown, Green & Longmans.

Horton, G. P. 1933. "A Quantitative Study of Hearing in Guinea Pigs," *Journal of Comparative Psychology* 15:59.

Karlberg, P. J. E. 1957. "Breathing and Its Control in Premature Infants." In *Physiology of Prematurity*, Transactions of the Second Conference, Josiah Macy Foundation.

Leger, J., and Masson, G. 1947. "Factors Influencing an Anaphylactic Reaction in the Rat," *Federation Proceedings* 6:150.

Liddell, H. 1951. "The Influence of Experimental Neuroses on Respiratory Function."

In H. A. Abramson, ed., *Treatment of Asthma*. Baltimore, Md.: Williams & Wilkins. Pp. 126–147.

Luparello, T. J. 1971. Personal communication.

Luparello, T. J., Lyons, H. A., Bleecker, E. R., and McFadden, E. R., Jr. 1968. "Influences of Suggestion on Airway Reactivity in Asthmatic Subjects," *Psychosomatic Medicine* 30:819.

Luparello, T. J., Stein, M., and Park, C. D. 1964. "Effect of Hypothalamic Lesions on Rat Anaphylaxis," *American Journal of Physiology* 207:911.

McFadden, E. R., Jr., Luparello, T. J., Lyons, H. A., and Bleecker, E. 1969. "The Mechanism of Action of Suggestion in the Induction of Acute Asthma Attacks," *Psychosomatic Medicine* 31:134.

McIlroy, M. B., and Marshall, R. 1956. "The Mechanical Properties of the Lungs in Asthma," *Clinical Science* 15:345.

MacKenzie, J. N. 1886. "The Production of 'Rose Asthma' by an Artificial Rose," *American Journal of Medical Science* 91:45.

MacLean, P. D. 1949. "Psychosomatic Disease and the 'Visceral Brain,' " *Psychosomatic Medicine* 11:338.

Macris, N. T., Schiavi, R. C., Camerino, M. S., and Stein, M. 1971. "The Effect of Hypothalamic Lesions on Immune Processes in the Guinea Pig," *American Journal of Physiology*.

Margolin, S. G. 1953. "Genetic and Dynamic Psychophysiological Determinants of Pathophysiological Processes." In F. Deutsch, ed., *The Psychosomatic Concept in Psychoanalysis*. New York: International Universities Press.

Masserman, J. H., and Pechtel, C. 1953. "Neurosis in Monkeys: A Preliminary Report of Experimental Observations," *Annals of the N.Y. Academy of Science* 56:253.

Mendelson, M., Hirsch, S., and Webber, C. S. 1956. "A Critical Examination of Some Recent Theoretical Models in Psychosomatic Medicine," *Psychosomatic Medicine* 18:363.

Mirsky, I. A. 1957. "The Psychosomatic Approach to the Etiology of Clinical Disorders," *Psychosomatic Medicine* 119:424.

Nilzen, A. 1954. "The Influence of the Thyroid Gland on Hypersensitivity Reactions in Animals: I," *Acta Allergica* (Copenhagen) 7:231.

Nilzen, A. 1955a. "The Influence of the Thyroid Gland on Hypersensitivity Reactions in Animals: II," *Acta Allergica* (Copenhagen) 8:57.

Nilzen, A. 1955b. "The Influence of the Thyroid Gland on Hypersensitivity Reactions in Animals: III," *Acta Allergica* (Copenhagen) 8:103.

Papez, J. W. 1937. "A Proposed Mechanism of Emotion," *Archives of Neurology and Psychiatry* 38:725.

Porter, J. C. 1963. "Secretion of Corticosterone in Rats with Anterior Hypothalamic Lesions," *American Journal of Physiology* 204:715.

Przybylski, A. 1962. "Effect of the Removal of the Cortex Cerebri and the Quadrigeminal Bodies Region on Histamine Susceptibility in Guinea Pigs," *Acta Physiol. Pol.* 13:535.

Przybylski, A. 1969. "Effect of Stimulation and Coagulation of the Midbrain Reticular Formation on the Bronchial Musculature. A Modification of Histamine Susceptibility," *Journal of Neuro-Visceral Relations* 31:171.

Rasmussen, A. F., Marsh, J. T., and Brill, N. Q. 1957. "Increased Susceptibility to Herpes Simplex in Mice Subjected to Avoidance-Learning or Restraint," *Proceedings of the Society for Experimental Biological Medicine* 96:183.

Rasmussen, A. F., Spencer, E. S., and Marsh, J. T. 1959. "Decrease in Susceptibility of Mice to Passive Anaphylaxis Following Avoidance-Learning Stress," *Proceedings of the Society for Experimental Biological Medicine* 100:878.

Rose, B. 1959. "Hormone and Allergic Responses." In J. H. Shaffer, G. A. LoGrippo, and M. W. Chase, eds., *International Symposium on Mechanisms of Hypersensitivity*. London: Churchill.

Schiavi, R. C., Adams, J., and Stein, M. 1966. "Effect of Hypothalamic Lesions on Histamine Toxicity in the Guinea Pig," *American Journal of Physiology* 211:1269.

Schiavi, R., Stein, M., and Sethi, B. B. 1961. "Respiratory Variables in Response to a

Pain-Fear Stimulus and in Experimental Asthma," *Psychosomatic Medicine* 23:485.

Seitz, P. F. 1959. "Infantile Experience and Adult Behavior in Animal Subjects," *Psychosomatic Medicine* 21:353.

Simonsson, B. G., Jacobs, F. M., and Nadel, J. A. 1967. "Role of Autonomic Nervous System and the Cough Reflex in the Increased Responsiveness of Airways in Patients with Obstructive Airway Disease," *Journal of Clinical Investment* 46:1812.

Stein, M. 1970. "Psychiatrist's Role in Psychiatric Research," *Archives of General Psychiatry* 22:481.

Stein, M., Schiavi, R. C., and Luparello, T. J. 1969. "The Hypothalamus and Immune Process," *Annals of the N.Y. Academy of Science* 164:464.

Stein, M., Schiavi, R. C., Ottenberg, P., and Hamilton, C. L. 1961. "The Mechanical Properties of the Lungs in Experimental Asthma in the Guinea Pig," *Journal of Allergy* 32:8.

Szasz, T. S. 1952. "Psychoanalysis and the Autonomic Nervous System," *Psychoanalytic Review* 39:115.

Szentivanyi, A. 1968. "The Beta Adrenergic Theory of the Atopic Abnormality in Bronchial Asthma," *Journal of Allergy* 42:203.

Szentivanyi, A., and Filipp, G. 1958. "Anaphylaxis and the Nervous System: Part II," *Annals of Allergy* 16:143.

Szentivanyi, A., and Szekely, J. 1958. "Anaphylaxis and the Nervous System: Part IV," *Annals of Allergy* 16:389.

Upton, M. 1929. "The Auditory Sensitivity of the Guinea Pig," *American Journal of Psychology* 41:42.

Vessey, S. H. 1964. "Effects of Grouping on Levels of Circulating Antibodies in Mice," *Proceedings of the Society for Experimental Biological Medicine* 115:252.

Wan, W. C., and Stein, M. 1968. "Effect of Vagal Stimulation on the Mechanical Properties of the Lungs in Guinea Pigs," *Psychosomatic Medicine* 30:846.

Weber, E. G. 1930. "The Upper Limit of Hearing in the Cat," *Journal of Comparative Psychology* 10:221.

Weiner, H. 1969. "Some Recent Neurophysiological Contributions to the Problem of Brain and Behavior," *Psychosomatic Medicine* 31:457.

Weiner, H., Thaler, M., Reiser, M. F., and Mirsky, I. A. 1957. "Etiology of Duodenal Ulcer: I. Relation of Specific Psychological Characteristics to Rate of Gastric Secretion (Serum Pepsinogen)," *Psychosomatic Medicine* 19:1.

Whittenberger, J. L. 1951. "Lung Volume and Airflow Characteristics." In H. A. Abramson, ed., *Treatment of Asthma*. Baltimore, Md.: Williams & Wilkins.

Wolff, H. G. 1950. "Life Stress and Bodily Disease: A Formulation," *Proceedings of the Association for Research in Nervous and Mental Disease* 29:1059.

The Royal Road to Anxiety: A Critical Review

I

Grinker's studies of anxiety are particularly fitting for review. Not only is anxiety a subject of the most central importance for psychiatry but it is one that has concerned him closely for many years.

To suggest that it is the importance of the problem that dictated his interest, however, would be an oversimplification. The choice of a subject for study by any scientist necessarily arises out of his individual background. If one seeks to achieve a meaningful understanding of an investigator's work, knowledge of that background is essential. Hence, perspective on Grinker's approach to the study of anxiety can best be achieved by taking note of his intellectual and philosophical roots. Three major separate but intertwined themes can be discerned in his career: the neurological scientist, the psychoanalyst, the psychotherapist.

THE NEUROLOGICAL SCIENTIST

Though Grinker is well known as a clinical neurologist through his textbook and early papers, this initial medical career role is apt to be misleading. Primarily, he always has been more a neurobiologist (or, biobehavioral scientist) than a clinical neurologist. His initial interest in the structure of the nervous system and its pathology soon shifted to a concern with its function, including its regulatory role in the bodily economy (Grinker, 1969, pp. 2–3). More influential than his European teachers were his colleagues and mentors in the group of neural scientists at the University of Chicago. His heritage stems from Coghill,

C. J. Herrick, and Percival Bailey, and from Sherrington and Hughlings
Jackson. From these sources derives his concern for elucidating the
basic principles of human behavior and homeostatic balance within the
perspective of the evolution of the structure-function of the nervous system. Thus, here one can see the origins and background of his application of field or general systems theory to human behavior and, specifically relevant to us, of the consideration of anxiety as a system, to be
discussed below.

THE PSYCHOANALYST

Anxiety was, of course, a pivotal concept for Freud. Indeed, anxiety
plays a nuclear role in almost every theory of personality. It is not sufficiently appreciated, however, that it was Freud who first provided a systematic consideration of the role of anxiety as part of a psychological
theory of human behavior. Moreover, Freud noted that anxiety stands in
a crucial position in psychosomatic relationships because of the physiological changes that are its concomitants. As a psychoanalyst, therefore,
Grinker's concentration on this topic is understandable. But the particular focus of his interest stems less from Freud, I believe, than from the
influence of Franz Alexander. As Grinker (1964) himself describes it:
"Alexander's pioneering psychosomatic studies opened for me a persistent interest in a field in which my neurological background could fuse
and steady my psychiatric present." And it is clear that Grinker's focus
on anxiety is as a psychosomatic phenomenon.

THE PSYCHOTHERAPIST

Here, again, the use of a specialty designation may be misleading. For
what is relevant is Grinker as a compassionate physician. (In this sense
his background as a clinical neurologist is relevant.) In whatever direction his interests have taken him, one clear theme in all his work is an
orientation toward obtaining useful information about processes responsible for "disease." Resonating with his personal attributes is his activist
therapeutic orientation. It is no surprise that his first major study of anxiety and stress occurred in the setting of a pressing concern about understanding and treating disease: war neuroses. At the heart of his interest in psychosomatic processes, including anxiety, is the wish to further
an understanding of psychosomatic disease. I make this point not in an
effort at praise but because of the insight it provides into his work.
Though much of what he has written is conceptual, the primary thrust
of these theoretical efforts is one of heuristic clarification directed toward realistic problem-solving via research. His orientation is intensely
pragmatic and rooted strongly in empiricism. The studies of anxiety are
certainly no exception.

II

Given the fact that anxiety is a major factor in psychological function, the reasons for its study seem self-evident. However, it is clear that Grinker has had certain additional purposes in mind. One of these is implicit in the title of his major theoretical paper in this area, "Anxiety as a Significant Variable for a Unified Theory of Human Behavior" (1959). In this (Grinker, 1959, p. 538), he makes clear his view that unified theory can best be achieved by "recognizing that there are some processes for which *general commonality* or *'systems'* properties may be traced from their biological roots to their social flowering," and by studying such a process "as it becomes a part of various systems of organizations and observ [ing] its relationship to the other parts of a system and to its whole. Finally, we may study how it is involved in functions of each system in relation to other systems." Obviously, he regards anxiety as such a process.

A careful reading of this paper, however, suggests that inversion of its title makes equal sense, that is, that it also uses unified theory as a way of casting fresh light on the phenomenon of anxiety itself. The overarching concepts of field theory are applied to develop a heuristic model for the practical scientific study of anxiety. Indeed, he has made clear that he views the verification of the utility of this model through its application to studies of anxiety as another of his major purposes in this research. For example, he states (Grinker, Korchin, Basowitz, Hamburg, Sabshin, Persky, Chevalier, and Board, 1956, p. 421): "In part, the present extensive research program was undertaken because we felt there was a need for a methodological model based on tenable theory, applicable to general problems of psychosomatic research." The empirical orientation is evident.

A detailed consideration of the model itself can be found elsewhere in this volume. It is useful, however, to emphasize a few features. The first follows from the fact that variables in multiple systems are to be studied. This requires a complex experimental design and a team approach. The necessity to measure these variables demands the expertise that only specialists in the applicable fields can bring. Thus, the team must be multidisciplinary, including not merely diverse types of professionals but specifically those who fall into each of the major levels of analysis involved, that is, biological, psychological, and social. It also requires that appropriate, specific methods be developed for objective quantification of each of the variables; and hence that each be operationally defined (Oken, 1965). The model also dictates the choice of variables. If these are to be representative of the central processes at each of the subsystem levels studied, they must be as close as possible to direct measures of these processes. Further, to understand relationships among systems that can be expected to have different temporal parameters, repeated simultaneous measures must be taken and special techniques applied to elicit the larger patterns of response.

It is important to clarify the two separate ways in which Grinker uses the term "anxiety." In the narrower sense, within the studies, he has applied a definition restricted to a conscious reportable affective state. Consistent with the point just made, this definition is an operational one. (Thus, for example, it excludes the construct "unconscious anxiety.") And it is limited entirely to the psychological subsystem, thus permitting study of its interrelationship with other subsystem variables. In the second, wider, sense, "anxiety" is used to refer to the entire system, including all its levels or subsystems, among them the affective, that is, anxiety in the narrower sense. As long as one keeps these distinct (and in his writings his referent is always clear from the context) no confusion need arise. Obviously, it is the broader system that is of primary interest to us here.

Finally, it is important to explicate a major premise that underlies application of the model. This is the assumption that conditions of emergency function provide a special view of processes central to organismic function. By heightening or disrupting these processes, an exploded view is obtained of relationships that, in the normal, are silent. This premise, of course, underlies much medical research, and its general correctness is affirmed by its excellent record of payoff. Whether it applies to any given instance, however, is problematical until empirical evidence of its specific validity can be established. The likelihood that it will apply to other abnormal situations is especially high, provided that the disturbed state has a specific relation to the abnormal conditions of interest. There are good grounds for assuming this relationship between anxiety and psychosomatic disorders, and the view has gained wide acceptance. This is the basis for the study of stress.

Up to this point, I have eschewed use of the term "stress" to prevent confusion. But it is essential now to clarify how the concept of stress relates to that of anxiety. To do so, it is best to go back to Selye's (1950) definition that stress represents "the sum of all the *non*-specific systemic reactions of the body"[1] to a simulus. Put into the terms of general systems theory, this could be rephrased as "the system of common processes set in motion in a variety of subsystems as the result of a stimulus which strains any given subsystem beyond its capacity to respond with processes located entirely within that subsystem." One can readily see that Grinker's definition of the system anxiety is essentially identical, that is, the two terms are "homomorphic" (Aikin, 1961). Within this anxiety/stress system, then, the overt affect of anxiety is the sign reveal-

---

[1] Italics mine—D. O.

ing that the psychological subsystem has been so strained, that is, it cannot deal with a stimulus solely through its intrasystemic (psychological) adaptive mechanisms.

III

In examining this theoretical model of anxiety it is important to recognize what it is not, as well as what it is. It is not really a psychological theory of anxiety. It provides no new psychological insight. With minor modifications, its psychological properties are merely those put forward by Freud in *The Problem of Anxiety* (1936). (Grinker does tend toward a monistic conception, in which signal and traumatic anxiety are on a single continuum, rather than Freud's dualism. But this modification, similar to that elaborated by Rangell [1968], is a minor one.) Though anchored in psychoanalytic theory, its specific feature is that it is a psychosomatic (or a psychosomatosocial) theory. Indeed, it is the only theory of anxiety that can be genuinely so classified. In making this distinction, I refer not only to its concern with processes at all three levels of organization, but also to the fact that it does not address itself to the issue of the genesis of anxiety. To the extent that other psychological theories[2] turn out to have greater validity than the psychoanalytic (though, for the most part, they are far from being mutually exclusive), only one aspect of Grinker's view will be diminished in value; and much of it could be integrated with several of these theories.

The theory of Schachter (1966) comes perhaps closest to Grinker's in its breadth, since it combines elements of social psychology and psychophysiology. This contemporary derivative of the James-Lange theory starts from the view that all emotions have a common physiological pattern of "arousal" (See below.), and postulates that this generic state is labeled as specific affect on the basis of cognitive cues defining the situation in which it arises. Its flaws (Lykken, 1967; Plutchik and Ax, 1967; Shapiro and Crider, 1969) aside, this theory cannot be put into the same category as Grinker's model. Despite its theoretical sweep, the focus within each area is narrow: Its physiological concerns are restricted to nervous system arousal, and its psychological elements to limited determinants of cognition. Though Schachter's experiments are ingeniously designed, data are limited to (well-chosen) measures of subjective affect. Finally, this theory neither attends to nor provides any basis for understanding psychosomatic processes, normal or pathological.

[2] For an excellent overview of the major current theories of anxiety the reader is referred to the book edited by Spielberger (1966). Though this provides no useful synthesis, each of the separate theoretical positions is described by its proponents; these descriptions are presented with clarity and effectiveness.

Despite its not being a psychological theory, one might hope that the research that Grinker's theory spawned might cast light on fundamental issues in the theory of affects. One such is the specificity question: Are the processes involved in states of anger, guilt, shame, and so on different from one another and from anxiety? Or is there but one generic unidimensional state of organismic arousal (Lindsley, 1951; Malmo, 1966) in which separate feeling states differ only in their cognitive properties, as Schachter has argued? Consonant with psychoanalytic theory as well as clinical experience, the Michael Reese group was strongly inclined toward the latter view. The early paratrooper study (Basowitz, Persky, Korchin, and Grinker, 1955) was encouraging in this regard, suggesting that there might be subclass differences in the pattern of responses even within anxiety, depending on whether its focus was harm or failure. Two later studies (Persky, Grinker, Hamburg, Sabshin, Korchin, Herz, Board, and Heath, 1958; Korchin and Herz, 1960) suggested that disintegrative anxiety was associated with greater adrenocortical responses than was anxiety of equal intensity that did not have this quality. Moreover, it was possible consistently to define and rate a variety of specific affect states separately (Hamburg, Sabshin, Board, Grinker, Korchin, Basowitz, Heath, and Persky, 1958; Oken, 1960). Considerable data also were added to the growing body of information that differentially patterned ("fractionated" [Lacey]) autonomic responses can occur. Among the determinants of response demonstrated in the studies were stimulus characteristics (Engel, 1960; Oken, Grinker, Heath, Herz, Korchin, Sabshin, and Schwartz, 1962), stable individual response propensities (Engel, 1960; Oken et al., 1962; Goldstein, Grinker, Heath, Oken, and Shipman, 1964), the tendency to suppress emotional expression (Oken, 1960; Oken et al., 1962), and the stability of the defensive organization (Oken, Grinker, Heath, Sabshin, and Schwartz, 1960); and there are others (for example, Lacey and Lacey, 1958; Lacey, Kagan, Lacey, and Moss, 1963). Such data raise grave doubts about the general applicability, if not the validity, of the arousal concept (Lacey, 1967). However, the data have been disappointing in demonstrating affect-specific patterning of autonomic responses. For example, the level of overall emotional response proved to correlate with adrenocortical response at least as well as did anxiety (Persky, et al., 1958).

This failure to demonstrate affect specificity may relate not to the validity of the concept but to the experimental design. Human beings rarely respond to stressors with pure affect states; nor did the research subjects. In the presence of complex admixtures of feelings, compounded by other factors such as those just noted, which enter in as determinants of response and which differ among subjects, eliciting a physiological pattern for any one affect may be impossible. The solution of the specificity question must await a better methodological approach that provides control for these interfering factors. Until then the issue remains unresolved.

The existence of this problem points up a deficiency in affect theory generally. There is simply no satisfactory comprehensive taxonomy of affects, though some efforts have been made in this direction (for example, Tomkins, 1962, 1963). This gap is particularly troublesome in psychoanalysis, as a dynamic, epigenetic theory of personality, where it is essential that the data and conceptualizations developed for the affects separately be integrated in hierarchical fashion into a general theory. (See Rapaport, 1953; Novey, 1961; Engel, 1963; Lofgren, 1968.) Along this line, it is interesting to consider the possible relationship between primal anxiety, existing prior to ego organization, and nonspecific arousal and to inquire into the processes by which affect differentiation becomes organized. Grinker has postulated that at higher, traumatic levels of anxiety, progressive dedifferentiation and disintegration supervene, producing a state of global excitement approaching that of early infancy. As subsystems become, in turn, unable to handle the loads placed on them, additional subsystems are involved: The boundaries among them disappear. This, of course, is linked to his aforementioned strategy of using stress states to uncover system relationships. Likewise it is central in his theory of psychosomatic disease (Grinker, 1953), which provides, as I have suggested, a major basis for his interest in research on anxiety.

IV

The primarily heuristic nature of Grinker's theoretical model has been pointed out. The value of such a view lies in setting directions for research more than in its powers of explanation. Correspondingly, its measure lies in the fruitfulness of the research it engenders more than in whether such studies validate the theory. The valuable substantive contributions of the anxiety studies are summarized elsewhere in this volume. (See Appendix.) Here, I would like to pursue two other matters. One has to do with the applicability of the model to the research, that is, the extent to which it has proved actually possible to implement the model; the other relates to the methodological contributions of the studies.

Though one can appropriately describe the studies of anxiety as psychosomatic research, the model itself purports to be more general, that is, psychosomatosocial. Closer scrutiny of the social aspects of the theory reveals, however, that this aspect is, by far, its least well developed. It includes only brief, vague statements about rumor formation and spread in social groups as a product of social unrest. But this is recognized by Grinker himself as merely an analogy in the social system to the signal of anxiety in the psychological. It is not like adrenocorticotrophic hormone secretion, somatically "of the same order of events" (Grinker, 1959, p. 541). There is no consideration of more relevant social science concepts that might have provided the basis for inclusion of useful experimental variables, for example, social status, role, values. Under these circumstances, and the corresponding persistent absence of a social scientist in the research group, it comes as no surprise that no sociological variables were included in the studies. One can only hazard a guess as to the basis for this deficiency. That Grinker has had no social science background does not seem a sufficient explanation. Perhaps the answer lies in the problems posed by complexities of interdisciplinary research and of the team that must carry it out (Ruesch, 1956; Luszki, 1958). It is difficult enough to manage these problems when several disciplines representing two major systems areas are involved. To add specialists and concepts from a third—and to retain a genuine unity to the approach, rather than mere parallelism—may be more than we can now achieve (Ruesch, 1961). The model does pose a great demand. It is true that psychosomatic studies that included social variables have been successful (for example, Reiser, Reeves, and Armington, 1955; Back, Oelske, Brehm, Bogdonoff, and Nowlin, 1970). But these have been few in number as well as far simpler in their number of variables and participants as well as in their design. In truth, a major program of psychosomatosocial research has never been carried out.

The degree to which variables can be chosen to provide an index of central processes is limited by technical and practical factors as well as the level of conceptual understanding of the subsystem under scrutiny. Though anxiety ratings would seem to be a measure of one central process, the same cannot be said for blood or urine levels of adrenocorticosteroids, nor measures of heart rate, blood pressure, galvanic skin response, and so on. Regarding the former, production rates would be more significant, and indirect measures of these began to be added in the course of the studies (Persky, 1957). But even though direct measures have since become available, this is not a truly central measure: For this subsystem, one would want measures of pituitary adrenocorticotrophic hormone output, for which no useful method is yet available. In this sense, the studies failed to live up to the demands of the model. Yet much is to be learned from peripheral measures. As a matter of fact, if the goal is increased leverage on psychosomatic disorders, these measures may be more relevant. Though peripheral responses may be merely the "final common pathway" of a complex of processes, the behavior of, for example, the heart is highly significant in understanding the pathogenesis of cardiac disease. What we learned in the studies that contributed to an understanding of the patterning of autonomic responses constitutes a real step in achieving this type of understanding.

The need for studying relationships over time has posed further problems. Though reasonable success was achieved in obtaining multiple measures simultaneously, the more formidable issue has been how these can be related. Correlational and multivariate (for example, factor analytic) techniques were used to relate variables at a given point in time. These methods could be modified to deal with systems with differing lags in responding by offsetting measures in time in the analyses, though the number of potential combinations when dealing with repeated measures of multiple variables can be astronomical, requiring a computer. But this deals with only one aspect of the relation of dynamic variables. How does one relate the pattern of response among variables. We do not yet know how to express dynamic sequences of response in mathematical terms for even a single variable. How then is one to examine transactions among groups of such dynamic responses? The future of the systems approach rests on the solution of this exceedingly difficult problem.

The second point for consideration in evaluating the studies lies in the area of its methodological contributions. There were, of course, a large number of new procedural, technical, and statistical methods developed which have proved of great value. But there is one that stands out above the others. This was the development of specific rating scales as an operational method for quantifying subjective psychological phenomena. The need for this approach is now so clear, and the value of the rating scales developed so obvious, that it is difficult to add anything to this simple statement. Moreover, the success of these has provided a stimulus for the development of analogous scales by a variety of subsequent researchers, so that one tends now to take their development for granted. Yet, I believe that if there is one methodological advance that has paved the way for psychiatric research to move from its primitive beginnings to achieve a solid scientific foundation, it is this one. Grinker's was neither the first nor the only research to take this step. But his was an influential one in demonstrating the feasibility of developing such scales for measuring a phenomenon of obvious importance and in providing a model of how to go about doing so, as well as in establishing their practical utility.

A weakness in one of the scales [3] that was developed for rating defenses, which in retrospect seems obvious, is instructive to examine. The defense scales did prove reliable, but they did not make a major contribution to the study in which they were used (Oken et al., 1962). The scale we had anticipated as most useful, defense intensity-primitivity, combined the effort involved in maintaining a defense with its nature,

[3] A deficiency for which I must take responsibility. A more detailed description of the scale to be discussed can be found in Oken (1962).

categorized along a continuum defined by traditional psychoanalytic concepts. The error in this lay, I believe, in failing to be consistent with our own theoretical model. From the standpoint of the system anxiety, internal subsystem characteristics are not relevant per se. The nature of defenses fits in this category. What is relevant are the signs of strain portending a breakdown of boundaries between subsystems. From the standpoint of defense, evidences of increased effort, captured by the intensity component of the scale, is such a sign. The ineffectiveness of defense in blocking anxiety or other dysphoric affect would be another such sign. This latter dimension was used efficaciously by other researchers, who found, as one might expect, that it correlated with adrenocortical stress response (Wolff, Friedman, Hofer, and Mason, 1964).

V

Anxiety continues to be a topic of major concern in the literature. Spielberger (1966) has documented the progressive, almost exponential, growth of reports in this area since 1950. However, this seeming progress requires a critical look. Though numerous articles on the subject are listed in the *Index Medicus,* many are concerned with psychopharmacology, some with the role of anxiety in one or another clinical syndrome, and a few with theoretical issues. But almost none represent bona fide research per se. Consistent with this, the citations in the *Psychological Abstracts* contain few listings of work done by psychiatrists; and recent volumes in the annual psychiatric yearbook on "progress in" reveal but a handful of indexed references to anxiety, save for psychopharmacological studies.

The psychological literature is full of research reports. Last year, in his review of personality research, Adelson (1969) noted that "Anxiety was the most popular single topic." Yet there are questions about this work. Even if one restricts the view to human studies, one gets a sense of triviality and irrelevancy. Only a small part of it provides any genuine insight into anxiety as a major phenomenon in integrated human behavior. Part of the problem lies in designs based on comparison of subjects preselected as having high and low anxiety. It is essential to make a distinction between such trait anxiety, as a stable dimension of personality, and the affect state, a point that has been clarified by the work of both Cattell (1966) and Spielberger, Lushene and McAdoo (1970). Moreover, much of the selection for trait anxiety is based on instruments such as the Taylor Manifest Anxiety Scale, a measure that has been demonstrated to correlate with defensive style (Golin, Herron, Lakota, and Reineck, 1967; Kimble et al., 1967), depression (Crumpton, Grayson and Keith-Lee, 1967), and other variables. As S. B. Sarason (1966) has put it, "the verbal response to our scales may be telling us more about the self than about the affect." Most of the studies that do seem to be focused on the affect state concentrate on its relationship with but one or two personality or social psychological variables. Often these are of a nature that, to the clinician, seem less than crucial, for example, the learning of nonsense syllables, achievement motivation, or conformity. (Of course, such work may make a useful contribution to psychological theory and to such fields as education.) Where physiological variables are included, they tend to be used naïvely as indices of anxiety, as if these were clearly correlated with the affect state, which we know to be false (Shapiro and Crider, 1969). And at times the researchers use simple stimuli which lack real-life properties or assume that anxiety is produced because a plausible anxiety stimulus has been introduced.

Perhaps the one area in which significant progress has been made is that of psychological stress. An excellent recent book (Appley and Trumbull, 1967) contains the proceedings of a conference on this topic, in which the contributors (including, among them, two former colleagues of Grinker) report studies using approaches ranging from the biological to the anthropological. Though, individually, these studies have far less breadth and complexity than Grinker's, they do provide much new data and useful experimental models.

One of the contributors, Lazarus, recently has published a book of his own (Lazarus, 1966) on psychological stress that provides a theoretical approach as well as reporting on a series of studies carried out by him and his collaborators. This research comes closer, perhaps, than any other to that of the Michael Reese group in careful attention to both psychological and physiological phenomena and their covariation over time. As with Grinker, social variables are missing. But biological measures are fewer, and restricted to autonomic indices. The emphasis also differs. Lazarus is concerned largely with cognitive mechanisms: the processing of internal and external information in the appraisal of threat. This appraisal process is the focus of the theoretical development. Though anxiety is considered as one aspect of appraisal and coping processes, it is not considered to play a universal nor an essential role. That role is taken over by the concept of threat, defined in terms of the prospect of future harm, the latter being the product of the thwarting of a motive. Hence, this concept comes close to the way we might define anxiety. One difference—for me a troubling one—lies in the absence of impelling motivational properties of this construct in contrast with anxiety. Moreover, because of the cognitive orientation, Lazarus unfortunately pays little attention to quantifying anxiety or other affect states in his research. On the whole, however, this book—the best work on the topic in some time—is thoughtful, instructive, and presents a broad, integrated, and fair summary of work done in the field. It makes a significant contribution especially to our understanding of the cognitive, defensive, and coping processes in stress at the psychological level, though because it is thus circumscribed, it cannot be considered truly comparable to Grinker's work.

Like Lazarus, Weiss et al. (1968) used a film to stimulate anxiety. Their design provided for comparison of subjects preselected as being high and low in sensitivity to the stressful film theme: death and illness. The resulting cognitive defect occurred only in those subjects sensitive to the theme, thus affirming Grinker's oft-made point that a stimulus cannot be assumed as stressful, but becomes so only if it is meaningful to the specific subject.

That not all stress research is of high quality is attested to by the recent study of life in an undersea environment, Sealab (Radloff and Helmreich, 1968). Though the design was restricted by the characteristics of the setting, the inconsequentiality of the findings in a situation of so great promise is almost astounding. In a negative way, perhaps, this points out how important is the use of a carefully prepared plan based on a meaningful conceptual model, as in Grinker's work.

Likely the most interesting recent work is that of Epstein (1967), in part because the stress situation involved was sports parachuting, so similar to the early Michael Reese paratrooper study (Basowitz, Persky, Korchin, and Grinker, 1955). Like that project, the studies are well designed, using several operational measures of anxiety as well as examining physiological (but again, no social) variables and following these sequentially over time, though like Lazarus there is considerable focus on cognitive variables (learning to notice and evaluate cues, planning for action, and so on). This work adds a further dimension in studying the adaptations that occur over extended time, comparing the responses of the novice to those of the expert. The novices showed a steep rise in fear and physiological arousal peaking at or just before the jump. With increasing experience, the pattern shifted to an inverted "V," whose peak (and onset) occurred progressively earlier, and there was a secondary rise after the jump. The novices, moreover, were noted to use drastic all or none type defenses and to have a tendency to respond to more diffuse cues, that is, those not very relevant to the dangers, though they also failed sometimes to pay sufficient heed to relevant cues. Since the time dynamics of the physiological indices differed from one another as well as from the fear (though with the same pattern) a concept of defense in depth is postulated. This is consistent with Grinker's consideration of hierarchies of response, but is a less-developed concept that fails to consider the relationships among the subsystems. The postjump rise is of particular interest since the same intriguing end phenomenon was found in the paratrooper study.

This research was summarized in an important recent article by Epstein (1967), who uses it as a basis for a unified theory of anxiety and derives what he considers to be a basic biological principle. The law of excitatory modulation states that a gradient of inhibition develops in response to a stimulus that, though it occurs after the excitation gradient, is steeper. The end phenomenon is interpreted as the result of the parallel greater steepness of the falloff of inhibition following termination of the stimulus. This highly compressed summary hardly does justice to the richness of the theory nor of the breadth of the supporting argument (the sources include Pavlov and Freud among others). It is, however, a theory of arousal rather than of anxiety. For example, anxiety is defined only as undirected arousal following perception of danger (in contrast to fear in which there is an avoidance motive and object); and there is no clear definition of danger. Moreover, the theory has a mechanistic quality. Epstein does not, for example, consider the end phenomenon in terms of the release of defenses occurring with the perception that it is safe to be anxious once the danger is over and mobilization of resources is no longer necessary. Still, this is a major contribution to the field.

VI

To evaluate Grinker's anxiety studies in fuller perspective, one must examine what has happened in the broad field of psychosomatic research. The comments of three recent presidents of the American Psychosomatic Society provide a guide. Greene (1968) has noted that the history of psychosomatic research can be divided into three phases: the anecdotal, the more controlled systematic clinical study, and the (present) phase of focused experimental investigations. This last phase is characterized by a notable improvement in the quality of methodology: precise definition and measurement of variables, technical refinements, careful attention to experimental controls, the application of statistical checks, and so on. With this, there has been a shift away from global questions of etiology to focus on explicating the precise mechanisms by which changes occur. By and large, this has been salutory. But it has not occurred without a price. Such research is inclined to focus on narrow, if not microscopic, portions of the field. As Greene described it, this has produced "bits of information which, pasteurized by the computer, have yielded results that are as clean and neat as they are sterile and lifeless." It is not just that more is known about less and less. The very validity of that knowledge is called into question since it is based on the study of isolated functions. It is doubtful that these data portray processes occurring within the complexity of the organism with its fields of transactional relationships among multiple systems and subsystems. As Reiser (1961) has indicated, the clinical and experimental data on human psychosomatic functioning "resist interpretation by linear stimulus-response models of the simple or multiple factor type." (In the same paper, Reiser also provides some excellent examples of the ways in which social-psychological variables may be determinants of somatic responses.) Similarly, Mason (1968), discussing the scope of psychoendocrine research, remarks critically that "there has been a tendency to focus upon a *single* endocrine system at a time" and goes on to note that "Several lines of evidence appeared to indicate that the study of endocrine regulation may uniquely require an approach which considers the *full scope* of activity of the many interdependent endocrine systems simultaneously."

It seems evident that these statements represent what Grinker has been saying for years, and what his studies of anxiety have attempted to achieve. The issue, of course, is not so simple. Precision in experimentation is essential, and it is far more difficult to achieve in research based on the systems model. Moreover, it is not easy to extract useful information from the complex data the model yields. These limitations are a consequence of the state of the art, that is, of the lack of development of operations and, especially, concepts for handling transacting multivariate data. We have a long, long way to go in this development. We also have far to go in developing technical procedures for obtaining measures of variables that are truly central for each subsystem. In this sense, one might characterize the anxiety studies as premature. Certainly they would have been better had they been postponed to a time when the state of the art was more refined. But to say this is to forget how science advances. Methodological progress does not occur de novo. It is out of studies such as these that the art becomes refined. The anxiety studies have provided both a model and a stimulus for the methodologic advances that the field has needed. This is a signal achievement. To leave it even at this, however, is to underestimate their full value. Their tangible yield should not be forgotten. From them has come substantial progress in our understanding of the psychophysiological and psychoendocrine processes involved in anxiety. From them also have come a variety of specific methods for eliciting, measuring, and examining psychosomatic data.

REFERENCES

Adelson, J. 1969. "Personality," *Annual Review of Psychology* 20:217–252.
Aikin, L. R. 1961. "Anxiety and Stress as Homomorphisms," *Psychological Record.* 11:365–372.
Appley, M. H., and Trumbull, R., eds. 1967. *Psychological Stress.* New York: Appleton-Century-Crofts.
Back, K. W., Oelske, S. R., Brehm, M. L., Bogdonoff, M. D., and Nowlin, G. B. 1970. "Physiological and Situational Factors in Psychopharmacological Experiments," *Psychophysiology* 6:749–760.

Basowitz, H. A., Persky, H., Korchin, S. J., and Grinker, R. R., Sr. 1955. *Anxiety and Stress.* New York: McGraw-Hill.

Cattell, R. B. 1966. "Anxiety and Motivation: Theory and Crucial Experiments." In C. D. Spielberger, ed., *Anxiety and Behavior.* New York: Academic Press. Pp. 23–62.

Crumpton, E., Grayson, I. M., and Keith-Lee, P. 1967. "What Kinds of Anxiety Does the Taylor MA Measure?," *Journal of Consulting Psychology* 31:324–326.

Engel, B. T. 1960. "Stimulus-Response and Individual-Response Specificity," *Archives of General Psychiatry* 2:305–313.

Engel, G. L. 1963. "Toward a Classification of Affects." In P. H. Knapp, ed., *Expression of the Emotions in Man.* New York: International Universities Press.

Epstein, S. 1967. "Toward a Unified Theory of Anxiety." In B. A. Maher, ed., *Progress in Experimental Personality Research,* vol. 4. New York: Academic Press. Pp. 1–89.

Freud, S. 1936. *The Problem of Anxiety.* New York: Norton.

Goldstein, J. B., Grinker, R. R., Sr., Heath, H. A., Oken, D., and Shipman, W. G. 1964. "Study in Psychophysiology of Muscle Tension: I. Response Specificity," *Archives of General Psychiatry* 11:322–330.

Golin, S., Herron, E. W., Lakota, R., and Reineck, L. 1967. "Factor Analytic Study of the Manifest Anxiety, Extraversion, and Repression-Sensitization Scales," *Journal of Consulting Psychology* 31:564–569.

Greene, W. A. 1968. "The Fallacy of Misplaced Concreteness," *Psychosomatic Medicine* 30:873–880.

Grinker, R. R., Sr. 1953. *Psychosomatic Research.* New York: Norton.

Grinker, R. R., Sr. 1959. "Anxiety as a Significant Variable for a Unified Theory of Human Behavior," *Archives of General Psychiatry* 1:537–546.

Grinker, R. R., Sr. 1964. "Psychoanalytic Theory and Psychosomatic Research." In J. Marmoston and E. Stainbrook, eds., *Psychoanalysis and the Human Situation.* New York: Vantage Press.

Grinker, R. R., Sr. 1969. "An Essay on Schizophrenia and Science," *Archives of General Psychiatry* 20:1–24.

Grinker, R. R., Sr., Korchin, S. J., Basowitz, H., Hamburg, D. A., Sabshin, M., Persky, H., Chevalier, J. A., and Board, F. 1956. "A Theoretical and Experimental Approach to Problems of Anxiety," *Archives of Neurology and Psychiatry* 76:420–431.

Hamburg, D., Sabshin, M., Board, F. A., Grinker, R. R., Sr., Korchin, S. J., Basowitz, H., Heath, H., and Persky, H. 1958. "Classification and Rating of Emotional Experiences," *Archives of Neurology and Psychiatry* 79:415–426.

Kimble, G. A., et al. 1967. "Anxiety?," *Journal of Personality and Social Psychology* 7:108–110.

Korchin, S. J., and Herz, M. 1960. "The Differential Effects of 'Shame' and Disintegrative Threats on Emotional and Adrenocortical Functioning," *Archives of General Psychiatry* 2:640–651.

Lacey, J. I. 1967. "Somatic Response Patterning and Stress: Some Revisions of Activation Theory." In M. H. Appley and R. Trumbull, eds., *Psychological Stress: Issues in Research.* New York: Appleton-Century-Crofts. Pp. 14–37.

Lacey, J. I., and Lacey, B. C. 1958. "The Relationship of Resting Autonomic Activity to Motor Impulsivity," *Research Publications of the Association for Nervous and Mental Disease* 36:144–209.

Lacey, J. I., Kagen, J., Lacey, B. C., and Moss, H. A. 1963. "The Visceral Level: Situational Determinants and Behavioral Correlates of Autonomic Response Patterns." In P. H. Knapp, ed., *Expression of the Emotions in Man.* New York: International Universities Press. Pp. 161–196.

Lazarus, R. S. 1966. *Psychological Stress and the Coping Process.* New York: McGraw-Hill.

Lindsley, D. B. 1951. "Emotion." In S. S. Stevens, ed., *Handbook of Experimental Psychology.* New York: Wiley. Pp. 473–516.

Lofgren, L. B. 1968. "The Psychoanalytic Theory of Affects," panel report, *Journal of the American Psychoanalytic Association* 16:638–650.

Luszki, M. B. 1958. *Interdisciplinary Team Research: Methods and Problems*. New York: New York University Press.

Lykken, D. 1967. "Valin's 'Emotionality and Autonomic Reactivity': An Appraisal," *Journal of Experimental Research in Personality* 2:46–55.

Malmo, R. B. 1966. "Studies of Anxiety: Some Clinical Origins of the Activation Concept." In C. D. Spielberger, ed., *Anxiety and Behavior*. New York: Academic Press. Pp. 157–177.

Mason, J. W. 1968. "The Scope of Psychoendocrine Research," *Psychosomatic Medicine* 30:565–575.

Novey, S. 1961. "Further Considerations on Affect Theory in Psychoanalysis," *International Journal of Psychoanalysis* 42:21–31.

Oken, D. 1960. "An Experimental Study of Suppressed Anger and Blood Pressure," *Archives of General Psychiatry* 2:441–456.

Oken, D. 1962. "The Role of Defense in Psychological Stress." In R. Roessler and N. S. Greenfield, eds., *Physiological Correlates of Psychological Disorder*. Madison: University of Wisconsin Press. Pp. 193–210.

Oken, D. 1965. "Operational Research Concepts and Psychoanalytic Theory." In N. S. Greenfield and W. C. Lewis, eds., *Psychoanalysis and Current Biological Thought*. Madison: University of Wisconsin Press. Pp. 181–200.

Oken, D., Grinker, R. R., Sr., Heath, H. A., Sabshin, M., and Schwartz, N. 1960. "Stress Response in a Group of Chronic Psychiatric Patients," *Archives of General Psychiatry* 3:451–466.

Oken, D., Grinker, R. R., Sr., Heath, H. A., Herz, M., Korchin, S. J., Sabshin, M., and Schwartz, N. B. 1962. "Relation of Physiological Response to Affect Expression," *Archives of General Psychiatry* 6:336–351.

Persky, H. 1957. "Adrenocortical Function in Anxious Human Subjects: The Disappearance of Hydrocortisone from Plasma and its Metabolic Fate," *Journal of Clinical Endocrinology and Metabolism* 17:760–765.

Persky, H., Hamburg, D. A., Grinker, R. R., Sr., Basowitz, H., Sabshin, M., Korchin, S. J., Herz, M., Board, F. A., and Heath, H. A. 1958. "Relation of Emotional Responses and Changes in Plasma Hydrocortisone Level after Stressful Interview," *Archives of Neurology and Psychiatry* 79:434–447.

Plutchik, R., and Ax, A. F. 1967. "A Critique of 'Determinants of Emotional State' by Schachter and Singer," *Psychophysiology* 4:79–82.

Radloff, R., and Helmreich, R. 1968. *Groups under Stress: Psychological Research in Sealab*, vol. 2. New York: Appleton-Century-Crofts.

Rangell, L. 1968. "A Further Attempt to Resolve the 'Problem of Anxiety,'" *Journal of the American Psychoanalytic Association* 16:371–404.

Rapaport, D. 1953. "On the Psychoanalytic Theory of Affects," *International Journal of Psychoanalysis* 34:177–198.

Reiser, M. F. 1961. "Reflections on Interpretation of Psychophysiologic Experiments," *Psychosomatic Medicine* 23:430–439.

Reiser, M. F., Reeves, R. B., and Armington, J. 1955. "Effect of Variations in Laboratory Procedures and Experimentor upon the Ballistocardiogram, Blood Pressure and Heart Rate in Healthy Young Men," *Psychosomatic Medicine* 17:185–199.

Ruesch, J. 1956. "Creation of a Multidisciplinary Team," *Psychosomatic Medicine* 18:105–112.

Ruesch, J. 1961. "Psychosomatic Medicine and the Behavioral Sciences," *Psychosomatic Medicine* 23:277–286.

Sarason, S. B. 1966. "The Measurement of Anxiety in Children: Some Questions and Problems." In C. D. Spielberger, ed., *Anxiety and Behavior*. New York: Academic Press. Pp. 63–79.

Schachter, S. 1966. "The Interaction of Cognitive and Physiological Determinants of Emotional State." In C. D. Spielberger, ed., *Anxiety and Behavior*. New York: Academic Press. Pp. 193–224.

Selye, H. 1950. *The Physiology and Pathology of Exposure to Stress*. Montreal: Acta.

Shapiro, D., and Crider, A. 1969. "Psychophysiological Approaches in Social Psychology." In G. Lindzey and E. Aronson, eds., *Handbook of Social Psychology*, 2d ed. Reading, Mass.: Addison-Wesley. Vol. 3, pp. 1–49.

Spielberger, C. D. 1966. "Theory and Research on Anxiety." In C. D. Spielberger, ed., *Anxiety and Behavior*. New York: Academic Press. Pp. 3–20.

Spielberger, C. D., ed. 1966. *Anxiety and Behavior*. New York: Academic Press.

Spielberger, C. D., Lushene, R. E., and McAdoo, W. G. 1970. "Theory and Measurement of Anxiety States." In R. B. Cattell, ed., *Handbook of Modern Personality Theory*. Chicago: Aldine.

Tomkins, S. S. 1962. *Affect, Imagery, Consciousness*. Vol. 1. *The Positive Affects*. New York: Springer.

Tomkins, S. S. 1963. *Affect, Imagery, Consciousness*. Vol. 2. *The Negative Affects*. New York: Springer.

Weiss, B. W., et al. 1968. "Relationship Between a Factor Analytically Derived Measure of a Specific Fear and Performance after Related Fear Induction," *Journal of Abnormal Psychology* 73:461–463.

Wolff, C. T., Friedman, S. B., Hofer, M. A., and Mason, J. W. 1964. "Relationship Between Psychological Defenses and Mean Urinary 17-Hydroxycorticosteroid Excretion Rates," *Psychosomatic Medicine* 26:576–609.

Biological Approach to
Stress and Behavior[*]

Research related to the correlation between affective states and the secretion of adrenocortical hormones has always posed frustrating questions. It was very obvious that these adrenocortical hormones were elaborated and secreted under a variety of psychological conditions. But of
what use was the output of these hormones; what purpose did they
serve? Grinker once remarked to me that he could not believe that evolution has created an organism in such a stupid fashion as to permit a
nondiscriminate output of pituitary-adrenal hormones which seemed to
have no relation to the tissue and metabolic needs of the system. Research in the past decade, directed toward the question of the functional
properties of these hormones in relationship to behavior, has highlighted
the fact that they indeed do play a role in behavior and behaviors that
may be important for the adaptive, coping mechanisms of living organisms.

However stress is defined, there is little question that many affective
states are accompanied by the physiological events that have been so
well described during the past two decades (Board, Persky, and Hamburg, 1956; Persky, Grinker, Hamburg, Sabshin, Korchin, Basowitz, and
Chevalier, 1956; Persky, Hamburg, Basowitz, Grinker, Sabshin, Korchin,
Herz, Board and Heath, 1958; Persky, Korchin, Basowitz, Board, Sabshin,
Hamburg, and Grinker, 1959; Mason, Brady, and Tolson, 1966). Selye
(1950) and numerous others have established that the pituitary-adrenal

    ° This study was supported by MH 12732, ONR Contract N00014-67-A-0112-
0009, and the Leslie Fund, Chicago and the U.S. Public Health Service research scientist award K3-MH-19,936 from the National Institute of Mental Health.

axis is activated by stimuli that have been categorized as neurogenic or psychic since they do not involve any overt tissue damage. However, in spite of the many studies that have measured adrenocortical function under psychological conditions that activate the pituitary-adrenal system, relatively little is known about the influence of these systems on behavior. In fact, what had been reported in the literature appeared to be, to say the least, discouraging. The many studies that have tested the behavioral effects of adrenalectomy have largely yielded negative results. In view of these data we are presented with a paradox. It is relatively easy to understand the necessity for the activation of the pituitary-adrenal system where there is tissue damage, insofar as the adrenal steroids appear to maintain tissue integrity under conditions of physiological insult. However, since many situations that do not involve tissue damage also activate these systems, one needs to come to the conclusion that either the evolution of the central nervous system has not reached the point where the organism is able to discriminate between potentially physiological damaging situations and actual conditions that involve tissue damage or that the hormones are involved in those behaviors concerned with stress.

It is important to note that much of the research that was done, both in humans and in animals, attempting to establish correlates between emotional states and pituitary-adrenal activity, was based primarily on the unstated assumption that there was a fervent hope that the physiological indices of emotional juice could be located. However, though this aspect of the enterprise was to prove unsuccessful, primarily because of the nonspecific nature of this pituitary-adrenal activation, the research did indeed lead to asking specific questions about the functional role of these secretory products on behavior.

That a relationship should exist between the hormones of the pituitary-adrenal system and behavior is not surprising. It has long been known that those steroids emanating from the gonads have a marked influence on behavior and, further, advances in neuroendocrinology have established an intimate relationship between the central nervous system and endocrine activity.

The purpose of this essay, therefore, is to review the results that have been obtained over the past decade. These results have demonstrated (1) that the hormones emanating from the anterior pituitary and the adrenal cortex do indeed play a significant role in behavior, (2) that these hormones appear to have essentially antagonistic actions, and (3) that there are behaviors that appear to be sensitive to adrenocorticotrophic hormone (ACTH) and others that appear to be principally regulated by the adrenocortical hormones. Another purpose of this essay is to present a theoretical model to hopefully integrate these findings into a motivational framework.

NEUROENDOCRINE INTEGRATION

Since it is a basic assumption of physiological psychologists that the central nervous system is in some way involved in behavior, the existence of a central nervous system component that regulates the synthesis and release of anterior pituitary hormones and that, in turn, is regulated by the level of circulating hormones makes the effects of circulating hormones on behavior much more explicable. Figure 6–1 presents a simplified illustration of some of the pathways involved in the control of glucocorticoid released by the adrenal cortex. To this schemata presented by the late Professor Ernst Scharrer (1966) we have added one additional pathway, namely, the pathway of the so-called short feedback loop in which it has been postulated that increased ACTH acts directly on the central nervous system to regulate the subsequent release of ACTH. The essentials of neuroendocrine integration have also been described by Scharrer (1960):

1. The central nervous system receives and integrates information and stimuli that affect endocrine functions. This input, in addition to hormonal feedback, originates in organs of special senses, *i.e.* chemoreceptors, photoreceptors and acoustic apparatus, or may be received from systems serving general visceral and somatic sensation, such as touch, pain and temperature. Other sources of stimuli are intrinsic rhythms, psychogenic factors and conditioned reflexes. Stimuli which result from changes of pH, temperature, osmolarity and chemical composition of the blood reach nerve centers by way of vascular channels. By the same route, drugs or toxins exert their effects.

2. The combined input must eventually reach a final common path which may be either nervous or hormonal.

3. The role of efferent nervous pathways in the control of endocrine functions is still uncertain. Most endocrine organs do not seem to be under direct nervous control; autonomic fibers entering them are thought to be vasomotor rather than secretomotor.

4. The hormonal final common path has been described; it acts either directly on effector organs or through endocrine glands that serve as amplifiers of signals from the central nervous system.

5. Hormonal and neural feedback from the target organs complete a neuroendocrine circuit.

6. Every animal species may, in accordance with its particular requirements, select combinations of pathways by which it accomplishes appropriate control of endocrine functions. These combinations are probably rarely as simple as they were previously envisaged when emphasis centered around feedback homeostasis. In the higher vertebrates they may include, for example, the circuits of the mesencephalic reticular formation and the limbic system, which provide the versatility and integrative capacity required for dealing with complex and changeable situations.

Thus, we can see that not only are the peripheral endocrine organs, in this instance the adrenal, the target organs for the trophic hormones released from the anterior pituitary, but further, that the central nervous

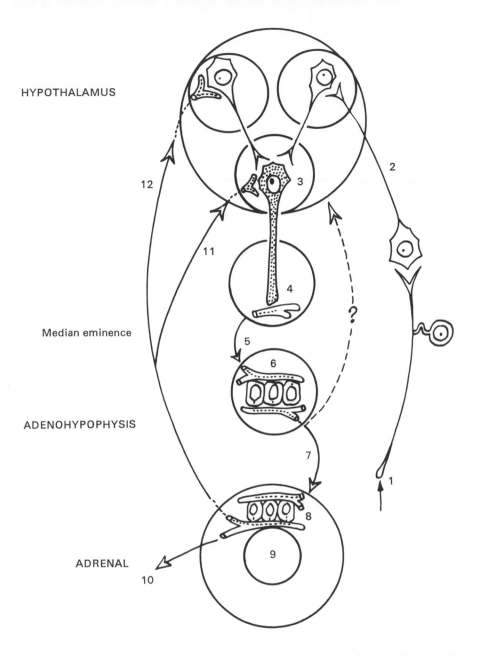

HYPOTHALAMUS

12

2

3

Median eminence

11

4

5

6

ADENOHYPOPHYSIS

7

?

1

8

ADRENAL

9

10

Figure 6–1. Simplified illustration of some of the pathways involved in the control of glucocorticoid release by the adrenal cortex. A stressful pain stimulus (1) via afferent nervous pathways (2) reaches neurosecretory cells (3), the axons of which end in the median eminence (4), where corticotropin-releasing factors are released. These are carried through hypothalamic-hypophyseal portal vessels (5) to the anterior pituitary (6), which in turn releases ACTH into the general circulation (7). ACTH acts on the adrenal cortex (8); the adrenal medulla (9) is presumably not involved. Glucocorticoids reach all tissues of the body by way of the circulating blood (10), including the neurosecretory cells in the hypothalamus (11) and nerve centers of higher order (12). Which of these represent the hormone-sensitive neurons is not decided at present. Further, ACTH may act on the neurosecretory cells (3) or nerve centers of higher order (12) to modulate ACTH release (?). (After Scharrer, 1966.)

system is also a target organ for the peripheral circulating hormones. Recently, by using a technique of implanting corticoids directly into the central nervous system, it has been demonstrated that the feedback loop postulated to exist between the peripheral target organ and the central nervous system can be interrupted by the continuing presence of steroids in various brain areas (Davidson, Jones, and Levine, 1968). It is not without interest that the brain areas that appear to be steroid-sensitive and those that are involved in mediating control of ACTH secretion are the same areas that appear to be involved in the mediation of much basic motivational and emotional behavior, such as rage, plasticity, fear, pleasure and pain, attention and startle, as well as various patterns of sexual behavior.

ACTH AND ACQUISITION OF A CONDITIONED AVOIDANCE RESPONSE

Though a number of studies by Moyer (1958a, 1958b) and his associates reported that adrenalectomy had little or no influence on the acquisition of a conditioned avoidance response using a runway, Applezweig and Baudry (1955) reported that surgical removal of the pituitary severely retarded the acquisition of an avoidance response in a shuttlebox, while the administration of adrenocorticotropin somewhat improved avoidance conditioning in hypophysectomized rats. However, these studies were open to a number of criticisms. (1) It could be argued that removal of the pituitary leads to a debilitation of the organism in view of the suppression of most endocrine activity. (2) Inasmuch as the whole pituitary was removed, there exists the possibility that the results obtained in completely hypophysectomized rats may not only be due to removal of the anterior pituitary, but may indeed be a result of the absence of the hormones emanating from the posterior pituitary. Subsequent investigations (de Wied, 1964), however, indicated that adenohypophysectomy (removal of only the anterior pituitary) also results in a marked suppression of the acquisition of an avoidance response in animals who were maintained on a regimen of hormone replacement to counteract the other effects of adenohypophysectomy. (3) Injection of a long-acting ACTH in adenohypophysectomized animals restored these animals to normal, whereas ACTH in the normal, intact animal had no effect. The fact that total bilateral adrenalectomy had no influence on avoidance conditioning in the same situation where hypophysectomy had a profound effect was one of the indications that there is a behavioral effect of ACTH that is independent of its steroidogenic activity.

It should be noted however that though the evidence does indicate that ACTH may be involved in the acquisition of a conditioned avoidance response, these results might in part be explained by a deficiency in the motor and sensory capacities of the animal as a function of hypo-physectomy. Thus de Wied found that the speed of escape in an electri-fied runway was markedly suppressed in adenohypophysectomized rats but that treatment with ACTH restored these runway speeds to normal. Furthermore, treatment of the animals with thyroxin, cortisone, and tes-tosterone also restored the animals' ability to learn a conditioned avoid-ance response. Thus it appeared that the effects of hypophysectomy and ACTH on acquisition of a conditioned avoidance response in a shuttle-box could most easily be explained in terms of the general impairment of sensory or motor functions in the animal when the anterior pituitary is removed.

However, recently the evidence that ACTH and other pituitary pep-tides may play a significant role in the acquisition of learned behavior has become more compelling. De Wied has shown that fragments of the ACTH molecule (see the next section), which have no effect on steroido-genesis and which do not alter the observed defects in the electrified run-way seen in hypophysectomized animals, completely restore the ability of hypophysectomized rats to learn a conditioned avoidance response. Evidence that ACTH is involved in the acquisition of a passive avoid-ance response has been obtained by Levine and Levin (1970). Further, Guth, Seward, and Levine (1971) have found that if ACTH is adminis-tered only during the trial on which the animals are shocked, passive avoidance in rats is markedly enhanced. In the experiment, rats were trained to obtain water in the larger of two chambers. On the test day a passive avoidance was established by shocking the animal during the approach to water. On subsequent days, the time to emerge from the small chamber into the larger chamber when water was available was recorded. Initially all animals showed a marked hesitancy to emerge. The control animals did, over days, emerge more rapidly. Those ani-mals that were given a single injection of ACTH only on the shocked trial exhibited significantly longer suppression of the emergence response. These data do indicate that ACTH is of importance in the acquisition of the passive avoidance behavior. Further, there are studies that report a correlation between pituitary-adrenal activity and avoidance behavior on a discriminative avoidance schedule (Bohus and Endoröczi, 1965) and on a Sidman avoidance schedule (Wertheim, Conner, and Levine, 1969).

ACTH AND EXTINCTION OF A CONDITIONED AVOIDANCE RESPONSE

In addition to the role that ACTH and other pituitary peptides may have in the acquisition of learned behaviors there is a great deal of evi-dence that these hormones are involved in the maintenance of a previously learned avoidance response. The influences of the peptides of the anterior pituitary and of adrenocortical hormones on extinction of a previously learned conditioned avoidance response are indeed profound, and it is in these studies on extinction that we see a clear differentiation of the behavioral effects of ACTH in opposition to the behavioral effects of adrenocortical hormones. The findings of Murphy and Miller (1955) were the first demonstration of the effects of ACTH on extinction of a conditioned avoidance response. These investigators trained rats in a shuttlebox with and without ACTH. It was found that ACTH did not affect the acquisition of the response. However, if ACTH was administered during extinction, extinction was markedly retarded. Miller and Ogawa (1962) showed that ACTH inhibits extinction in the adrenalectomized rat. However, the extensive and elegant series of studies by de Wied (1969) have yielded the most conclusive evidence of the effects of ACTH and related ACTH-like peptides on resistance to extinction. The results in general indicate that ACTH retards extinction of a conditioned avoidance response, whereas glucocorticoids facilitate extinction of the same response.

To investigate the role of endogenous ACTH during extinction, de Wied (1967) studied animals that had been chronically adrenalectomized. It had been previously demonstrated (Cox, Hodges, and Vernikos, 1958) that chronic adrenalectomy leads to a chronic increase in circulating levels of adrenocorticotropic hormone. Hypophysectomized animals were studied in the same experiment. It was observed that adrenalectomy resulted in a marked inhibition of extinction, whereas hypophysectomy restored extinction to normal. Thus, an excess of circulating ACTH is reflected in inhibition of extinction, whereas the absence of this hormone is reflected in a normal extinction of the conditioned avoidance response. In order to further demonstrate the influence of ACTH on extinction, a synthetic glucocorticoid, dexamethasone, was administered to animals during extinction. In addition to its peripheral glucocorticoid effects, dexamethasone has been shown to be one of the most potent inhibitors of ACTH. Animals treated with dexamethasone showed a marked facilitation of extinction which was dose-dependent with increasing doses of dexamethasone leading to more rapid extinction. Corticosterone, the principal natural glucocorticoid of the rat adrenal cortex, also has a facilitatory effect on extinction of a conditioned avoidance response. It might be argued, of course, that the action of glucocorticoids on extinction is due primarily to the extent to which these corticoids inhibit ACTH secretion. However, de Wied (1966) has shown that both dexamethasone and corticosterone facilitate extinction in a hypophysectomized animal, thus indicating that there are effects of the glucocorticoids that are independent of their effects on ACTH, as there are effects of ACTH that are independent of their influence on adrenocorticoids.

Van Wimersma Greidanus (1970) has also reported that extinction of a conditioned pole-jumping response was facilitated by pregnene-type steroids, that is, steroids containing twenty-one carbon atoms and a double bond. Thus, corticosterone, dexamethasone, pregnenolone, and progesterone resulted in more rapid extinction of this avoidance response. The fact that progesterone does not appear to have the capacity to block ACTH, as is the case with the other steroids in this study, gives further evidence of an independent, non-ACTH dependent, role of steroids on behavior.

Furthermore, the extra adrenal effects of ACTH and other pituitary polypeptides on behavior have been conclusively demonstrated. The ACTH molecule consists of a sequence of thirty-nine amino acids (Turner, 1960, p. 66). The presence of the first sequence of twenty-four amino acids is necessary for the ACTH molecule to exert an influence on the adrenal cortex with the resultant increase in secretion of adrenocortical hormone. De Wied (1966) and Greven and de Wied (1967) have studied fractions of the ACTH molecule and have demonstrated that the molecule containing the 1–10 sequence of amino acids, though having no influence on the adrenal cortex, inhibits the extinction of a pole-jumping response. The first ten amino acids of the ACTH molecule also inhibit extinction of the shuttlebox avoidance response (Bohus and de Wied, 1966). However, if the phenylalanine in the seventh position of this peptide is replaced by its dextorotatory form, extinction is facilitated. These results occur both in sham-operated and hypophysectomized animals. It is indeed surprising that the ACTH 1–10 (7dPhe) is not simply an inactivated molecule, but the dextorotatory form of the ACTH actually has a facilitatory influence on extinction.

ACTH AND PASSIVE AVOIDANCE

Though the studies reviewed above seem to show the clearest demonstration of the independent action of ACTH and corticosterone on behavior, there is another series of studies that appear to implicate ACTH in the regulation and modulation of aversive behavior. Lissak and Endröczi (1961) have shown that the degree of passive avoidance learning was positively correlated with responsiveness to stress. Rats that showed little passive avoidance learning exhibited significantly less adrenal ascorbic acid depletion to a subsequent stress of unilateral adrenalectomy than did those rats that showed greater passive avoidance conditioning. Insofar as adrenal ascorbic acid is a reflection of circulating levels of ACTH, it would appear that the higher the level of ACTH, the better the acquisition of passive avoidance response. Levine and Jones (1965) tested the effects of ACTH in a passive avoidance situation. In this study rats were trained to press a lever in order to obtain water on a continuous reinforcement schedule. After the behavior had been well stabilized, the animals were given a response-contingent electric shock. Following two such electric shocks, no subsequent aversive stimuli were presented to the animal, and the rate of recovery of the lever-pressing response was observed. In the control animals, either injected with a placebo or not injected, a bimodal distribution was observed, similar to that obtained by Lissak and Endröczi. Those animals that failed to return to the bar press after seven days of testing also had significantly heavier adrenals than the animals that did resume the operant response. However, complete suppression of the operant response was observed in all animals if ACTH was administered during the seven-day test period. However, if ACTH was administered only during that period of time prior to and terminating with the end of the response-contingent shock, these animals, instead of failing to return, showed a much more rapid return to operant responding. Inasmuch as it has been demonstrated that, following treatment with ACTH, ACTH release is subsequently suppressed, it could be argued that this suppression of ACTH facilitated extinction of the passive avoidance response, whereas continual administration inhibited the extinction of the passive avoidance behavior.

In a recent replication of the above study, Anderson, Winn, and Tam (1968) demonstrated a similar effect of ACTH on passive avoidance behavior in hypophysectomized animals. Administration of hydrocortisone had no effect on the length of time it took animals to resume responding following the establishment of passive avoidance with electric shock. Further evidence that ACTH is implicated in passive avoidance comes from a recent study by Levine and Levin (1970).

STEROIDS AND BEHAVIOR

Whereas thus far we have emphasized the effects of ACTH and its molecular fragments on behavior, it should be noted that steroids were also implicated in the extinction process and facilitated extinction. This action appeared to be independent of ACTH inasmuch as extinction was facilitated by steroids in the hypophysectomized animal.

Recently it has been reported (Wertheim, Conner, and Levine, 1967) that treatment of rats with ACTH or dexamethasone was accompanied by changes in performance on a free-operant avoidance (Sidman) schedule. These changes can be best characterized as producing more efficient avoidance behavior. The following changes were noted: (1) There was a higher frequency of long duration interresponse times and greater stability among these IRT's. (2) There were fewer short interresponse times, total responses, and shocks. In addition, the period of high-density shocks (warm-up), which is usually observed at the beginning of the Sidman avoidance schedule, was suppressed during those sessions when the animal was treated either with ACTH or dexamethasone. The similarity of the behavior modification by both ACTH and dexamethasone suggests that there is an interaction of Sidman avoidance behavior with the effects attributable to increased plasma corticosteroid levels and not to ACTH per se. If the role of ACTH were other than its glucocorticoid releasing effects, then opposing behavioral changes would have been expected following the administration of these two hormones, since dexamethasone markedly suppresses endogenous ACTH levels. The nature of the behavior modification observed, presumably as a function of the increase in circulating steroids either by ACTH or dexamethasone, might best be characterized as a response suppression which results in better timing behavior. This characteristic of response suppression, which is being attributed to glucocorticoids, could also account for the facilitation of the extinction response observed in the experiments reported in the previous section.

As a direct result of these studies, the question was raised as to whether the effects of steroids on temporal discrimination were restricted to those situations in which there was an aversive component or whether a more generalized effect on timing behavior under appetitive schedules could also be observed.

In experiments in our laboratory (Levine, 1968) using a reinforcement schedule that requires the animal to withhold a response until a given amount of time has passed in order to be reinforced, it was found that during those sessions in which the organism was treated either with dexamethasone or ACTH, as in the previous experiment on free-operant avoidance behavior, delayed performance was improved so that the animals were able to obtain more reinforcements under these conditions. Further, with monkeys on a fixed-interval, variable-interval, multiple schedule, it was observed that during those sessions in which the animal was administered exogenous ACTH, the postreinforcement pause on the fixed-interval schedule was significantly longer. Thus, whereas the animal obtained equivalent numbers of reinforcements, the response rate was markedly reduced, indicating again a suppression of irrelevant responses, which can be interpreted as better timing behavior. It should be noted that ACTH had no influence on the response rates under the variable-interval schedule. Though this experiment has not as yet been extended to examine corticosteroids, in view of the data on other free-operant performances we have every reason to believe that this again is a steroid-dependent behavior.

There is one final series of experiments that also demonstrates the behavioral effects of pituitary-adrenal hormones. Relearning of an avoidance response in the shuttlebox has been studied in several experiments in which randomly selected groups of rats were given a number of avoidance-training trials before being returned to their home cage for an intersession interval, after which avoidance training was then resumed.

In the original experiment by Kamin (1963), the number of avoidance responses made during the relearning session was a function of the intersession interval. Brush (1962) found that the slowest relearning occurred after a one-hour intersession interval. Since the original experiment by Kamin, a number of investigators have confirmed and extended these results. These time relationships are of particular significance in examining the physiological bases for this phenomenon. Rarely do neurophysiological events occur within the time span observed for the maximum effect of these experiments. The time relationships intuitively resemble the temporal relationships between the onset of noxious stimuli and the pattern of the stress response observed in humoral events. With the intuitive hypothesis that the Kamin effects may indeed be due to alterations in pituitary-adrenal activity, the following studies were conducted. In the initial experiment (Brush and Levine, 1966) animals were given twenty-five CS-UCS pairings, and blood samples were obtained at those intersession intervals when the animal would normally have been returned to the shuttlebox. The data indicated that the descending limb of the behavioral function was found to be correlated with a corresponding decrease in plasma concentrations of corticosterone. Thus, during the period when relearning is rapid, these animals also appear to have markedly elevated corticosterone. The suppression of relearning is also correlated with a suppression of plasma levels of corticoids. In order to determine whether or not this correlation represented a causal relationship, a subsequent experiment was conducted (Levine and Brush, 1967) in which animals were given either ACTH or hydrocortisone immediately following original avoidance training. When high corticosteroid levels are maintained by exogenous ACTH administration, high levels of performance are maintained. Once again, inasmuch as increased steroids and adrenalectomy produce opposing results, it seems reasonable to hypothesize that the effects of the pituitary-adrenal system on the Kamin effect are due primarily to alterations in circulating levels of glucocorticoids.

In a recent series of studies Johnson and Levine (1971) demonstrated that adrenalectomy results in an almost complete suppression of relearning. In contrast, an implant of crystalline hydrocortisone placed in the median eminence of the hypothalamus caused a significant increment in the retention of the partially learned avoidance response. At first glance it would appear that the diametrically opposed results obtained following adrenalectomy, which elevates ACTH, and steroid implants that suppress ACTH, could be interpreted as a function of ACTH rather than the presence or absence of steroid. However, since ACTH administration has the same effects as hydrocortisone injections or implantation, it would appear that the suppression of relearning by adrenalectomy is caused by the lack of available glucocorticoids needed to act on the central nervous system.

Though it is possible to interpret these data in terms of some hormonal effect on the memory process, the fact that ACTH facilitates avoidance learning in a group run seven days after original avoidance training tends to refute such an interpretation. It is unlikely that the steroid elevation produced by ACTH administration seven days after learning could preserve an already decayed memory trace. Rather it is more likely that ACTH treatment or corticosteroid replacement reinstates motivational cues associated with the avoidance training situation, therefore facilitating performance.

Because of the nature of the interactions of ACTH and adrenocortical steroids, it is difficult to determine with certainty whether a particular behavioral effect is due to ACTH or to corticosteroids. Since ACTH secretion directly causes adrenocortical activity, it is often impossible to ascertain whether the behavioral effect caused by ACTH injection is due to the ACTH itself or to the effects of corticosteroids. Both injection and median eminence implantation of hydrocortisone suppress ACTH response. Consequently, it is necessary to discover whether any behavior that these techniques affect is altered by the steroid itself or by its negative feedback control of ACTH secretion.

The current data on pituitary-adrenal effects support three independent positions on what aspect of the system really affects behavior. The first position is that all effects are caused by glucocorticoids and is supported by experiments that show that ACTH and glucocorticoid injections have the same behavioral effects. De Wied's (1964) finding that both ACTH injections and steroid-thyroxine "cocktail" had the same effect on improvement of acquisition of a conditioned avoidance response in hypophysectomized animals supports this notion. Brush and Levine (1966) found that both ACTH and hydrocortisone injections facilitated performance in the second Kamin session. Also Wertheim et al. (1967) found the same improvement of Sidman performance with ACTH and dexamethasone injections.

A second position would be that all pituitary-adrenal effects on behavior are simply due to ACTH. This position would explain any effects shown with glucocorticoid manipulations as due to the suppression of ACTH secretion. Data that support this idea include that of Anderson et al. (1968) showing that ACTH and hydrocortisone have different effects on extinction of a passive avoidance response. Miller and Ogawa (1962) found an effect of ACTH on inhibition of extinction of a shuttlebox avoidance response in adrenalectomized rats, a preparation where

ACTH could not act by eliciting corticosteroid secretion. Further, de Wied (1969) found that ACTH 4–10 with no apparent biologic—that is, adrenocorticotrophic—effects, improved acquisition in hypophysectomized animals.

It is also possible that both ACTH and glucocorticoids affect behavior. The two classes of hormones could either behave in an independent and opposing fashion or be effective in different situations with different parameters of time and/or response. For instance, de Wied (1964) found that both hydrocortisone cocktail and ACTH 4–10 improve acquisition of hurdle response in hypophysectomized animals. In neither situation does injection of one of the compounds affect the level of its complement. Further, de Wied found that ACTH retards extinction and hydrocortisone facilitates it. However, it can be argued that hydrocortisone administration works by depressing all secretion of ACTH. Perhaps the most convincing argument for the position that both ACTH and glucocorticoids affect behavior is that there exist data in support of each of the other two positions.

THEORETICAL CONSIDERATIONS

On the assumption that there do exist two independent hormonal actions, one for ACTH and the other for steroids, there still exists the need for theoretical integration of much of the data presented.

The concept of inhibition has taken a central place in almost every theory of brain function. Pavlov (1960) utilized the concept of inhibition extensively to explain many of the processes occurring in the brain during conditioning and extinction. Pavlov postulated two kinds of inhibition, internal and external. It is within the framework of the concept of internal inhibition that the data presented in this essay with regard to the effects of hormones on conditioning are most explicable. On internal inhibition Pavlov states:

Conditioned stimuli, acting as they undoubtedly do through the intermediation of definite cortical cells, provide the obvious means whereby the physiological characteristics of these cortical cells can be studied. One of the most important of these properties is that under the influence of conditioned stimuli they pass, sooner or later, into inhibition. In the previous lectures upon internal inhibition it was shown that in all cases when a positive conditioned stimulus repeatedly remains unreinforced, it acquires inhibitory properties, *i.e.* the corresponding cortical cells enter under its influence into a state of inhibition. The present lecture will be devoted to the study of the intimate mechanism of this phenomenon, and of the part played therein by the unconditioned reflex and by other conditions which retard or accelerate the development of this inhibitory state.

The transition of the cortical cells into an inhibitory state is of much more general significance than could be inferred from the facts which have been discussed up to the present, concerning the development of internal inhibition. The development of inhibition in the base of conditioned reflexes which remain without reinforcement must be considered only as a special instance of a more general case, since a state of inhibition can develop also when the conditioned reflexes are reinforced. The cortical cells under the influence of the conditioned stimulus always tend to pass, though sometimes very slowly, into a state of inhibition. The function performed by the unconditioned reflex after the conditioned reflex has become established is merely to retard the development of inhibition.

In order to test more directly the implication of the possible involvement of adrenocortical hormones in the development of internal inhibition, the effects of adrenalectomy and central implants of corticosteroids were studied in a simple habituation experiment (Brown, Kalish, and Farber, 1951). Habituation of a skeletal-motor response to the sudden onset of white noise occurred more rapidly in the animals with central hydrocortisone implants and was markedly retarded in the adrenalectomized animals. Insofar as habituation is considered to be a process of self-inhibition (Pribram, 1967), either glucocorticoids or the absence of ACTH facilitate inhibition of the startle response and high ACTH, and the absence of glucocorticoids prevents this inhibitory process.

Insofar as the theory of extinction critically depends on internal inhibition, the effects glucocorticoids have on extinction would lead to the hypothesis that the action of the glucocorticoids is primarily to facilitate the development of internal inhibition. This concept, of course, is entirely consistent with the action of glucocorticoids on the central nervous system since it has already been demonstrated that glucocorticoids have anesthetic properties, and further, within the negative feedback system of regulation of ACTH and glucocorticoids, the primary function of glucocorticoids on the brain is to inhibit further elaboration of corticotropin-releasing factors and ACTH. However, since ACTH apparently has a paradoxical effect on extinction, that is, it retards rather than facilitates, one must postulate an opposing role of ACTH with regard to the development of internal inhibition, that is, ACTH disinhibits the inhibitory process. If one assumes that the passive avoidance situation, as described previously in this chapter, is one that requires extinction of the fear components related to electric shock, the action of ACTH in that particular study is also to retard extinction of those fear components that would permit the animal to return to bar-pressing in the absence of any further noxious stimulation.

The data concerning both Sidman avoidance and other temporally based schedules are also consistent with the hypothesis that glucocorticoids act to facilitate the inhibitory process. Animals working on a Sidman schedule show many characteristic behaviors. Among these are the typical warm-up observed in the Sidman avoidance and bursting behavior, that is, rapid trains of responses following the presentation of electric shock when the animal fails to avoid. Under the influence of glucocorticoids there exists a considerable shortening of the warm-up period, and bursting behavior tends to be suppressed. Further, there is evidence that if animals are given a period of free shock or other stress procedures, for example, ether, immobilization (which would presumably raise glucocorticoids), there is a significant reduction in the warm-up period.

Since "timing behavior" per se involves an inhibitory process, and insofar as glucocorticoids enhance performance on time-based schedules, in addition to suppressing those aspects of behavior (for example, bursting) which interfere with the performance of a temporal response, it seems reasonable to assume that once again in the Sidman avoidance, and other temporally based schedules, the glucocorticoids are exerting a facilitatory effect on the inhibitory process.

Finally, with regard to those experiments concerned with the Kamin effect, one can use the theoretical notions concerning internal inhibition again as an explanatory device. Since the avoidance behavior paradigm requires the inhibition of irrelevant responses in order for efficient avoidance performance to occur, the elevation of glucocorticoids would possibly facilitate the suppression or inhibition of these random irrelevant responses.

The purpose of any theory is both to explain and to generate further research. The present theory of the influence of hormones, of the glucocorticoids, and of ACTH on the processes of internal inhibition meets the requirements of these elements of a theory. The number of experiments generated by the hypothesis that glucocorticoids facilitate internal inhibition are legion. The concept of internal inhibition has recently received a great deal of attention both in the explanation of the effects of various brain lesions (Kimble, 1968) and as a central concept in many learning theories (Walker, 1958). It is hoped that the utilization of these theoretical concepts will facilitate explanation and further research in this area.

In addition to the number of other effects we have discussed, of the pituitary peptides and adrenocortical hormones on behavior, there has now been an extensive series of reports that indicate that sensory detection and perception are also influenced by pituitary-adrenal activity. Patients with hypoadrenocorticalism show a raised threshold for detection of sensory information in the sense modalities of taste, audition, olfaction, and proprioception. Patients suffering from Cushing's disease show a lowered threshold for these same sensory systems. However, though sensory detection is greater, perception of the sensory information appears to be impaired under conditions of hypoadrenocorticalism.

The action of all hormones is based on the property that there are receptor sites for hormones. It is an amazing process that certain tissues of the body are able to selectively extract from a mass of compounds existing in the blood, specific compounds acting on specific tissues. Thus, though the blood contains innumerable compounds, the uterus, for example, is quite capable of extracting only estrogen and progesterone, which act on uterine function. The seminal vesticles and prostate gland of the male are capable of extracting and responding only to testosterone. There are numerous examples of this receptor site hormone relationship. The brain appears also to be a specific target organ and receptor site for hormones. There is abundant evidence that hormones are taken up by the brain and certain brain regions selectively take up some hormones and other brain regions selectively take up others. Recent research by McEwen and Weiss (1970) has indicated that the hippocampus appears to be a specific receptor site for adrenocortical hormones. Other evidence indicates that the lateral hypothalamus may be a receptor site for gonadal hormones. However, almost nothing is known of how these hormones act on the cells of these receptor sites. Which particular areas of the brain are involved still needs extensive exploration. What is of importance is that these hormones do influence behavior and in general appear to have adaptive significance for the living organism. This adaptive significance is seen in many ways. Adequate sensation and perception are important for organisms' capacity to integrate the universe around them. Habituation to novel stimuli is a basic physiological and behavioral mechanism in order for organisms to function in their environment and certainly the ability to learn is so basic for survival that organisms incapable of some form of learning are unable to survive.

REFERENCES

Anderson, D. C., Winn, W., and Tam, T. 1968. "Adrenocorticotrophic Hormone and Acquistion of a Passive Avoidance Response: A Replication and Extension," *Journal of Comparative Physiological Psychology* 66:497–499.

Applezweig, M. H., and Baudry, F. D. 1955. "The Pituitary-Adrenocortical System in Avoidance Learning," *Psychological Reports* 1:417–420.

Board, F., Persky, H., and Hamburg, D. A. 1956. "Psychological Stress and Endocrine Functions: Blood Levels of Adrenocortical and Thyroid Hormones in Acutely Disturbed Patients," *Psychosomatic Medicine* 18:324–333.

Bohus, B., and de Wied, D. 1966. "Inhibitory and Facilitatory Effect of Two Related Peptides on Extinction of Avoidance Behavior," *Science* 153:318–320.

Bohus, B., and Endröczi, E. 1965. "The Influence of Pituitary-Adrenocortical Function on the Avoiding Conditioned Reflex Activity in Rats," *Acta Physiologica Hungarica* 26:183–189.

Brown, J. S., Kalish, H. I., and Farber, I. E. 1951. "Conditioned Fear as Revealed by Magnitude of Startle Response to an Auditory Stimulus," *Journal of Experimental Psychology* 41:317–328.

Brush, F. R. 1962. "The Effects of Intertrial Interval on Avoidance Learning in the Rat," *Journal of Comparative Physiological Psychology* 55:888–892.

Brush, F. R., and Levine, S. 1966. "Adrenocortical Activity and Avoidance Learning as a Function of Time after Fear Conditioning," *Physiological Behavior* 1:309–311.

Cox, G. S., Hodges, J. R., and Vernikos, J. 1958. "The Effect of Adrenalectomy on the Circulating Level of Adrenocorticotrophic Hormones in the Rat," *Journal of Endocrinology* 17:177–181.

Davidson, J. M., Jones, L. E., and Levine, S. 1968. "Feedback Regulation of Adrenocorticotropin Secretion in 'Basal' and 'Stress' Conditions: Acute and Chronic Effects of Intrahypothalamic Corticoid Implantation," *Endocrinology* 82:655–663.

de Wied, D. 1964. "Influence of Anterior Pituitary on Avoidance Learning and Escape Behavior," *American Journal of Physiology* 207:255–259.

de Wied, D. 1966. "Inhibitory Effect of ACTH and Related Peptides on Extinction of Conditioned Avoidance Behavior in Rats," *Proceedings of the Society for Experimental Biological Medicine* 122:28–32.

de Wied, D. 1967. "Opposite Effects of ACTH and Glucocorticosteroids on Extinction of Conditioned Avoidance Behavior." In L. Martini, F. Fraschini, and M. Motta, eds., *Hormonal Sterioids*. Netherlands: Mouton. Pp. 945–951.

de Wied, D. 1969. "Effects of Peptide Hormones on Behavior." In W. F. Ganong and L. Martini, eds., *Frontiers in Neuroendocrinology*. New York: Oxford University Press. Pp. 97–140.

Greven, H. M., and de Wied, D. 1967. "The Active Sequence in the ACTH Molecule Responsible for Inhibition of the Extinction of Conditioned Avoidance Behaviour in Rats," *European Journal of Pharmacology* 2:14–16.

Guth, S., Seward, J. P., and Levine, S. 1971. "Differential Manipulation of Passive Avoidance by Exogenous ACTH," *Hormones and Behavior* 2:127–128.

Johnson, J. T., and Levine, S. 1971. "Avoidance Learning as a Function of Time after Avoidance Training in Adrenalectomized and Hydrocortisone-Implanted Rats. Submitted for publication.

Kamin, L. J. 1963. "Retention of an Incompletely Learned Avoidance Response: Some Further Analyses," *Journal of Comparative Physiological Psychology* 56:713–718.

Kimble, D. P. 1968. "The Hippocampus and Internal Inhibition," *Psychological Bulletin* 70:285–295.

Levine, S. 1968. "Hormones and Conditioning." In W. J. Arnold, ed., *Nebraska Symposium on Motivation, 1968*. Lincoln, Nebraska: University of Nebraska Press. Pp. 85–101.

Levine, S., and Brush, F. R. 1967. "Adrenocortical Activity and Avoidance Learning as a Function of Time after Avoidance Training," *Physiology and Behavior* 2:385–388.

Levine, S., and Jones, L. E. 1965. "Adrenocorticotrophic Hormone (ACTH) and Passive Avoidance Learning," *Journal of Comparative Physiological Psychology* 59:357–360.

Levine, S. and Levin, R. 1970. "Pituitary-Adrenal Influences on Passive Avoidance in Two Inbred Strains of Mice," *Hormones and Behavior* 1:105–110.

Lissak, K., and Endröczi, E. 1961. "Neurohumoral Factors in the Control of Animal Behaviour." In A. Fessard, R. W. Gerard, and J. Konorski, eds., *UNESCO Symposium: Brain Mechanisms and Learning*. Oxford: Blackwell. Pp. 293–308.

Mason, J. W., Brady, J. V., and Tolson, W. W. 1966. "Behavioral Adaptations and Endocrine Activity: Psychoendocrine Differentiation of Emotional States." In Rachmiel Levine, ed., *Endocrines and the Central Nervous System*. Baltimore: Williams & Wilkins. Pp. 227–250.

McEwen, B. S. and Weiss, J. M. "The uptake and action of corticosterone: regional and subcellar studies on rat brain. In D. de Wied and J.A.W.M. Weijnen, eds., *Pituitary, Adrenal and the Brain*. Amsterdam: Elsevier. Pp. 200–212.

Miller, R. E., and Ogawa, N. 1962. "The Effect of Adrenocorticotrophic Hormone (ACTH) on Avoidance Conditioning in the Adrenalectomized Rat," *Journal of Comparative Physiological Psychology* 55:211–213.

Moyer, K. E. 1958a. "The Effect of Adrenalectomy on Anxiety Motivated Behavior," *Journal of Genetic Psychology* 92:11–16.

Moyer, K. E. 1958b. "Effect of Adrenalectomy on Emotional Elimination," *Journal of Genetic Psychology* 92:17–21.

Murphy, J. V., and Miller, R. E. 1955. "The Effect of Adrenocorticotrophic Hormone (ACTH) on Avoidance Conditioning in the Rat," *Journal of Comparative Physiological Psychology* 48:47–49.

Pavlov, I. P. 1960. *Conditioned Reflexes*. New York: Dover.

Persky, H., Grinker, R. R., Sr., Hamburg, D. A., Sabshin, M., Korchin, S. J., Basowitz, H., and Chevalier, J. A. 1956. "Adrenal Cortical Function in Anxious Human Subjects: Plasma Level and Urinary Excretion of Hydrocortisone," *Archives of Neurology and Psychiatry* 76:549–558.

Persky, H., Hamburg, D. A., Basowitz, H., Grinker, R. R., Sr., Sabshin, M., Korchin, S. J., Herz, M., Board, F. A., and Heath, H. A. 1958. "Relation of Emotional Responses and Changes in Plasma Hydrocortisone Level after Stressful Interview," *Archives of Neurology and Psychiatry* 79:434–447.

Persky, H., Korchin, S., Basowitz, H., Board, F., Sabshin, M., Hamburg, D. A., and Grinker, R. R., Sr., 1959. "Effect of Two Psychological Stresses on Adrenocortical Function," *Archives of Neurology and Psychiatry* 81:210–226.

Pribram, K. H. 1967. "Emotion: Steps Toward a Neuropsychological Theory." In D. C. Glass, ed., *Neurophysiology and Emotion*. New York: Rockefeller University Press and Russell Sage Foundation. Pp. 3–39.

Scharrer, E. 1966. "Principles of Neuroendocrine Integration." In Rachmiel Levine, ed., *Endocrines and the Central Nervous System*. Baltimore: Williams & Wilkins. Pp. 1–35.

Selye, H. 1950. *Stress*. Montreal: Acta.

Turner, C. D. 1960. *General Endocrinology*. Philadelphia: Saunders.

van Wimersma Greidanus, Tj. B. 1970. "Effects of Steroids on Extinction of an Avoidance Response in Rats: A Structure-Activity Relationship Study." In D. de Wied and J.A.W.M. Weijnen, eds., *Pituitary, Adrenal and the Brain*. Amsterdam: Elsevier. Pp. 185–191.

Walker, E. L. 1958. "Action Decrement and Its Relation to Learning," *Psychological Review* 65:129–142.

Wertheim, G. A., Conner, R. L., and Levine, S. 1967. "Adrenocortical Influences on Free-Operant Avoidance Behavior," *Journal of the Experimental Analysis of Behavior* 10:555–563.

Wertheim, G. A., Conner, R. L., and Levine, S. 1969. "Avoidance Conditioning and Adrenocortical Function in the Rat," *Physiological Behavior* 4:41–44.

Twenty-five Years after
Men under Stress

Twenty-five years have elapsed since the publication of *Men under Stress* (Grinker and Spiegel, 1945), but it is still too early to assess its ultimate impact. Such a conclusion may seem strange in describing a book that has had such a remarkable influence on psychiatry and on fields not directly related to mental health. The treatment of war neuroses has been modified radically as a consequence of Grinker and Spiegel's experiences and recommendations. Psychosomatic medicine achieved a mid-century renaissance as a product of the book and related work. Furthermore, a generation of physicians entered psychiatry by stepping on the rungs of a career ladder erected, in part, out of the wartime experiences.

Yet, as significant as these developments have been, the implications of *Men under Stress* transcend the shorter pragmatic considerations. The book points toward a transactional theory of human adaptation and maladaptation. With national concern for health delivery, for community mental health as well as mental illness, there is now renewal of psychiatric interest in stress research. This has occurred under the rubric of crisis theory during the 1960's. There is increased attention given to the processes of human adaptation by a variety of behavioral scientists. Accordingly, there is reason to hope that stress and crisis theory will develop substantially—and relevantly—during the next decade. If these expectations are correct, *Men under Stress* should have a significant impact in such an evolution of theory.

The first part of the chapter will focus on the book and its subsequent impact. After this discussion there will be an overview of current issues relevant to crises and critical transitions with an attempt at integration of *Men under Stress* with the current approaches to crisis theory.

THE STUDIES OF *MEN UNDER STRESS*

One of the most important aspects to *Men under Stress* has been its universal appeal to psychiatrists of diverse interests and styles. For clinicians, the book proved to be a gold mine, and they could easily become engrossed in the remarkable richness of the case histories, the clarity of the expository formulations, and the specificity of the therapeutic interventions. In reviewing the book, Myerson (1946) illustrated this reaction when he stated:

This volume, *Men under Stress*, makes the finest of impressions. It is a distinguished book, both in format and in the essential brilliancy of its English. From the classical beginning wherein the flying man is linked to his great prototype, Icarus, to the solemn warning in the final paragraphs of the book, there is nothing lacking, and there is everything which such a book should have. There are descriptions justifiably to be described as classical, because there is a living mixture of fine emotion, calm objectivity and keen insight, a combination which represents an ideal in the semantics of psychiatry. The language is always plain and frequently bold. It pulls no punches, and yet it is subtle with an underlying irony and, I believe, is always permeated by a wholehearted sympathy for the sufferings and trials which the "men under stress" have undergone.

Though a major theme of this chapter involves the richness of the theoretical yield inherent in *Men under Stress*, it is significant to note the excellent quality of clinicoempirical observations throughout the book. Utilizing a parsimonious but elegant style that is all too rare in clinical description today, sixty-five cases were described in enough detail for the reader to understand the major issues. Each case was introduced by a heading that stated the critical point, for example, "hostility reactive to a homosexual conflict accentuated by the symbolic significance of enemy attacks"; "stable individual with strong ego-ideals, who experienced gradually increasing free anxiety, eventuating in combat failure in spite of over-compensation"; "pentothal-induced abreaction, indicating the relation of guilt to unconscious current and past hostility." In the narrative subsequent to the heading the clinical facts were outlined perceptively and convincingly. Indeed *Men under Stress* was a remarkable teaching volume for a generation of psychiatrists, and it could still serve as a basic primer for the current group of trainees.

Most young psychiatrists do not use the term "narcosynthesis," but more importantly they do not understand the significance of the concepts underlying the term. In describing their treatment techniques, Grinker and Spiegel stated:

The capacity of the ego to synthesize the material (which is more important than the abreaction), released from its isolation, is not determined by the patient's ability to verbalize insight or to repeat the material in a conscious state. The ego often synthesizes and learns unconsciously; hence the term "narcosynthesis." Many symptoms disappear spontaneously and recovery occurs directly under pentothal. Yet even then a subsequent working through of the material is helpful in strengthening the ego's grasp on the emotional drives.

This critical concept has great value for the current therapies including crisis intervention, and *Men under Stress* should be on the relevant bibliography for those who attempt to engage in such intervention.

To some extent, the authors anticipated developments that would take place over the twenty-five years following their book:

As the public and medical men become increasingly aware of the need for psychiatric help and the goals of rational psychotherapy, the paucity of well trained psychiatrists becomes increasingly apparent. It is estimated that for future needs of peacetime America, 10,000 to 17,000 additional psychiatrists will be necessary. Because of the length of time necessary for training, the relatively few teaching institutions and the limited number of aspirants, it is obvious that decades will elapse before psychiatric services to the people of the United States will even approach their needs, which will steadily increase in the interim.

The decades have elapsed and the manpower has begun to be available but still in quantity insufficient for the predicted steady increase of needs. Unfortunately, some of the therapeutic practices developed as a product of World War II receded into the background during the 1950's when commitment by many psychiatrists to long-term psychotherapy was at its apogee. There is reason to conclude that the therapeutic practices initiated by Grinker and Spiegel will have a resurgence during the next decade as more systematic attention is given to the clinical aspects of crisis intervention in community mental health.

The theoretical yield from *Men under Stress* has already been striking but in all likelihood the yield will be augmented over the next decades. The implications for psychosomatic medicine have been discussed elsewhere in this book (see Chapter 4) in regard to Grinker's consistent contributions to the transactional approach to psychosomatic processes (Grinker, 1953). As the reductionism inherent in specificity theory has become more apparent as a constraining force, field theory has emerged during the 1960's as a dominant model in current psychosomatic medicine. On a broader plane, *Men under Stress* has had direct impact on the entire field of stress and coping. The excellent overview of this subject by Lazarus (1966) explicitly demonstrates this impact by numerous references to Grinker. For example, in discussing appraisal, threat, and stimulus factors in coping, Lazarus states:

Grinker and Spiegel's clinical description illustrates a number of things. For one thing, changes in the appraisal of the viability of certain types of coping lead to changes in the coping strategy and the affective state, as when depression turns into anxiety with the return of hope. Similarly, feelings of invulnerability tend to disappear when they can no longer be sustained in the face of the evidence. Some kinds of appraisal and their consequent coping processes are a function not only of the stimulus configuration, but of personality dispositions. These dispositions account for individual differences in the reaction.

Throughout the book Lazarus returns to *Men under Stress* as the forerunner of key concepts. Similarly, Mechanic (1962) utilizes Grinker's concepts in his social psychological studies of adaptation. Though Mechanic emphasizes the importance of group interaction in stress somewhat more than Grinker, he acknowledges that his concept of group influence had been anticipated in several of the authors' interpretations in *Men under Stress*. It is impressive to note how often Mechanic compares subjects in his study with case examples cited by Grinker and Spiegel. It seems highly likely that those who will attempt to formulate new models of human coping to stress will continue to utilize *Men under Stress* as a basic building block. Indeed, Grinker's concepts as they were enunciated in 1945 afford an excellent starting point for studies of adaptation as well as maladaptation even though the pathological responses received more explicit attention by Grinker and Spiegel.

MEN UNDER STRESS AND CRISIS THEORY

One of the by-products of psychiatry's heterogeneity and increased size is the possibility that analogous concepts may develop somewhat independently in diverse segments of the field. To a significant extent the concepts of crisis and of stress seem to illustrate the pattern of separate and even isolated growth. The concept of crisis has evolved rapidly during the 1960's, though its origins are most often traced back to Lindemann's pioneering work (1944). The Laboratory of Community Psychiatry in the Department of Psychiatry and the School of Public Health at Harvard under Caplan's leadership has played a key role in developing crisis theory. Caplan (Caplan, 1961; Caplan, Mason, and Kaplan, 1965) has contributed to the definition of crisis and colleagues at the laboratory have broadened the concept in their investigations of critical transitions (Rapoport, 1967). The impetus for crisis theory has been identical to several of the major forces resulting in the community psychiatry "movement" of the 1960's (Sabshin, 1966a, 1966b). On the one hand there has been the need to begin scientific approaches to primary prevention (Caplan, 1964; Bloom, 1968), including the identification of high-risk groups prior to their entrance into crisis periods. There has also been belated recognition that increased services provided for those who were experiencing crises might afford an excellent opportunity to reduce the incidence and prevalence of mental disorders. A variety of human experiences have been investigated in this context, though the

primary emphasis has been placed on acute situations that produce turbulent reactions in a large segment of the population undergoing such
experiences (Miller and Iscoe, 1963). These have included natural disas-
ters (Tyhurst, 1957), the birth of a premature child (Caplan, 1960), reactions to forthcoming surgery (Janis, 1958), sudden bereavement (Lindemann, 1944) and mourning (Pollock, 1961), and severe physical illness
(Visotsky, Hamburg, Goss, and Lebovits, 1961). Simultaneous with the
increasing attention to these situations as special foci for the delivery of
improved mental health services, there has developed increasing conviction that during periods of crisis many people are more amenable to significant personality change. Though the scientific basis for such an hypothesis is meager at the present time, the assumptions of greater
plasticity of human behavior during crises have provided additional motivation for so-called crisis intervention (Rapoport, 1962; Harris, Kalis,
and Freeman, 1963; Parad, 1965). Community mental health clinics
have adopted such intervention as their major mode of therapy with the
goal of facilitating individuals and groups in coping with a variety of
threatening stimuli. By providing services close to where people live
and work, it is hoped that there will be a decrease in the need for psychiatric hospitalization and a reduction in the likelihood for protracted
disability.

The similarity of crisis intervention to the techniques of treatment described in *Men under Stress* is striking. Indeed, the history of military
psychiatry since World War II (Glass, 1955) indicates that psychiatric
leaders in the Department of Defense have taken full advantage of the
concepts introduced by Grinker and Spiegel and have developed model
programs of primary as well as secondary prevention. What is surprising
is the apparent overlooking of *Men under Stress* by many of those who
are now advocating crisis intervention in the community mental health
centers throughout the country. In part this lack of awareness reflects
the fact that crisis theory seems to have evolved independently of stress
theory and the cross references are surprisingly sparse. Grinker has had
remarkable influence on the stress literature in the field of psychological
research and in psychosomatic medicine. The direct applicability of
many of his concepts to crisis theory has yet to be discovered, but there
is every reason to predict that *Men under Stress* and Grinker's models of
transactional processes will have substantial influence on the scientific
evolution of crisis intervention.

Efforts to develop a conceptual integration between stress and crisis
theories are highly likely to occur during the 1970's. Indeed, such efforts
have already begun without being identified explicitly as intending to
achieve such an integration. Our current typologies of crises are crude,
and, in the absence of adequate empirical testing, there is as yet no consensus regarding the most useful classificatory system. Erikson's admittedly oversimplified dichotomization between developmental and acci-

dental crises (Erikson, 1959) has been a helpful starting point for a few investigators. Though some writers (Davis, 1970) have pointed out that such dichotomization appears to be arbitrary and inapplicable under certain circumstances, there has been insufficient attention given to empirical investigations of the similarities and the differences between the methods utilized by various individuals in coping with simultaneous developmental and accidental crises. Even rarer are research data related to the sequential transactions between an individual's coping with an accidental crisis and the resultant positive impact on one of his later developmental stages (for example, in certain cases parents who have coped with a fatally ill child may improve their competence or adaptive capacity to subsequent vicissitudes in their lives) (White, 1959; Chodoff, Friedman, and Hamburg, 1964; Sabshin, Futterman, and Hoffman, 1971). Though much has been written regarding maladaptive sequences (for example, in many cases failure to solve an adolescent identity crisis may lead to subsequent difficulties in dealing with occupational stresses), there have been few studies testing specific hypotheses in this area. An obvious consequence of meager anterospective predictions has been the lack of attention to the negative cases where the predictions have proven to be inaccurate. More subtly, the gaps in empirical investigations of adaptation in addition to maladaptation in specific populations have helped to sustain distortions in conceptualizing human behavior resulting in a pronounced bias toward pathology within psychiatry and in other mental health fields (Offer and Sabshin, 1966).

Several recent developments in psychiatry may contribute simultaneously to a reduction of this bias and a serious commitment to empirical investigations of both accidental and developmental crises. Community psychiatry, despite its internal conceptual weaknesses and the plethora of its critics (Sabshin, 1969), affords much greater opportunities for empirical investigations of adapting and maladapting populations than has been available heretofore (Sabshin, 1968). Though crisis intervention has evolved pragmatically in the community setting, it will be possible and, indeed, necessary to develop a more adequate crisis theory if community psychiatry is to succeed in secondary prevention and to evolve a more rigorous primary preventive approach. Despite excellent efforts to collect and to interpret data (Visotsky et al., 1961; Coehlo, Hamburg, and Murphey, 1963; Smith, 1966; Hamburg and Adams, 1967), we are still lacking a cohesive theory to explain why certain groups and individuals, purportedly of high risk, adapt to unexpected and life-cycle stresses with adequate or even superior coping. Our understanding of crisis and of stress will be grossly inadequate until much more data is available about coping to an even wider variety of crises. When coupled systematically with increased specific information about maladaptation to stress, a broader, more integrated model will emerge by a process in which the stress and the crisis fields will be incorporated into a coherent entity. Grinker's specific contributions to stress theory and his more generalized model of transactional fields will play an important part in this unifying process.

Men under Stress can stand on its own merit as a fundamental contribution to our understanding of psychological stress and psychopathology. Grinker's (1956) later attempt to conceptualize a unified theory of human behavior had its antecedents in *Men under Stress* and his subsequent formulations about psychosomatic processes (Grinker, 1953). The need for a better understanding of human crises and stresses should impel a new generation of mental health workers to rediscover *Men under Stress*. Hopefully, Grinker's twenty-five year progression from studies of the wartime stresses to transactionalism and field theory will have begun to be repeated in psychiatry as a whole when a new conceptual model is developed by the integration of the stress and the crisis fields. Such integration, in turn, will help to stimulate further interest in an even broader based unified theory of human adaptation and maladaptation, which will borrow heavily on Grinker's past work and on his future contributions.

REFERENCES

Bloom, B. L. 1968. The Evaluation of Primary Prevention Programs. In L. M. Roberts, N. S. Greenfield, and M. H. Miller, eds., *Comprehensive Mental Health: The Challenge of Evaluation*. Madison, Wisc.: University of Wisconsin Press. Pp. 117–135.
Caplan, G. 1960. "Patterns of Parental Response to the Crisis of Premature Birth: A Preliminary Approach to Modifying Mental Health Outcome," *Psychiatry* 23:365–374.
Caplan, G. 1961. *An Approach to Community Mental Health*. New York: Grune and Stratton.
Caplan, G. 1967. *Principles of Preventive Psychiatry*. New York: Basic Books.
Caplan, G., Mason, E., and Kaplan, D. 1965. "Four Studies of Crisis in Parents of Prematures," *Community Mental Health Journal* 1:149–161.
Chodoff, P., Friedman, S., and Hamburg, D. A. 1964. "Stress, Defenses, and Coping Behavior: Observations in Parents of Children with Malignant Disease," *American Journal of Psychiatry* 120:743–749.
Coelho, G., Hamburg, D. A., and Murphey, E. 1963. "Coping Strategies in a New Learning Environment," *Archives of General Psychiatry* 9:433–443.
Davis, D. R. 1970. "Depression as Adaptation to Crisis," *British Journal of Medical Psychology* 43:109–116.
Erikson, E. H. 1959. *Identity and the Life Cycle*. New York: International Universities Press.
Glass, A. J. 1955. "Principles of Combat Psychiatry," *Military Medicine* 117:27–33.
Grinker, R. R., Sr., 1953. *Psychosomatic Research*. New York: Norton.
Grinker, R. R., Sr., ed. 1967. *Toward a Unified Theory of Human Behavior* (1956), 2d ed. New York: Basic Books.
Grinker, R. R., Sr., and Spiegel, J. P. 1945. *Men under Stress*. Philadelphia: Blakiston.
Hamburg, D. A., and Adams, J. E. 1967. "A Perspective on Coping Behavior: Seek-

ing and Utilizing Information in Majc Transitions," *Archives of General Psychiatry* 17:277–284.

Harris, M. R., Kalis, B., and Freeman, E. 1963. "Precipitating Stress: An Approach to Brief Therapy," *American Journal of Psychotherapy* 17:465–471.

Janis, I. 1958. *Psychological Stress*. New York: Wiley.

Lazarus, R. S. 1966. *Psychological Stress and the Coping Process*. New York: McGraw-Hill.

Lindemann, E. 1944. "Symptomatology and Management of Acute Grief," *American Journal of Psychiatry* 101:141–148.

Mechanic, D. 1962. *Students under Stress: A Study in the Social Psychology of Adaptation*. Glencoe, Ill.: The Free Press.

Miller, K. S., and Iscoe, I. 1963. "The Concept of Crisis: Current Status and Mental Health Implications," *Human Organization* 22:195–201.

Myerson, A. 1946. "Review of *Men under Stress*, by R. R. Grinker., Sr., and J. P. Spiegel," *American Journal of Psychiatry* 103:138–139.

Offer, D., and Sabshin, M. 1966. *Normality: Theoretical and Clinical Concepts of Mental Health*. New York: Basic Books.

Parad, H. J., ed. 1965. *Crisis Intervention: Selected Readings*. New York: Family Association of America.

Pollock, G. H. 1961. "Mourning and Adaptation," *International Journal of Psychoanalysis* 42:341–362.

Rapoport, L. 1962. "Working with Families in Crisis: An Exploration in Preventive Intervention," *Social Work* 7:48–56.

Rapoport, R. 1967. "The Study of Marriage as a Critical Transition for Personality and Family Development. In P. Lomas, *The Predicament of the Family*. New York: International Universities Press. Pp. 169–205.

Sabshin, M. 1966a. "The Boundaries of Community Psychiatry," *Social Service Review* 40:245–282.

Sabshin, M. 1966b. "Theoretical Models in Community and Social Psychiatry." In L. M. Roberts, S. Halleck, and M. Loeb. eds., *Community Psychiatry*. Madison, Wisc.: University of Wisconsin Press. Pp. 15–30.

Sabshin, M. 1968. "Toward More Rigorous Definitions of Mental Health." In L. M. Roberts, N. S. Greenfield, and M. H. Miller, eds., *Comprehensive Mental Health: The Challenge of Evaluation*. Madison, Wisc.: University of Wisconsin Press. Pp. 15–27.

Sabshin, M. 1969. "The Anti-community Mental Health 'Movement,'" *American Journal of Psychiatry* 125:471–477.

Sabshin, M., Futterman, E. H., and Hoffman, I. 1971. Empirical Studies of Healthy Adaptation. In preparation.

Smith, M. B. 1966. "Explorations in Competence: A Study of Peace Corps Teachers in Ghana," *American Psychologist* 21:555–566.

Tyhurst, J. S. 1957. "The Role of Transition States: Including Disaster in Mental Illness," *Symposium on Preventive and Social Psychiatry*, Walter Reed Institute of Research. Washington, D.C.: Government Printing Office.

Visotsky, H. M., Hamburg, D. A., Goss, M. E., and Lebovits, B. Z. 1961. "Coping Behavior under Extreme Stress," *Archives of General Psychiatry* 5:423–448.

White, R. B. 1959. "Motivation Reconsidered: The Concept of Competence," *Psychological Review* 66:297–333.

Normality:
For an Abnormal Age

Psychiatry and clinical psychology need a theoretical and empirical grasp on normality for several reasons. For these clinical disciplines, the abnormal, the personally and socially deviant and problematic, is home base. Yet, any conception of the abnormal requires, at least implicitly, a view of what is normal. If the view remains implicit, it may be based on assumptions that embody factual error and covert values; far better that the standard of comparison by which abnormality is identified be explicit and subject to empirical correction.

If the clinician needs a conception of normality as counterpoise to his preoccupation with the abnormal, as a comparative basis for establishing the meaning of abnormality, he has a distinguishable though related reason for concern with normality in his need to clarify his therapeutic goals. These goals will depend on many considerations: on his theories about psychological functioning, on the social context of his practice (for example, mental hospital vs. consulting room vs. community [see Smith, 1968]), on his own values and ideals, on pragmatic judgments about the kinds and degree of psychological change that can be achieved and sustained. What he thinks and knows about normality will surely affect how his therapeutic goals are formulated.

Still a third reason for concern with normality is more theoretical. The broad impact of psychoanalysis on psychology and other sciences of human behavior highlights a methodological assumption that has gained wide acceptance—that our understanding of the normal can be greatly advanced by a close look at the deviant, in whom processes that are hidden from observation in the normal person are prominently displayed for analytic dissection. Productive as this strategy has undoubtedly been, it has not been immune to challenge on principled grounds as potentially misleading (for example, Asch, 1952; Allport, 1961). Certainly, normal functioning deserves scrutiny in its own right, and, on the face of it, it is likely that the complementary methodological principle is equally justified: Good understanding of normal functioning should usefully modify our thinking about the abnormal. Such a principle, indeed, follows from the general systems approach that Grinker (1967a) espouses. Traffic in this methodological direction has dealt primarily in the currency of general and developmental psychology and of single-variable research; the potential contribution of holistic personality studies (paralleling the attention clinical research gives to abnormal "cases") and of multivariate research has been richly exemplified (for example, White, 1952; Heath, 1965) but far from realized.

But the meanings of normality slip around elusively as we encounter the term in these different contexts. A useful catalog of current usage has recently been provided by Offer and Sabshin under Grinker's sponsorship (Offer and Sabshin, 1966; Sabshin, 1967). They distinguish four functional perspectives on normality: (1) normality as health; (2) normality as utopia; (3) normality as average; and (4) normality as process.

1. Normality as health in effect treats it as a residual category left over after identified cases of illness or malfunction are excluded. People who are not sick, and that is most of us, are normal. This, as Sabshin (1967) observes, is the traditional medical-psychiatric approach. It fits our accustomed patterns of thought and practice and research and so remains widely prevalent in spite of telling objections that have been raised by many recent critics. Thus, Szasz (1961) and other more temperate critics among whom I count myself (Smith, 1968) object to the medical characterization of much psychological malfunctioning, rejecting the term "illness" as metaphoric and unwarranted in this application. Other critics, notably Clausen (1968), point to the slippery criteria by which the psychiatric case is identified in practice and in epidemiological research. The seeming solidity of the cases to which normality is residual in this approach dissolves on close inspection. Jahoda (1958), in her classic review of concepts of positive mental health, rejects "absence of mental disease" as a criterion of mental health (read "normality" in the present context) on the grounds that what is regarded as mental disease is subject to wide cultural variation and also that our conceptual framework should be kept open to the possibility that man's resources of strength and areas of vulnerability and malfunction may vary in some independence of one another. That is, we should not decide arbitrarily that health and illness can usefully be represented as opposite ends of a single dimension. For all these reasons I will henceforth ignore this first perspective on normality as an obstacle to clear thinking. What is pragmatically useful in the approach can be reformulated in one or another of the three remaining ones.

2. By normality as utopia Offer and Sabshin designate the approaches to normality that treat it as an ideal, or set of ideals, of optimal functioning. In this usage, nobody is completely normal. The facets of ideal

normality that have been proposed in thoughtful discussions of psychotherapeutic goals by psychoanalysts of various persuasions, by
humanistic-existential therapists, and others have been well cata-
loged by Jahoda (1958), whose book is firmly planted in this perspective and demonstrates that to be utopian in the sense of positing
idealized evaluative criteria is fully compatible with an empirical orientation. Elsewhere (Smith, 1969b), I have argued that clarity about the
frankly evaluative nature of such utopian criteria has the merit of avoiding the surreptitious advocacy of values in scientific-professional disguise. The conditions and consequences of placement on any particular
evaluative dimension, however, are entirely an empirical matter, knowledge about which may affect the choices we make as we assign relative
priority to competing values.

3. Normality as average is the straightforward statistical notion, useful, indeed unavoidable, for descriptive purposes. When, as in the case
of psychiatric research, the tedious work of conducting normative studies to define the normal in this sense has largely been skimped, pioneer
efforts such as Grinker's (1963) description of the homoclites and Offer's
of suburban adolescent boys (Offer, 1969) can provide a useful corrective to untested assumptions about what is in fact prevalent in human
dispositions and behavior. More systematic studies in this vein are being
initiated by the National Health Survey of the National Center for
Health Statistics (see Sells, 1968).

The trouble with this approach to normality lies not in its proper descriptive use but in its ready misuse, its confusion with the other meanings of normality. What is prevalent need not be biologically healthy or
desirable in terms of other evaluative criteria. All too often it is not. And
depending on the intrinsic nature of the dimensions or variables on
which one is averaging, the ideal, utopian norm may lie near the mid-
point of the scale (for example, flexibility as intermediate between rigid
overcontrol and impulsive undercontrol) or it may just as well be located near one end or the other (for example, intelligence, capacity for
intimacy, mendacity). The judgment as to what is healthy or desirable
—or important—does not follow from the descriptive statistics.

4. Normality as process is the most elusive category because, more
than with any of the other meanings, this one hinges especially on the
development of our scientific theories, which are still primitive. Offer
and Sabshin classify here conceptions of normality as processes tending
toward adaptation, whether in the perspective of individual development or of evolutionary or sociocultural processes. As Grinker has long
argued (Grinker, 1967a, 1967b), general systems theory provides a
widely applicable abstract framework within which, in principle, it is
possible to identify processes that promote the maintenance and growth
of the system (organism, personality—or society; the framework is very
abstract) and processes that comprise decompensation and regression.

To the extent that our empirically based theory can be carried forward to fill in this abstract paradigm, we will have attained evaluative criteria of normality that rest on a solid empirical footing. They will still be evaluative, and the value choice involved will still remain optional, like all value choices (the saint or martyr or rebel may set other values higher than those of system growth and maintenance). But evaluative criteria rooted in a well-matured systems theory should have an intrinsic connection with empirical fact that our present criteria largely lack (see Smith, 1969c).

For the present, the process approach to normality is best represented by the various recent attempts to conceptualize and study effective coping behavior (for example, Murphy, 1962; Haan, 1963; Hamburg and Adams, 1967). Short of the full development of a systems theory of personality, there will be wide agreement that active coping processes will in one way or another be part of an adequate formulation. The more we understand about them the better.

Grinker, in his collaborative study of mentally healthy young males half facetiously, half seriously, gave them the neologistic diagnostic label "homoclites" to legitimize them for psychiatric inquiry and put aside, for the nonce, intrusive problems of values (Grinker, Grinker, and Timberlake, 1962). The primary value of this characterization of very ordinary young men is as a benchmark for comparison, not only with the abnormal seen in clinical practice but also with the idealizations of normality held by highly educated and culturally sophisticated mental health professionals. The fact that Grinker's homoclites, undergraduates at the YMCA-related George Williams College with its tradition of muscular Christianity, were not statistically representative of the general population of young American men but were rather a fairly homogeneous group suitable for composite description adds to their stimulus value in provoking us to rethink our conceptions of normality.

Though the design of the study is rather primitive, Grinker's commentary on his findings is, as we might expect, thoroughly sophisticated. In offering the homoclites as a version of average normality, he wastes little time in wonder at the contrasts they provide with the normal abnormal experience of psychiatry and psychoanalysis, though he vividly conveys the sense that acquaintance with them was eye-opening. From a critical standpoint, Grinker employs his homoclites to question the relevance of current utopian views of normality, which may have arisen from a narrow cultural and experiential basis. He states (Grinker, 1963, pp. 128–129):

The psychiatrist is educated, trained, and experienced in psychopathology and the treatment thereof. . . . In addition to knowing only patients, by virtue of his geographical location he also sees mostly those engaged in the rat-race of city life. Finally, he is caught in his own middle-class perspective.

The ordinary person has simple and reasonable values. He wants to feel good, work well, love and be loved, play and enjoy life occasionally and have hope for the future. There are wide ranges and many permutations of these values. He settles or is willing to settle for less than he originally hoped for, holding to Freud's adage that life is difficult but it is all we have.

What we as psychiatrists see to be the goal of American families includes: upward mobility regardless of intellectual, aptitudinal or social fit; doing and becoming which is operationally goal-changing rather than goal-seeking; permissiveness rather than boundary fixing of behavior, work, strict religious belief, and discipline; and child-rearing according to the latest fad based on current theory.

The difficulty is that these cultural values which Spiegel showed not to be entirely held by Irish-Americans or Italian-Americans are also not cherished by upper-lower or lower-middle class and later Protestant-American main-streeters in Kansas, Minnesota, or Illinois or in America as a whole.

Thus what is normality and what is mental illness is confused because of the value discrepancies among psychiatrists, people, and cultures.

In a word, psychiatrists and other mental health professionals may have applied out of their appropriate context utopian versions of normality that are specific to particular subcultures, with resulting confusion.

In broadening the sociocultural base in which behavior is evaluated, Grinker's study contributes to a more adequate conception of "normality as average." But Grinker is admirably clear that the homoclites, widely distributed throughout America as they undoubtedly are, represent only one kind of average normality, a relatively complacent type with limited ambitions better suited to maintaining a stable society behind the leaders than to giving leadership or to adapting in rapidly changing times. Within a broader adaptational frame of reference—normality as process —Grinker (1963, p. 131) speculates that the homoclites may have achieved their version of normality at the cost of "the limitation of patterned behavior suitable for a few roles and a restricted range of environments. . . . What prepares the developing boy for multiple roles and a wide range of environments *and* the behavioral and psychodynamic criteria of health, this study cannot answer."

Grinker's homoclites—and their conceptual descendants, Offer's modal adolescent boys in the suburbs (Offer, 1969)—are, one suspects, the psychiatric equivalent of President Nixon's "silent majority." I would not want to align myself with those observers who, in Grinker's words, "have developed a kind of moral judgment signifying intense anxiety when they imply the question, 'Should the homoclite exist at all?'" (Grinker, 1963, p. 129). Clearly homoclites do exist, and their version of the human condition also deserves respect. But just as I question the long-run viability of the prejudices and narrow goals that Mr. Nixon finds and cultivates in his silent constituency, so I would underline Grinker's doubts about the adaptive limitations of his homoclites. They will no doubt always be with us, and they contribute stability and their own version of sanity to the social mix. But they are likely to find the modern flux increasingly baffling. They will make little contribution to solving our urgent social problems, which their reactions to bafflement are already complicating. We badly need to rear citizens to other, more flexible and autonomous forms of normality. I am calling, let it be plain, for a particular utopian version of normality as especially relevant to our present predicament. In the remainder of this essay, I shall try to explicate these utopian criteria and sketch some of the present basis for thinking that they can be grounded empirically—in my judgment, a major task of the coming years for research on psychological development and functioning in social contexts.

Grinker provides me with an advantageous point of departure in his presidential address from which I have been quoting. The passage is one in which he lays the basis for confronting his homoclites from the adaptational perspective. Grinker (1963, pp. 130–131) writes:

The success of prediction regarding health depends on the possible relationship between person and environment. *The important question in the current fast-moving and changing social and cultural world is what stresses are, or will be, impinging on the individual.*[1] With environments no longer stable even in the previously primitive cultures, the individual is required to make extremely rapid changes. Mental health thus depends less on stability but more on the flexibility of the individual.

This is indeed an important question, one that as we have seen raises serious questions about the viability of the homoclitic adaptation. Heinz Hartmann's "average expectable environment" (Hartmann, 1958) is becoming a will-o'-the-wisp—or a nightmare. But in the present, newly salient perspective of warranted concern over sheer human survival— surely the ultimate adaptational context—we are now compelled to frame an even more important set of questions: In the current fast-moving and changing social and cultural world, which in so many respects is becoming manifestly more stressful, less livable, and may be speeding toward irretrievable disaster, how can individuals organized in society gain control over these stresses? How can they divert the ominous trends of population, pollution, escalating armament, and dehumanization of technology run rampant? How can they avert the disaster that looms so threateningly? And how—here the special competences of psychology and psychiatry become involved—how can we rear and educate people who are capable of reconstructing their human situation, rather than flexibly accommodating to the inevitable? How can we foster autonomy and realistic self-direction and political efficacy among people who have been reared to adjust—the socially oppressed and excluded, even the homoclites?

[1] Italics mine—M. B. S.

In one of my earlier attempts to grapple with the relation of human values to concepts of mental health (Smith, 1969b), I argued that mental health is not a theoretical concept but merely a rubric for evaluations of human personality and that lists of mental health criteria are inherently arbitrary, reflecting standards of evaluation that we hold to on essentially nonscientific grounds, though we may revise our standards in the light of empirical relationships. Subsequent consideration of the intrinsic requirements of different major contexts of mental health practice—institutional psychiatry, the psychoanalytic consulting room, and most recently, the community as seen in the now-aborted War on Poverty—led me to begin to back away from this rather unsatisfying relativism (Smith, 1968). From the point of view of the public interest inherent in community psychology, the appropriate criterion for emphasis seemed to me to be one of human effectiveness vs. ineffectiveness. This was at least a relativism in a broader societal framework. I would now go a step further. Just as the newly emerging sense of urgent human priorities is giving rise to an insistent demand for a rescaling of national priorities, so it seems to me to cut through academic relativism and point unequivocally to new priorities among utopian criteria of normality, priorities that are called for by the hard adaptational facts. Whether it makes sense for psychologists and psychiatrists to respond to the new urgencies depends, of course, on whether our conceptual equipment, our research and technical knowledge, can in fact be brought cogently to bear. My intent in what follows is to pursue related themes in recent research and theorizing far enough to suggest that such is indeed the case.

I draw the text for my utopian message from Jacqueline Grennan Wexler—formerly Sister Jacqueline—who recently wrote (1969): "The question haunting society is whether or not the individual at all controls his own destiny." This is the question that underlies the adaptational issues I have just raised: Can man gain control over the societal processes that threaten to overwhelm him? The question is likely to evoke ambivalent feelings of hope and despair. In the light of our vastly increased knowledge and technological competence, what were once accepted as acts of God—fated natural calamities beyond the reach of human choice or control—are now displayed as failures of man, susceptible in principle to rational, planful attack. Yet the complexity of the problems that beset us and of the crowded urban society in which we live tends to leave us feeling more powerless than ever. The new potentialities for choice, for control over our destiny, do not produce the actuality.

It is also the question on which a remarkable diversity of contemporary reform movements converge. Whether the context be the new nationalism of the former colonial world, the strident claims for self-determination symbolized by Black Power and Student Power, or the efforts to break through the custodial tradition in prisons and mental hospitals, the focus now is on augmenting people's capacities to take charge of their own lives. Authoritarianism is in ill repute, and paternalism no longer seems a benign and acceptable policy.

If whether the individual can control his own destiny is indeed the haunting question of our times, is it one to which psychological science can contribute? In spite of recurrent discussions of autonomy as a psychological condition or variable (see, especially, Angyal, 1941; Riesman, Denney, and Glazer, 1950), psychology has largely been tongue-tied about this problem. Our metatheories and paradigms have not equipped us to deal with it coherently. We spot the time-worn paradoxes of free will lurking to bemuse us, and we throw up our hands in retreat.

In launching American psychology with its greatest classic, William James (1890) faced the problem directly, but left it dramatically unresolved. His psychology is at once explicitly deterministic and voluntaristic. In the spirit of science, James adopted its deterministic methodological premise—the commitment to push to the limits the search for causes. But this did not shake his belief in human freedom, and his psychology could still include the will as a traditional chapter, long since read out of respectable academic psychology where it is replaced by motivation.

Modern psychologies have typically chosen one or the other horn of James's dilemma. Both for behaviorism in its various versions and for psychoanalysis, free will is an illusion. Our experience of choice as human actors is denied validity. Models of man emerge that appear to be radically incompatible with the assumptions about human nature that underlie democratic political institutions. A reductionistic, mechanistic version of determinism—a determinism that becomes a dogmatic principle rather than a methodological commitment—can undergird programs for the manipulation and control of behavior; it does not provide a language in which we can talk meaningfully about personal or political freedom, about ethical or political responsibility, about personal or democratic choice.

The existential and humanistic doctrines that Maslow (1966) calls "third force psychology" react against what they regard as the dehumanizing tenets of both behaviorism and psychoanalysis. They pick the other horn of the dilemma and opt for human freedom as a dogmatically given absolute. But in the process, they abandon the hard-won gains of deterministic science (Smith, 1966). Grinker (1970) characterizes "this kind of existentialism"—the concern of "third force psychology" with an idealized, finalistic view of human potentiality—as "our modern delusionary system devised to alleviate the pain associated with the abandonment of certainty and meaning." I would not go so far; it seems to me that Maslow, Rogers, May, and Fromm have suggestive contributions to make on matters of human import that have been neglected by behavioristic psychology and by psychoanalysis. But they are too ready to give up the advantages of inquiry that is governed by the rules of the scientific game—corrigibility and cumulativeness. Their view of man as a free agent is essentially nonempirical and therefore vulnerable.

If this must indeed be our choice, between a scientific deterministic psychology that disparages man and a humanistic voluntarism that idealizes him at the cost of discarding science, it is easy to see why so many of our students are fleeing from scientific psychology. But it is a Hobson's choice and, as I have become convinced, an unnecessary one.

The convergent lines of thought and investigation that I will be discussing lead to a view of free will not as an illusory paradox ("we have to act as though we have it, but we don't") or as a metaphysical postulate but as an empirical variable. In this view some people have more free will than others. The extent to which they enjoy freedom is subject to causal analysis. It can be changed, increased, or decreased. And it has important consequences, as well as causal antecedents. Our orienting question, of whether the individual at all controls his destiny, becomes an empirical one, within the bounds of psychological science. We will see that there are conceptual handholds for coming to grips with it, and also relevant though imperfect measures that are beginning to generate usable data. In other words, I am not conjuring up a speculative solution to a philosophical impasse; I am attempting rather to explicate and advertise some new ways of thinking, the full import of which has not been adequately recognized.

The empirical free will that I am talking about is not a matter of arbitrary fiat or chance, of course. It does not hinge on a layman's naïve interpretation of Heisenberg's principle of indeterminacy. The indeterminate is not what people have meant by freedom. It certainly provides no basis for a conception of ethical responsibility. What we mean by freedom, rather, is personal causation or self-determination, causal processes with ascertainable antecedents in which the self figures as an agent. The antonym of freedom is not determination but constraint. Freedom is limited by causal processes that bypass the self or constrain its options of choice; it is enhanced by processes that increase one's range of choice and one's resources for attaining what one has chosen.

But it is time to begin fleshing in these abstractions. The best case study I know to exhibit self-determination as an empirical variable and to suggest some of the considerations that may bear on its deliberate nurture is Claude Brown's contemporary classic, *Manchild in the Promised Land* (1965). Most often read for its shocking and realistic account of ghetto life, it is also a sophisticated psychological drama, the story of Brown's dawning realization that he could be the architect of his life.

Growing up in the Harlem culture of poverty, the young Claude as we encounter him is caught in a vicious circle. Life is a hopeless jungle. The most one can aspire to is successful predation in the street life and an early death. The self-sustaining dynamics of the culture of poverty have been described and illustrated by Oscar Lewis (1959, 1966) and by

Chilman (1966); they appear vividly in the Harlem world of the young Claude Brown. The critical missing ingredient is hope. In a realistically hopeless situation, people dependably reinvent a defensive fatalism, time and again the world over. To hope is to be disappointed. Fatalism is at least a tenable posture toward a life situation that would otherwise be insupportable. One lives in a narrowly restricted present, and one gets one's kicks where and when one can. But the fatalistic adjustment is not just a private invention; it is a culturally transmitted solution to life's problems, a solution that does not solve the problems, but makes life possible. Storefront religion, cynical predation, and drug-induced oblivion are part of the culture; they do not have to be reinvented.

The culture of poverty is a trap, a vicious circle, because its hopeless prophecies are self-confirming. If you do not hope, you will not try, and if you do not try you do not acquire the skills or take advantage of the opportunities that might make hope warranted, even under these difficult conditions. People enmeshed in the culture of poverty are very low in the variable that we are interested in. They have little capacity for self-determination, little effective free will.

In Claude's case, there were mitigating features from the start that laid an essential basis for his subsequent escape. For reasons that he cannot explain, Claude was no ordinary street boy; at an early age he set out to become a first-class scrapper and hustler. Constitution and early experience had clearly endowed him with an ample supply of the motivation to have effects on one's environment that Robert White (1959) has analyzed in his discussion of the concept of competence. Because of his underlying hopelessness, however, Claude's deviantly channeled competence motivation seems heedless and self-destructive. During his early years Claude could realistically aspire to be a big man on the street, which he indeed became. But though he knew the street life as "nasty, brutish, and short," it did not occur to him that he could escape it to a better one. His beliefs about self and world drastically limited his actual freedom of choice, beyond the severe limitations imposed by social realities.

The drama that unfolds concerns Claude's realization, at first faltering, then progressively firmer, that escape was possible, that there were steps he could take to bring it about, and that he could commit himself to take those steps. It is a story of many backslidings and of ultimate success—of a working through in real life of changes in his self-concept that leave him, at the end, able to choose his life, not merely to adjust to the miserable realities that were initially dealt out to him. This change in Claude identifies the variable I am talking about.

As a case history, *Manchild in the Promised Land* is very suggestive about causal factors in this progression toward greater self-determination and freedom. Close study would identify many clues about such factors; here I will only suggest a few that stood out in my reading.

Preeminent is the effect on Claude's self-concept of receiving the full respect and trust of impressive adult figures whom he could idealize. Two
such figures play prominent roles in his story: Ernst Papanek of the Wilt-
wyck School, and the Rev. William James. Brown makes it explicit that
his experience of their seemingly unwarranted but also toughminded acceptance of him, respect for him, and confidence in him made a crucial
difference in his feelings about himself; his account also illustrates how
tentative and uncertain the process of change can be. The causal mechanism here might be labeled the "Quaker principle": Treat a person with
respect to make him worthy of respect.

One can also find evidence in Claude's story for the importance of
successes—"reinforcements" in the current jargon—that added to his resources for further coping. Claude's widely recognized accomplishments
as a hustler were surely important in sustaining him through difficult
times, and in making it possible for him to revise his conception of what
he could do with his life. When, finally, he could reject the street as a
bad life in which he had nevertheless been successful on the street's own
terms, neither his delinquent friends nor he himself could accuse him of
copping out from weakness or failure.

Fortuitous traumatic experiences may also have played their constructive part in saving Claude from two extreme hazards of the street life
that could easily have undercut all the positive influences—heroin and
guns. After a particularly nasty initial experience with heroin, Claude
had no further truck with it. It was no temptation, and his revulsion
against the junkies' way of life, as it engulfed his brother and his close
boyhood friends, helped to propel him from the streets where its use was
so common and so visibly destructive. Perhaps his distaste for guns had
a similar traumatic basis. At all events, we learn that the necessity to
carry a gun and to be prepared to use it was a main reason for Claude's
deliberate withdrawal from the life of a big-time hustler, just as he was
rising to a level of conspicuous success.

The positive and negative import of social support also comes through
clearly. To choose a new life meant leaving Harlem, disengaging from
old friends, finding new ones. No more than anyone else was Claude independent of his social environment, either early or late in the story.
His achievement, a major one, was to become capable of choosing his
environment. He could not have rebuilt his life within the old environment. "Autonomy" is never absolute.

And, finally, one gets from Claude's life story a feeling for the spiral,
cumulative nature of the processes of change as they involve the self.
When one is locked into a vicious circle of hopelessness and failure, the
first steps out are very difficult, and backsliding is certain. That is what
it means to be "locked in." But successes, insecure at first, cumulate in
their effects. One gains in confidence. Eventually one comes to the point

where one can profit even from failures, not be bowled over by them. Near the end of Claude's story, the reader has confidence that, come what may, Claude is in command of his life. The vicious circle has been transformed into a benign one, in which the consequences of his new-found self-direction will predictably tend to sustain him in his revised conception of himself and of what life can be. Life may still contain neurotic conflicts, and even tragedies, but it will be his.

This sketchy account of a rich personal document makes it plain, I hope, how self-determination can be conceived as a complex empirical variable that falls within the causal framework of a deterministic psychology. I turn now to some convergent strands of recent research that suggest its fruitfulness in more systematic inquiry. They hinge on the suggestion that what a person thinks and feels about himself makes a crucial difference to his effective freedom.

A first approximation to the variable that concerns us was presented not long ago by Rotter (1966; see also Lefcourt, 1966) under the ponderous heading, "generalized expectancies for internal versus external control of reinforcement." Perhaps Rotter's most important contribution was a simple pencil-and-paper scale that has lent itself to wide and, thus far, productive use. I now think that the scale captures the distinction among people that I have been trying to draw rather imperfectly. But it undoubtedly hits the general target area, so the relationships that it has turned up are interesting.

Rotter came to devise his measure via a circuitous route. In the framework of his social learning theory of personality, he and his students had been doing experimental studies of the very different dynamics of task performance when the performing person believes that his outcomes result from his own skill and effort as compared with when he believes that they are the product of fate or chance. Reinforcement is under internal control in the former instance, under external control in the latter —in the eyes of the experimental subject. It occurred to Rotter that he could supplement his experimental manipulations of internal vs. external control to gain greater predictability if he were able to take into account individual differences in his subjects' expectancies about whether, in general, their outcomes were the product of their own skill and abilities or of fate and chance. Hence the I–E scale, made up of items paired for choice in which the respondent has to choose between skill and ability vs. fate and chance as the source of major kinds of outcomes in his life. The scale, once launched, has had a life of its own. Standardized on Ohio State students (who are more like homoclites, I imagine, than are University of Chicago students), the measure yielded a single general factor.

As Rotter (1966, p. 25) summarized the findings that bear on the validity of the scale and hence on its interest for us, there is

strong support for the hypotheses that the individual who has a strong belief that he can control his own destiny is likely to (a) be more alert to those
aspects of the environment which provide useful information for his future behavior; (b) take steps to improve his environmental position; (c) place greater
value on skill or achievement reinforcements and be generally more concerned
with his ability, particularly his failures; and (d) be resistive to subtle attempts
to influence him.

These are surely earmarks of a coping orientation.

The defects of the scale in its present form arise from its relation to
Rotter's experimental setting, in which ability, skill, and effort are contrasted with fate and chance. It will be remembered that the items in
the scale pose the same forced choice. This might be an adequate choice
if fatalists from the culture of poverty were to be contrasted, in a more
innocent society than that of today, with believers in the Horatio Alger
myth. But for alienated or dissident youth and for blacks imbued with
the ideology of the Black Power movement, these are not the only alternatives. Perceived external control may not rest with fate but with "the
Man" or the system. And the perceived resources for personal control
may reside not only in the individual himself, but also, vicariously, in
his identification with a charismatic leader or a powerful movement and,
quite realistically, in the possibilities of joint action. Gurin, Gurin, Lao,
and Beattie (1969) have shown that distinctions of this kind have to be
drawn when the I–E scale is applied to Negro samples. Not surprisingly, Negroes who blame the system are more likely to favor collective action. But in their measurements, these investigators have not
freed themselves from Rotter's paired-choice format, and the job of reconstructing an instrument that can map out what now appears to be a
complex of core beliefs about self and world still remains to be done.
Meanwhile, the generally intelligible relationships that Rotter's I–E
scale has produced give much support to the promise of this approach.
They also indicate that people may be more able to give honest answers
to these questions than to questions that bear on the delicate, ambivalent, and therefore highly elusive matter of self-esteem (see Wylie
[1968] for the difficulties encountered with self-esteem measures).

Meanwhile, Richard de Charms (1968, pp.273–274), whose theoretically elaborated account of personal causation is akin to my own, had
arrived at a distinction quite similar to Rotter's, between man as Origin
and as Pawn.

That man is the origin of his behavior means that he is constantly struggling
against being confined and constrained by external forces, against being moved
like a pawn into situations not of his own choosing. . . . Play that is forced becomes work; if one can choose his work without regard to external pressures
and necessity, it takes on many of the aspects of play. . . . An Origin is a person who perceives his behavior as determined by his own choosing; a Pawn is
a person who perceives his behavior as determined by external forces beyond
his control. . . .

The personal aspect is more important motivationally than objective facts. If the person feels he is an Origin, that is more important in predicting his behavior than any objective indications of coercion. Conversely, if he considers himself a Pawn, his behavior will be strongly influenced, despite any objective evidence that he is free. An Origin has strong feelings of personal causation, a feeling that the locus for causation of effects in his environment lies within himself. The feedback that reinforces this feeling comes from changes in his environment that are attributable to personal behavior. This is the crux of personal causation, and it is a powerful motivational force directing future behavior. A Pawn has a feeling that causal forces beyond his control, or personal forces residing in others, or in the physical environment, determine his behavior. This constitutes a strong feeling of powerlessness or ineffectiveness.

The research that de Charms reports creates Origin-like and Pawn-like situations experimentally. He has not as yet published a measure of individual differences in the Origin-Pawn variable, but rather assimilates Rotter's I–E data to his purposes.

My own entry to this area came from attempting to understand observations I had made of the experience and performance of Peace Corps volunteers. In working over my data and those of my colleague Raphael Ezekiel (1968) I found myself turning to Robert White's (1959) concept of competence to formulate what made the difference between able young men and women who rose to the challenge with full commitment and performed with corresponding effectiveness, and equally able young people who responded to the same objective situations as frustrating, not challenging, and spent their energies on adjusting, not coping (see Smith, 1968). White's proposal of a biologically intrinsic motive to have effects on one's environment—effectance, he called it—is more modest and, I think, more tenable than de Charms's postulate (1968, p. 269) that "Man's primary motivational propensity is to be effective in producing changes in his environment," but they are talking about the same human characteristic. Competence in dealing with the environment, in White's view, rests on this motivational foundation.

The more effective volunteers among a generally impressive group, who had joined the Peace Corps when it was still a great unknown, seemed more than the others to have preserved this responsiveness to challenge with which most infants start life amply endowed. By this point in their young adult lives, of course, it was no longer the somewhat randomly directed effectance of infancy. Their readiness for engagement and commitment was very much integrated into their selves. I came to think that generalized attitudes toward the self, viewing the self as worthy of being taken seriously (self-respect) and capable of producing desired effects, lay at the core of the motivational complex that involved them in benign circles of challenge, coping, accomplishment, and hope rather than vicious ones of passivity or frustration, defense, failure, and fatalism. This cluster of self-attitudes seemed to me the common thread linking White's competence, Rotter's internal control of reinforcement, and de Charms's sense of being an Origin. I now propose it as a utopian version of normality that is especially cogent to our present adaptive predicament. It permits us to translate Jacqueline Wexler's (1969) question into psychological terms that we are beginning to know how to work with.

Bits of evidence, often collected under seemingly unrelated labels, are beginning to emerge and to fall into place concerning the family relations in childhood, the educational practices, the kinds of life situations that contribute to people's ability to take charge of their own lives. I am not prepared to review them here, nor is there space for such a review. Rather, I would call attention to a consequence that follows theoretically, I think, from the fact that the dispositions we are concerned with are reflexive self-attitudes. To an important degree, the self is or becomes what one thinks it is; the self is the prime domain of the self-fulfilling prophecy (Merton, 1957). That is, what one believes about one's causal efficacy affects what one tries and what one does. We have seen evidence for this in the story of Claude Brown, in the correlates of Rotter's I–E measure, and in de Charms's observation that the feeling of being an Origin creates the fact of acting like an origin. In the sphere of the self, the self-fulfilling prophecy is a prime causal mechanism. This is the basis, I think, for the Quaker principle as illustrated in the transformation of Claude.

Of course there are many problems in this sketchy formulation that badly need to be clarified. What, for example, of the constraints of reality? Can illusory vicarious power create a sense of efficacy that lays the psychological basis for making it more realistic later on? What is the role of myth in compensating for but perpetuating the actual status of Pawn ("opium of the people")? In helping Pawns become Origins? The heated discussions among liberal psychologists that are predictably evoked by the more romantic reconstructions of Black history and the more unrestrained expressions of Black Power ideology make it evident that we are not clear about the answers.

To avoid misunderstanding, I must close by making explicit the limited intent of this essay. I decided to highlight a utopian version of normality that contrasts strikingly with Grinker's homoclites because I believe that better understanding of how to foster this particular brand of normality is attainable and important. But I still agree with Jahoda (1958) that to characterize positive mental health—utopian normality—requires multiple criteria. And in a day of Women's Liberation, I cannot help recognizing that there is a male bias to the criterion that I have selected for emphasis. The homoclites were males, Offer's adolescents are males, and my Promethean phrasing of competence is probably a masculine-slanted version. Within the female role as modern culture presently defines it, there are ways in which women, too, can be Origins, though the culture builds in highly probable Pawn-like ingredients. We

clearly need a new psychology of women, which encompasses the present cultural definitions—so different from those that prevailed in
Freud's day—and also transcends them in seeking to identify enduringly
distinctive directions of fulfillment that arise from the biology of sexual
differentiation.

I think it a good guess that, culture aside, men will more frequently
be moved to produce large and satisfying environmental effects—from a
feminine point of view, often absurd ones, no doubt—and women more
often predisposed to cultivate the values of human responsiveness and
love that make the enterprise humane and worth the candle. But even
this remains in doubt, and the range of individual differences in each
sex is surely large. At any rate, the expressive-compassionate component
of the feminine role and perhaps of the female biological bent need not
impose Pawnhood. As our present imprecise ideas about effectance and
self-determination become elaborated in research, we may hope that the
versions that apply to the two sexes will receive equal attention. We risk
fatuity if we neglect sex differences in developing our utopian criteria of
normality, and we do injustice if we unthinkingly apply male-oriented
standards to both sexes.[2]

[2] Since completing this essay, I have been impressed and influenced by Steiner's
(1971) integrative review of a quite different, mainly experimental, literature of social
psychological research under the rubric of "perceived freedom." There is much in
his contribution that complements mine.

REFERENCES

Allport, G. W. 1961. *Pattern and Growth in Personality*. New York: Holt, Rinehart
& Winston.
Angyal, A. 1941. *Foundations for a Science of Personality*. New York: Common-
wealth Fund.
Asch, S. E. 1952. *Social Psychology*. Englewood Cliffs, N.J.: Prentice-Hall.
Brown, C. 1965. *Manchild in the Promised Land*. New York: Macmillan.
Chilman, C. S. 1966. *Growing Up Poor. An Over-View and Analysis of Child-Rear-
ing and Family Life Patterns Associated with Poverty*. Washington, D.C.: Division
of Research, Welfare Administration, U.S. Department of Health, Education and
Welfare.
Clausen, J. A. 1968. "Values, Norms, and the Health Called 'Mental': Purposes and
Feasibility of Assessment." In S. B. Sells, ed., *The Definition and Measurement of
Mental Health*, U.S. Public Health Service Publication, no. 1873. Washington,
D.C.: National Center for Health Statistics. Pp. 116–134.
de Charms, R. 1968. *Personal Causation: The Internal Affective Determinants of Be-
havior*. New York: Academic Press.
Ezekiel, R. S. 1968. "The Personal Future and Peace Corps Competence," *Journal of
Personality and Social Psychology*, 8 (no. 2), monogr. suppl., pt. 2.
Grinker, R. R., Sr. 1963. "A Dynamic Story of the 'Homoclite.'" In J. H. Masserman,
ed., *Science and Psychoanalysis*, vol. 6. New York: Grune & Stratton. Pp.
115–134.
Grinker, R. R., Sr. 1967a. "Normality Viewed as a System," *Archives of General Psychiatry* 17:320–324.

Grinker, R. R., Sr. 1967b. *Toward a Unified Theory of Human Behavior* (1956), 2d ed. New York: Basic Books.

Grinker, R. R., Sr. 1970. "The Continuing Search for Meaning." *American Journal of Psychiatry* 127:25–31.

Grinker, R. R., Sr., with the collaboration of Grinker, R. R., Jr., and Timberlake, J. 1962. "'Mentally Healthy' Young Males (Homoclites)," *Archives of General Psychiatry* 6:405–453.

Gurin, P., Gurin, G., Lao, R. C., and Beattie, M. 1969. "Internal-External Control in the Motivational Dynamics of Negro Youth," *Journal of Social Issues* 25 (no. 3):29–53.

Haan, N. 1963. "A Proposed Model of Ego Functioning: Coping and Defense Mechanisms in Relation to IQ Change," *Psychological Monographs* 77 (no. 8, whole no. 571).

Hamburg, D. A., and Adams, J. E. 1967. "A Perspective on Coping Behavior: Seeking and Utilizing Information in Major Transitions," *Archives of General Psychiatry* 17:277–284.

Hartmann, H. 1958. *Ego Psychology and the Problem of Adaptation.* New York: International Universities Press.

Heath, D. H. 1965. *Explorations of Maturity.* New York: Appleton-Century-Crofts.

Jahoda, M. 1958. *Current Concepts of Positive Mental Health.* New York: Basic Books.

James, W. 1890. *The Principles of Psychology,* 2 vols. New York: Henry Holt.

Lefcourt, H. M. 1966. "Internal versus External Control of Reinforcement," *Psychological Bulletin* 65:206–220.

Lewis, O. 1959. *Five Families: Mexican Case Studies in the Culture of Poverty.* New York: Basic Books.

Lewis, O. 1966. *La Vida: A Puerto Rican Family in the Culture of Poverty—San Juan and New York.* New York: Random House.

Maslow, A. H. 1966. *The Psychology of Science: A Reconnaissance.* New York: Harper & Row.

Merton, R. K. 1957. "The Self-Fulfilling Prophecy." In R. K. Merton, *Social Theory and Social Structure,* rev. ed. Glencoe, Ill.: The Free Press. Pp. 421–436.

Murphy, L. 1962. *The Widening World of Childhood: Paths toward Mastery.* New York: Basic Books.

Offer, D. 1969. *The Psychological World of the Teenager: A Study of Normal Adolescent Boys.* New York: Basic Books.

Offer, D., and Sabshin, M. 1966. *Normality: Theoretical and Clinical Concepts of Mental Health.* New York: Basic Books.

Riesman, D., in collaboration with Denney, R., and Glazer, N. 1950. *The Lonely Crowd: A Study of the Changing American Character.* New Haven: Yale University Press.

Rotter, J. B. 1966. "Generalized Expectancies for Internal versus External Control of Reinforcement," *Psychological Monographs* 80 (no. 1, whole no. 609).

Sabshin, M. 1967. "Psychiatric Perspectives on Normality," *Archives of General Psychiatry* 17:258–264.

Sells, S. B., ed. 1968. *The Definition and Measurement of Mental Health.* U.S. Public Health Service Publication, no. 1873. Washington, D.C.: National Center for Health Statistics.

Smith, M. B. 1966. Review of A. H. Maslow, *The Psychology of Science: A Reconnaissance, Science* 153:284–285.

Smith, M. B. 1968. "Competence and 'Mental Health': Problems in Conceptualizing Human Effectiveness." In S. B. Sells, ed., *The Definition and Measurement of Mental Health,* U.S. Public Health Service Publication, no. 1873. Washington, D.C.: National Center for Health Statistics. Pp. 99–114.

Smith, M. B. 1969a. "Competence and Socialization" (1968). In M. B. Smith, *Social Psychology and Human Values: Selected Essays.* Chicago: Aldine. Pp. 210–250.

Smith, M. B. 1969b. "'Mental Health' Reconsidered: A Special Case of the Problem of Values in Psychology" (1961). In M. B. Smith, *Social Psychology and Human Values: Selected Essays.* Chicago: Aldine. Pp. 179–190.

Smith, M. B. 1969c. "Research Strategies toward a Conception of Positive Mental Health" (1959). In M. B. Smith, *Social Psychology and Human Values: Selected Essays*. Chicago: Aldine. Pp. 165–178.

Smith, M. B. 1969d. *Social Psychology and Human Values: Selected Essays*. Chicago: Aldine.

Steiner, I. D. 1971. "Perceived Freedom." In L. Berkowitz, ed., *Advances in Experimental Social Psychology*, Vol. 5. New York and London: Academic Press. Pp. 187–248.

Szasz, T. S. 1961. *The Myth of Mental Illness: Foundations of a Theory of Personal Conduct*. New York: Harper-Hoeber.

Wexler, J. G. 1969. "Campus Revolution: Social and Political Responsibilities," *Illinois School Journal* 49:171–179.

White, R. W. 1952. *Lives in Progress: A Study of the Natural Growth of Personality*. New York: Dryden.

White, R. W. 1959. "Motivation Reconsidered: The Concept of Competence," *Psychological Review* 66:297–333.

Wylie, R. C. 1968. "The Present Status of Self Theory." In E. F. Borgatta and W. W. Lambert, eds., *Handbook of Personality Theory and Research*. Chicago: Rand McNally. Pp. 728–787.

Transactional Psychotherapy

The invitation to write a chapter in a *Festschrift* volume designed to honor the many-sided contributions of a figure of such significant and diverse impact on the field of clinical psychiatry as Roy R. Grinker, Sr., can be variously responded to. I have chosen to discharge this pleasant task in my assigned area of transactional psychotherapy by tracing out his position in and influence on the field through me as I discern it (and to the extent that I properly reflect it), in a manner that is personal as much as it is general and that is characterized by the spirit of friendly disputation that can be in the best service of the scientific goals, the accretion of empiric knowledge, and the forging of useful theory, to which we both adhere. I will attempt this by showing how despite (or perhaps more appropriately, by way of) the argument of differences, of disagreement in vantage point or in conceptualization, my own professional development has been significantly affected by some of the major interests and influences that Grinker has so cogently and persuasively represented. In this sense I have chosen to make this a personal account; where I do this by reference to my own contributions, it is to show in them the further extension or my particular way of fulfillment of ideas and trends that Grinker has represented and expressed in his life's work. At the same time I will do this in the spirit stated, of facing differences squarely, not glossing them over in encomiums that are anyway unnecessary to one who properly warrants the honor of a *Festschrift* volume in the first place, as indeed Roy Grinker so richly does.

The task of assessing Grinker's place in the field of psychotherapy as a whole and in relation especially to that variant he designated transactional psychotherapy is, even when properly circumscribed, a significant one. In order to accomplish it in a focused way, it needs to be delimited. I will therefore not discuss the general theory of transactionalism as it defines an approach to the understanding of organismic functioning (and within that, of psychological functioning) advocated by Grinker and characterized by fields of multiple transactions, without operative levels or hierarchies, marked by multicausal approaches to multivariable problems (Grinker, 1965),[1] but I will discuss only what Grinker calls the transactional variant of psychotherapy, thus avoiding the argument, unnecessary in this context, as to whether the two usages of "transactional" connote sufficient identity of meaning to justify the same rubric. Nor will I discuss Grinker's excursionary polemics into the (political) espousal of the goal of self-conscious eclecticism as a corrective that he feels necessary against what he deems to be unwarranted "over-emphasis on the principles and methods of psychoanalytic psychiatry" (Grinker, 1964) in the field of clinical psychiatry, since (to anticipate the argument I will develop at some length) I think Grinker actually has deviated more in name than in substance from those self-same principles and methods of psychoanalytic psychiatry (and pari passu is less "eclectic" in his theoretic stance than he fancies). Nor will I discuss another such straw man that I think Grinker (1966) has confusingly adduced that it is the so-called superseding of "libido theory (a closed system) . . . by adaptational theory (an open system)" that has brought "biology and sociology back into focus," since Grinker's avowed transactional approach to psychotherapy does not, I think, necessarily rest on the adaptational psychodynamic viewpoint (of Rado) nor is it generally agreed that adaptational theory provides a necessarily more congenial framework for including biological and sociological considerations than does classical analytical theory.

To turn now to my central task, within the caveats thus declared, to delineate the nature and the significance of Grinker's contribution to the field of psychotherapy (its practice, its teaching, and its research) and to

[1] Stated summarily by Grinker (1965) as follows: "In my opinion we should approach living human beings as if they existed in a total field of multiple transactions without connotations of significance, hierarchical importance, or conceptual devices called levels. Thereby we avoid the dichotomies of nature vs. nurture, organic vs. functional, lower vs. higher, or reduction vs. extension. Furthermore, we can operationally behave in dealing with multivariable problems as if we really believed in multicausality of both healthy and disordered function. This is *transactionalism* in a total field whose constituents range from physiochemical to symbolic foci" (italics mine—R. S. W.). And psychiatry, therefore, "as a science is a composite field in which several disciplines using various methods from appropriate frames of reference are concerned with transactions (reciprocal underlining interrelationships). Each frame of reference and method is based on theoretical constructs somewhat different from all others." Here he means psychoanalysis, along with constitution, biogenetics, culture, etc., "all components of an extended field of which intrapsychic processes are but a part." And it follows that "the neurological or social sciences cannot explain behavior nor can the psychiatric behaviorist do without them. How should we consider the *operational* relationship of these other approaches of bordering fields to our own? Do these many techniques study hierarchical series of levels each with its own system or variables? Do we 'borrow' information from other sciences or extrapolate from them by analogy? Or is there a more fruitful way of viewing relationships?" Relevant to this inquiry is the assertion made in another paper (Grinker, 1966) stating, "I believe that behavioral transactional researches constitute a valuable bridge between internal psychological and biological processes on the one hand, and social, cultural, and economic (ecological) factors on the other hand."

spell out the very real and very significant issues in the field to which he has addressed himself and toward the resolution of which he has formulated the approach designated "transactional psychotherapy." These considerations can be set forth in summary form as follows (not necessarily in his—or my—order of importance or logical sequencing), and with Grinker's viewpoint I trust always clearly implicit in the assertion through which the issue is posited:

1. Psychoanalytic psychology and psychoanalytic psychotherapy as a conceptual explanatory framework for (a) the training of psychiatrists (and other mental health clinical professionals) so that they can be understanding of and psychologically helpful to individuals suffering emotional and behavioral distress and for (b) the treating of the patients who come to psychiatric clinics, social agencies, and general psychiatric practices cannot be simply conceptualized as a diluted direct application of psychoanalysis, a wholesale "baby psychoanalysis," or an analyzing (insofar as possible) sitting up and at lesser frequencies and intensities than classical psychoanalysis. Rather, the theory and the practice of dynamic psychotherapy, though rooted in, and in complex relationship to, its psychoanalytic parentage had had (in America at least) a substantial, vigorous, and disputatious (see: Grinker's own work) development of its own. Already more than a decade and half ago, in the years 1952 to 1954, within the official structure of the American Psychoanalytic Association a sequence of four full-day panels (English, 1953; Ludwig, 1954; Rangell, 1954; Chassel, 1955) was devoted to the comprehensive discussion of the range of issues—and of controversies—both conceptual and technical, comprised in the varying understandings of the complex relationship between psychoanalysis and the so-called dynamic psychotherapies. In 1966 I essayed to delineate the current state of all these issues (in the same spirit that has always moved Grinker's concerns) in a long book essay that reviewed eight major books in the field of psychotherapy—its theory, its practice, its research—in the context of the historical development of the field out of its origins in the scientific psychology innovated by Freud. Later (Wallerstein, 1969) I systematized this survey in a broadened context in a statement of the major scientific issues within this area of the relationship of psychoanalysis and psychotherapy as a sequence of questions with consideration of the major, often quite sharply dichotomized, positions that have been taken in relation to each of them by typical outstanding proponents of the various viewpoints. Grinker's views have always been represented in this dialogue, though not always under his personal banner of transactional psychotherapy.

2. There is need for what Grinker calls an "operational approach" to the teaching of dynamic psychotherapy (or the variant he calls "transactional psychotherapy") that renders the therapeutic activity appropriate

to the structural requirements and possibilities within the clinical setting, to the needs of the kinds of patients being served, to the teaching
opportunities and limitations, and to the realities of trainee and staff
life, including deployment, turnover, and the like. Most psychotherapy
teaching and supervision he has felt to be considerably overfocused on
but one aspect of the interacting field of forces, the specific psychological sequences and responses within the patient. By contrast, Grinker has
persistently underscored the dyadic and transactional nature of the clinical educational enterprise. This is precisely the spirit that came to in-
fuse the book by Ekstein and myself (1958) on the teaching and learning
of psychotherapy, built around the parallel and the complementary
psychological processes to be contemporaneously studied in the interacting systems of patient, student-therapist, supervisor, and administrator.

3. A proper major focus for the psychotherapeutic endeavor as distinct from psychoanalysis proper is on current rather than in any significant way on remote (infantile) conflict, what Grinker (and others) have
called primary involvement with the "here and now." Grinker (1961)
says of this, in describing transactional psychotherapy, "We do not emphasize the so-called genetic processes or the past experiences of childhood . . . past experiences of dissatisfaction. These form the neurotic
core of the personality and will persist." But rather, "We are content to
work with what the psychoanalysts call derivative conflicts. . . ." Gill
has given fullest expression to the theoretical underpinning of this view
that there can be sufficient resolution of such derivative conflict without
resolution of the related basic conflict (which presumably would require
full psychoanalysis?). Gill (1954) stated, "I would still like to hold open
the question that even though the basic conflict is unsolved and under
sufficient stress can once again reactivate the derivative conflicts, the derivative conflicts develop a degree of autonomy and exist in a form
which allows a relatively firm resolution even under psychotherapeutic
techniques. . . ." Unless this last statement is substantially true, much of
the argument advanced by Grinker and all others on the possibility and
the sufficiency of dynamic psychotherapy as an enterprise distinct from
psychoanalysis becomes untenable. This same position has actuated our
Psychotherapy Research Project at the Menninger Foundation (Wallerstein and Robbins, 1956) in its delineation of differential indications for
the deployment of a range of psychotherapeutic modalities geared to the
differing character and conflict structures of patients, each individually
considered in terms of what is necessary and what is possible of change.
And within this, of course, goals can be set both minimally and maxi-
mally. To quote from Gill (1951), "the choice of therapy may be divided
into that which determines the minimum necessary to restore the ego to
functioning, and that which strives for the maximum change that is possible."

4. The single case, carefully and intensively scrutinized, still serves as

the prime exemplar of our clinical method, our major source for the ac-
cumulation of clinical knowledge, the chief vehicle of our training enterprise (the transmission of accumulated clinical lore and wisdom to new
generations of students) and the wellspring as well of the scientific hy-
pothesis-generating activity in the clinical field. Yet an *n* of 1 or of very
few poses fundamental problems for generalization in empirical science,
generalization built ordinarily across a sampling of many. Nonetheless,
arguments have been made among both statisticians and clinicians in
support of the position that for many clinical research purposes more
can perhaps be learned from a smaller than from a larger number of
cases; and even in the experimental literature of general psychology, instances have been adduced under which the even more limiting condition of $n=1$ can still mark an appropriate and useful, and even the only
possible, research strategy. This supporting literature has been recently
summarized (Wallerstein and Sampson, 1971) in a long review of issues
in research in the psychoanalytic process. And this viewpoint, which
Grinker champions, is integral to the philosophical and methodological
underpinning of the Psychotherapy Research Project of the Menninger
Foundation, which no matter what its number (an *n* of 42 actually) is
not a group-averaging phenomenon but rather a succession of single
cases, built around individual prediction studies, and so on (Wallerstein,
Robbins, Sargent, and Luborsky, 1956).[2]

5. There is a paucity of adequately conceived evaluation and assessment research in the clinical field in general and in relation to the effectiveness of psychotherapy in particular. This makes it particularly difficult to measure in any, even imprecise sense, the usefulness of Grinker's,
or any other, model of the psychotherapeutic process, of how character
and illness structure are formed, and how change in them is induced.
Eysenck's oversimplistic thinking about the issues—and the solutions—
involved in clinical evaluation has been decisively faulted by an array
of clinician and researcher respondents.[3] But that Eysenck has failed to

---

[2] There are, of course, aspects of the data analysis of the project that rest on the
grouping of data and on group differences.

[3] These have been summarized very briefly in my review (1966) of the *Current
State of Psychotherapy* as follows: "The many ways in which the cogency of
Eysenck's data and conclusions can be faulted have already been amply documented
in the public aspects of the rancorous exchanges that subsequently ensued: that the
groups compared were not comparably ill and that it was not a matter of *random*
self-selection as to which kind of group an individual turned up in, hospitalized, outpatient analytic, eclectic therapeutic, insurance disability claimant, etc.; that the different groups in different initial illness states, starting at vastly different levels of
functioning had therefore widely different margins for improvement; that the therapies employed differed widely encompassing almost every known theoretical viewpoint, every kind of treatment setting, every variation in concomitant use of adjunctive agents such as drugs, etc.; that each of the therapeutic approaches was taken to
be unitary in nature and uniform in application, without regard to the discipline,
training, or experience of the therapists; that there were no agreed-upon criteria as
to what constituted improvement or cure—in fact, no statement other than the ac-

establish that psychotherapy has no demonstrated effect does not by it-self prove the converse, that therefore it does. The issue still remains an empirical one. The Psychotherapy Research Project of the Menninger Foundation is the most ambitious among those dedicated to its system-atic study and resolution (Wallerstein et al., 1956). The various concep-tual and methodological difficulties in the way of this empirical resolu-tion have been delineated in great detail (Wallerstein and Sampson, 1971). An emergent in that review is that Grinker's impassioned plea on behalf of the research effort is today being heeded by more and more—and even more important, in increasingly sophisticated ways.

6. And lastly in this catalog, Grinker has called attention as insis-tently as anyone to the language and terminology confusions that beset our field, confusions, for example, between the scientific language of the-oretical explanation and the clinical language of therapeutic inter-change. Waelder (1962) spoke of six levels of conceptualization in psy-choanalysis from the most specific, tied to observables, the level of observation, to the most abstruse and metaphysical, the level of philoso-phy. Each has a unique language appropriate to itself. Dealing on one level in the language of another is one of the central confusions rampant in clinical work; an egregious example is talking the language of ab-stract theory rather than the language of direct experience with patients. Nonetheless, the levels and the phenomena particular to each must be congruent with one another and (here I would differ with Grinker) we need not turn our back on psychoanalytic thinking (theorizing) appro-priate to the theoretic context, as Grinker I think mistakenly posits we need to, in order to be experientially relevant in the clinical context of therapist-patient interchanges. In his effort to create the theory for the transactional psychotherapy approach, Grinker feels he has had to cir-cumscribe the use of psychodynamic (psychoanalytically based) theory to the understanding of the "underlying motivations, conflicts, and de-fenses" alone, as if otherwise there would necessarily ensue the confus-ing use of the language of metapsychological theory to explain (and to substitute for) the concrete interactional exchanges of the therapy (truly a straw man as Waelder so clearly pointed out), or the equally confus-ing inappropriate use of the techniques of psychoanalysis where the techniques of psychotherapy are indicated—just because the theoretical language of both analysis and therapy is the same, the language of psy-choanalysis (another straw man, I think, as I will try to spell out at length in what follows). But, on the fundamental point of the differing

ceptance of what each reporter used (in his private judgment) as his simple (or sim-ple-minded) unitary and global improvement criterion; that differences in goals striven for that would underlie the standard against which improvements would be assessed were not considered; and that in numerous ways, the control groups hardly adequately qualified as such, including even the unaccounted-for elements of treat-ment that pervaded, to a variable and unmeasured degree, the so-called untreated groups."

languages appropriate to the differing conceptual contexts, one can only hold with Grinker's thrust.

In all then, Grinker has stood clearly athwart the major progressive trends in the evolution of the theory and practice of modern-day psychotherapy—and in the training for it and the research on it. My personal notations have demonstrated, I trust, the reflections in the one instance of the manner in which the influences represented by a figure such as Grinker have pervaded an entire field of endeavor. What then is the quarrel alluded to in the opening phrase concerning the spirit of "friendly disputation" within which this critical essay is offered? Essentially, it has to do with what Grinker takes to be the core of his contribution to the development of psychotherapy theory, the effort to eschew the psychoanalytic theory of personality and of therapy (as the guiding theory underlying the psychotherapy endeavor) and the postulated need to replace it with that presumed blend of field and role and communication theory that Grinker calls the "theory of transactional therapy." To make my counterargument that Grinker not only need not make this switch (which would I feel, if it were possible—as a substitution—be a retrogressive step) but that he, in fact, has not made it and is, despite his protestations to the contrary using the very same psychoanalytic theory that he feels he has surmounted—to make this argument, I will need to offer extended examples from the two published versions of Grinker's major presentation in this area, each entitled "A Transactional Model of Psychotherapy" (Grinker, 1959, 1961).

Grinker (1961) begins his exposition quoting (apparently approvingly) Fromm-Reichmann's contention that there is no valid therapy other than psychoanalysis or that based on psychoanalytic principles. He says in seeming extension of this, that "of all the theoretical systems in psychiatry, psychoanalytic psychodynamics yields most satisfaction because of its completeness, its sense of closure and its analogical fit" (Grinker, 1961) and that therefore "the basic core of our pedagogical processes in psychiatric training is the psychodynamics of Freudian psychoanalysis" (Grinker, 1959), and moreover "American psychotherapy has imitated and closely approached the psychoanalytic model" (Grinker, 1961). But then he introduces a presumed conceptual difference in the requirements of theory and the requirements of practice when he says "This theoretical system, in general, I also consider to be the essence of the best modern *psychodynamic theory*. However, I contend that the theory has a *minor place in the operational procedures of psychotherapy*" (Grinker, 1961), and extends this further as follows, "Actually, psychoanalytic theory, laudable as it may be as a system fruitful for viewing the psychological processes of humans, places a *heavy load of interference* on the psychotherapist who attempts to understand and to help a sick person through psychotherapy." [4] The basic flaw appears to Grinker to be the prejudicing perspective created by an overarching conceptual framework within which the data of observation are to be viewed and organized into meaningful configurations. Presumably this operates especially within the psychoanalytic theoretical framework because of the central significance of interpretation as a therapeutic intervention (change agent) in psychoanalytically based and psychoanalytically understood psychotherapeutic modalities. Grinker (1961) says, "It is exactly in this operation (conveying a hypothesis or an inference, i.e. an interpretation) where understanding is *preconceived* rather than *learned in the therapeutic situation,* that psychoanalytic theory enters into the interpretation of symbolic content of communications and the choice of therapeutic interferences." [5] These are therefore "loaded operations" (Grinker, 1959). (It would be a digression here from my task of developing Grinker's viewpoint to raise the question, so fully discussed in the literature of the philosophy of science—see Polanyi [1959]—as to whether it is ever possible or would be even desirable to be free from theoretical predilection in the empirical pursuit in science, or, put differently, whether the problem is not one of being able to transcend the constraints of theory while still being supported by it and bound within the predispositions with which it colors our observing and our thinking.)

Grinker (1959) then asks, "Can we set up a model of treatment based on an operational theory of psychotherapy, not on a theory of psychodynamics or diagnosis? If so, it should be derived from the empirical operations involved in psychotherapy." Here he relies in the first instance on role theory, since, after all, the parties to psychotherapy enact a variety of roles within a "nondimensional space" traversed by verbal, nonverbal, and paralingual communications. He distinguishes explicit (instrumental) roles related to problem-solving as the psychotherapeutic task and implicit (expressive) roles, the latter based on complex early internalizations and identifications. Needless to say, these are automatic patterns and do not imply artificiality. The "explicit roles structure the relationship; the content is furnished by the implicit roles . . . for a therapeutic effect, the to-and-fro play of implicit roles furnishes the content of the relationship derived from each actor's personality, healthy or sick" (Grinker, 1961). When there is complementarity of roles between the two actors in the psychotherapeutic drama, "few decisions are made, there is little strain." Conversely, when role complementarity gets disturbed, a disequilibrium ensues within which is the opportunity for significant therapeutic gain. This is stated as follows, "therapy involves behavior of a patient in terms of his role in transaction with the therapist. Complementarity of roles when established represents stability and har-

[4] Italics mine—R. S. W.
[5] Italics mine—R. S. W.

mony and is conducive to *communication of information*. Disequilibrium because of non-complementarity eventually results in re-equilibrium, representing the disruption of an old repetitive process and the establishment of a new system. When this is attained by modification mutually achieved, there has occurred a learning process" (Grinker, 1959). To this spelling out of his incorporation of elements of role theory, Grinker adds aspects of field theory ("We adopt field theory to emphasize the extent of influences surrounding the two-person system of therapist and patient.") and communication theory (since "messages are received, acknowledged, and corrected in a cyclic transaction, which changes in time by virtue of the communications within the transaction that express forms of role performance having explicit and implicit meanings indicating past learning and identifications, as well as current relearnings, which we term therapy"). Thus, "In summary, I envisage a model for psychotherapy that encompasses aspects of several modern major theories of human behavior." This is, all told, "transactional . . . the process is reciprocal and cyclical . . . and ever changing."

How then does the psychotherapy look that is conducted within this avowedly different conceptual framework? It will be simplest to follow the discussion of the elements in the sequence in which Grinker presents them. He talks first (Grinker, 1961) of repetitive behavior patterns displayed within the therapy.

Activities within a transactional process, although they deal with current reality and start with well-defined explicit roles, expose the repetitive nature of the patient's unadaptive behavior and stimulate his recall of past experiences. Some of these are preconscious, and some are unconscious, but the orientation of the therapist remains in the present transaction within the field in which both members of the transaction find themselves. It can be stated quite certainly that this transactional approach evokes implicit expressive or emotional roles and incites repetition of old transactions and illuminates the genetic source of the current behavior. However, it does not require focusing on and interpretation of a vast, uncharted area of unconscious. On the contrary, it enables the therapist to orient himself in a special situation with a specific person in a transactional relationship which can be understood by common-sense evaluation of ordinary modes of communication.°

To anticipate a little my overall conclusion, here is encapsulated both the operating strength and the conceptual weakness of the stance that Grinker essays. He deals clearly with the incontrovertible phenomena of repetition and transference (though not by those names), starts with the model and vocabulary of role theory and works back to a description of good dynamic psychotherapy (an expressive psychotherapy, psychoanalytically based, but not psychoanalysis, avoiding the "vast, uncharted area of unconscious"), no doubt appropriate to that particular clinical

---

° Reprinted with permission from R. R. Grinker, Sr., "A Transactional Model for Psychotherapy," in Morriss I. Stein, ed., *Contemporary Psychotherapies*, (New York: Free Press, 1961).

context, which is however not described. He then (in my view) seriously undersells his effort by calling it the application of a commonsense psychology instead of the uncommon sense that is psychoanalytic psychology, devoted to rationally comprehending the irrational.

Somewhat further on, Grinker (1959) does acknowledge specifically that he has dealt in other terms with the analytic concept of transference, "Although we do not invite, and we certainly avoid, the development of a transference neurosis whenever possible, we are obviously still dealing with transference phenomena, which is another way of saying that we are dealing with back-and-forth implicit communications between therapist and patient in which the present is colored by the past." Incidentally, in connection with this discussion of transference manifestations, and with an eye no doubt, to the possible workings of the countertransference, Grinker then goes on to describe the therapist's (ideal) attributes in the therapeutic transaction in a manner that for the most part cannot be faulted: He describes the desire to help a suffering individual, the feeling that the patient is worthwhile and the optimism that the patient can change and grow, the respect for the patient, and so on. He is obviously on more controversial ground, as much within psychoanalysis as in his own chosen framework, when he states that "Respect for the patient also involves honesty on the part of the therapist in admitting his positive and negative feelings as they develop within the therapeutic transference. . . . [To do otherwise] is to block or distort the communication process."

Grinker then goes on to discuss the goals of psychotherapy. Here he tries to draw a distinction between the transactional approach and the more customary psychodynamic approach. He contrasts the more usual setting of long-range goals that "establish a theoretical bias" and a set of "formulated psychodynamics" that lead us then to fit the patient's communications into "the preconceived theoretical plan" with his own ostensibly more flexible and short-range approach where goals are not preset, cover only "successive short spans which involve specific therapeutic foci" (Grinker, 1959) and mostly are stated to be "a *result* of the transactional experiences between the therapist and patient. Slowly but surely, the motivation, the degree of illness, the capacity to endure the suffering of therapy, the ability to learn, as well as the special efficacy of the therapist in a particular case, becomes clear" (Grinker, 1961). Surely here Grinker describes primarily the difference between bad (rigid, dogmatic) therapy and good therapy; in good therapy, whatever the putative theoretical framework, there will be constant oscillation between, on the one hand, the overview and the long-range plan within which the elements as they appear are conceptualized and ordered, and, on the other hand, the openness and the readiness to be surprised (Reik, 1936) that will render the therapist available to new insights and revised formulations.

Grinker discusses the activity of the therapist and the technical use of dreams in therapy in like manner; that is, he contrasts the transactional approach and (his version of) the more customary psychodynamic approach. The transactional approach is declared to be "considerably more active than those workers who model their therapy after the psychoanalytic pattern. . . . We are active in that we choose the focus, we communicate adequately with the patient, avoiding long silences and the impassivity of a non-participant. Not only do we choose the focus for the subject of communication but we decide when a transaction should be dropped and a new one adopted" (Grinker, 1959). Here too, I feel the distinction in part specious. For again, in a good (that is, competent) psychotherapy the degree of therapist activity will be appropriately (and flexibly) tailored to the ever shifting psychic situation and therapeutic needs of the patient. To the extent that Grinker claims for the transactional approach a systematically greater degree of activity in the sense indicated ("we choose the focus . . . we decide when a transaction should be dropped and a new one adopted"), he is actually imposing the very preconceptions and prior formulations on the treatment that in his discussion of goal-setting he declared a special fault of psychoanalytically based therapies that the transactional approach would properly correct.

And as for dreams, again in presumed contrast to psychoanalytic psychotherapies, in the transactional approach, "We do not invite dreams, nor when they are recounted, do we interpret them. . . . As a result of the absence of emphasis on dream material, the patient soon learns not to use the dream as a means of avoiding direct communication. He does not practice remembering them, nor does he assume that without them communication is not possible" (Grinker, 1959). Here the proper contrast of course is between psychoanalysis as a specific therapy where certainly dream interpretation plays a central, or at least, a very significant, technical role, and all other psychotherapies (including the psychoanalytically based) where dreams are worked with much less, if at all, and usually in relation to the fit of their manifest content into the psychic situation and the transferences of the dreamer, rather than being "interpreted" (the translation of the manifest content into latent contents via undoing the dream work by dream interpretation based on following the dreamer's associations). By this point, a general trend seems to have emerged in this review of Grinker's transactional approach. The contrasts he draws (and very properly so) are often with psychoanalysis as therapy rather than with psychoanalysis as theory, and the differing therapeutic stance he adduces is by and large not different from that taken in psychodynamic (psychoanalytically based) psychotherapy which rests for its understanding on principles derived from the theory of psychoanalysis albeit extrapolating different kinds of technical interventions for the differing kinds of patients (not amenable to classical psychoanalysis) dealt with. (For a comprehensive statement of the differing kinds of technical interventions appropriate to expressive and to supportive psychotherapies as psychoanalytically conceived, see the classical articles by Gill [1951] and Bibring [1954].)

A central technical activity in all dynamic psychotherapies is interpretation, though it of course will be differently emphasized and used in relation to differing psychic contents in therapies that are more ego-supportive in contrast to those that are more uncovering and insight-aiming. This is clearly true too within the transactional approach, though Grinker tries to avoid the use of the word "interpretation." "The therapist primarily searches for information; and his interventions consist of making clear to the patient the information he has obtained about his implicit roles" (Grinker, 1961), or "Since the patient cannot achieve insight into his own neurotic processes, he depends upon information from the therapist in the two-person system as to how he seeks out stimuli to elicit old learned responses and hence perpetuate his neurotic behaviors" (Grinker, 1959). This is the Trojan horse where, under the rubric of conveying of information, interpretation has been brought back into the therapeutic enterprise out of the technical operating tenets of the theory Grinker decries as hampering therapeutic operations.

Grinker (1959) describes this actual process in considerable detail.

As the patient and the therapist approach each other, it is as though they are represented by arcs of two incomplete circles across which distorted, misunderstood, and incomplete messages traverse in both directions. As they come closer together and finally unite, complementarity of explicit roles is achieved. Then there is no gap and messages are not distorted. . . . At this point or "set," the therapist communicates his understanding of the patient's *implicit* role and the role the patient is attempting to ascribe or induce in him. . . . By recognizing these implicit role relations, the therapist . . . exerts pressure for understanding and the search for meaning . . . new solutions are sought. Some of these ways may be equally unrealistic, but finally a fit between implicit and explicit roles is achieved, repeated in form with different contents in "working through," and learning is consolidated. The therapist then has a sense of closure and decides that the therapeutic unit has been concluded.*

The therapist has thus fulfilled his task, "the function of making explicit by communication to the patient what his implicit roles really are." (Note, incidentally, that the interpretive effort is not presumed to exert a one-shot effect, but must be attended repetitively and in varying guises in order to accomplish the "working through" of its impact—the technically correct application of a specifically psychoanalytic concept, called by its unaltered name.)

Grinker (1959) then tries by reference to specific case material to "illuminate the differences between a modified psychoanalytic and a trans-

* Reprinted with permission from R. R. Grinker, Sr., "A Transactional Model for Psychotherapy," *Archives of General Psychiatry* (1959): 132–148.

actional approach" in the interpretive sphere. Thus one could (psychoanalytic approach) "consider that everything that the subject was saying had reference, in a disguised symbolic way, to her relationship with the therapist;" an overstatement, to be sure, of a well-known psychoanalytic truism concerning the nature of the transference. By contrast, it is stated that "the transactional concept utilizes the relationship between therapist and patient for the purpose of facilitating and permitting the patient to ventilate her feelings regarding all relationships." If one looks past the minor differing semantic shadings to the understanding that it is precisely because transference pervades *all* relationships to a variable and undetermined degree that we can always look for the manner in which, and the extent to which, it pervades the communications in this relationship in the therapy, whatever other meanings the therapeutic communication may have, we can then see more an identity than a distinction in *this* side-by-side comparison that Grinker offers of interpretation in the two therapeutic approaches.

And to pursue the matter still further, interpretation within the transactional approach is declared to lead to insights that ramify and uncover the past. In reference to another aspect of the same clinical example, it is thus stated (Grinker, 1959):

Although the transaction with the therapist was not made explicit, nevertheless, as it continued on a level at which anxiety was held to an optimum, memories were revealed regarding past transactions. In this sense the transaction was productive and will apparently continue at this level until it is stalemated. . . . As the memory of transactions outside the therapeutic relationship was revived, the patient truly ventilated her intense painful humiliation that she endured from men. Her final statements indicated almost a nonsequential jump into the *far distant past,* which indicates to us the effectiveness of this technique not only in evoking intellectual memories but also in reviving emotional experiences.[6]

Here the therapeutic transaction has dipped far back from the emphasis just on the "here and now"; it has in fact traversed the same range of temporal concerns from the immediate experience of the moment to whatever aspects of the revived past are available and relevant that is comprised in the gamut of psychoanalytically based psychotherapeutic approaches from the most supportive buttressing of current functioning derived in terms of the experiences of the present to the furthest reaching back in time that an expressive psychotherapy that does not have the full regression-inducing thrust of a psychoanalysis can reach.

To underline further my main thesis that the psychotherapeutic approach described by Grinker differs from psychoanalysis proper in just the ways that psychoanalytically oriented psychotherapies are classically marked off from psychoanalysis, let me as a final quotation from the statement of differences offer the following from Grinker (1959):

[6] Italics mine—R. S. W.

We attempt to avoid a transference neurosis and are helped by the fact that psychotherapeutic interviews are not conducted with greater frequency than once or twice a week. We thereby avoid regression and the development of highly infantile dependent relationships with the therapist. We do not call for free associations but ask for responses within the form of our communications. Nevertheless, without the invitation, we find that patients spontaneously communicate freely and in doing so are often able to relinquish conscious controls and liberate their preconscious processes of thinking, thereby exposing the distorted and repetitive unconscious controls. . . . Although we do not invite, and we certainly avoid, the transference neurosis whenever possible, we are obviously still dealing with transference phenomena, which is another way of saying that we are dealing with back-and-forth implicit communications between therapist and patient in which the present is colored by the past.

There could hardly be a clearer statement of the central technical distinctions between the psychotherapies based on analysis and psychoanalysis as therapy itself, and here Grinker has used the word "transference" and has carefully distinguished between working with the transference manifestations in psychotherapy and the unfolding and subsequent analytic resolution of the regressive transference neurosis in psychoanalysis.

Here we have come full circle in this extended description by quotation of Grinker's transactional psychotherapeutic approach. What he feels he has done, he states in his summary (Grinker, 1959) as follows,

In summary, what I have proposed is a method of psychotherapy based on the operations derived from field, role, and communications theories, rather than on a (psychoanalytic) theory of personality. This facilitates a vivid, current understanding of the patient without recourse to reified variables of unconscious, transference, countertransference, resistance, topological foci, processes involving energy, or any part-functions of the human being in behavior.

From the differing perspective I have tried to adumbrate in this review of Grinker's approach I feel that he has rather *been* describing psychoanalytically based dynamic psychotherapy, has in fact rendered an excellent representation of it, has systematically explored its reach and its limitations, and has set forth, though always only by implication, the nature of the similarities and differences between dynamic psychotherapy and psychoanalysis in a manner quite in accord with the main body of psychoanalytic conceptualization (Bibring, 1954; Gill, 1954; Rangell, 1954; Wallerstein, 1969). And though his vocabulary is in places somewhat different, delineating the concepts and the operations as much as possible in the vocabulary of role and field theory in the service of his avowed "struggle for eclecticism," Grinker is in fact more solidly within the psychoanalytic tradition and its framework than he willingly acknowledges. As much as any of us he has absorbed it and ordered his thinking by it. What he describes *is* dynamic psychotherapy within the commonly shared understanding of that word, and that is not what he somewhat depreciatingly calls just a euphemism for psychoanalysis but rather a separate body of precept and of technique, sharing with psychoanalysis its psychological understanding of man and dividing up with psychoanalysis its field of application to the psychological troubles of man (with dynamic psychotherapy falling heir, incidentally, to by far the overwhelmingly larger segment).

In this sense, I feel Grinker's terminal caveat in his exposition of the transactional approach to be somewhat gratuitously misleading. He states (1961), "The more we understand theoretical psychodynamics and the less we are influenced by it operationally, the better we may understand our patients and ourselves." The statement carefully does not quite say what the tenor of the argument throughout might imply that it was meant to say, that the theory (of analysis) should not influence or illuminate the operation (of therapy) because it is theory that is outmoded or inappropriate to the context. It says rather that good as the theory may be for understanding the mind of man in the large, that we must as far as possible free ourselves from the predilections it imposes on our thought processes when trying to understand the mind of any individual man in the particular within the specific therapy encounter. And read this way, no one should quarrel. We must indeed both remember and forget, operate within the framework the theory gives us to fit our observations together into comprehensive and comprehensible wholes, and at the same time transcend it in order to be open to unexpected observations and novel conceptualizations (Reik, 1936).

My own summation, and with it a final question to Grinker, is a different one. Grinker has played a signal role as one of the developers of the still rudimentary and sorely needed theory of psychotherapy technique. And it will only be the further development of a much more solid body of the theory of its technique that will make of dynamic psychotherapy the more systematic and scientific enterprise that will take it past the confusing present-day tower of Babel of competing voices in the marketplace of contending ideas amidst which we now labor. In the pursuit of this goal which we so thoroughly share, where is the gain in understanding that comes from the translation into a differing conceptual language of what is nonetheless recognizably the same (still useful) theory? The price I feel has been the introduction of needless (political) battle and the clouding of the perception of Grinker's major role as a proponent, a promoter, and a developer of the burgeoning field of psychotherapy (call it dynamic or psychoanalytic or transactional) in both its theory and its technique. In this sense I have welcomed the opportunity to place a psychoanalyst's appreciation of Roy Grinker's large contribution to the development of the field of psychotherapy (call it whatever) on record as part of this volume of salute to his influence on us all.

REFERENCES

Bribring, E. 1954. "Psychoanalysis and the Dynamic Psychotherapies," *Journal of the American Psychoanalytic Association* 2:745–770.

Chassel, J. O. 1955. "Panel Report: Psychoanalysis and Psychotherapy," *Journal of the American Psychoanalytic Association* 3:528–533.

Ekstein, R., and Wallerstein, R. S. 1958. *The Teaching and Learning of Psychotherapy.* New York: Basic Books.

English, O. S. 1953. "Panel Report: The Essentials of Psychotherapy as Viewed by the Psychoanalyst," *Journal of the American Psychoanalytic Association* 1:550–561.

Gill, M. M. 1951. "Ego Psychology and Psychotherapy," *Psychoanalytic Quarterly* 20:62–71.

Gill, M. M. 1954. "Psychoanalysis and Exploratory Psychotherapy," *Journal of the American Psychoanalytic Association* 2:771–797.

Grinker, R. R., Sr. 1959. "A Transactional Model for Psychotherapy," *Archives of General Psychiatry* 1:132–148.

Grinker, R. R., Sr. 1961. "A Transactional Model for Psychotherapy." In M. I. Stein, ed., *Contemporary Psychotherapies.* New York: Free Press of Glencoe. Pp. 190–213.

Grinker, R. R., Sr. 1964. "A Struggle for Eclecticism," *American Journal of Psychiatry* 121:451–457.

Grinker, R. R., Sr. 1965. "Identity or Regression in Psychoanalysis," *Archives of General Psychiatry* 12:113–125.

Grinker, R. R., Sr. 1966. " 'Open-System' Psychiatry," *American Journal of Psychoanalysis* 26:115–128.

Ludwig, A. O. 1954. "Panel Report: Psychoanalysis and Psychotherapy: Dynamic Criteria for Treatment Choice," *Journal of the American Psychoanalytic Association* 2:346–350.

Polanyi, M. 1959. *The Study of Man.* Chicago: University of Chicago Press.

Rangell, L. 1954. "Panel Report: Psychoanalysis and Dynamic Psychotherapy— Similarities and Differences," *Journal of the American Psychoanalytic Association* 2:152–166.

Reik, T. 1936. *Surprise and the Psychoanalyst.* London: Paul, Trench, Trubner.

Waelder, R. 1962. "Psychoanalysis, Scientific Method, and Philosophy," *Journal of the American Psychoanalytic Association* 10:617–637.

Wallerstein, R. S. 1966. "The Current State of Psychotherapy: Theory, Practice, Research," *Journal of the American Psychoanalytic Association* 14:183–225.

Wallerstein, R. S. 1969. "The Relationship of Psychoanalysis to Psychotherapy— Current Issues," *International Journal of Psychoanalysis* 50:117–126.

Wallerstein, R. S., and Robbins, L. L. 1956. "The Psychotherapy Research Project of the Menninger Foundation: IV. Concepts," *Bulletin of the Menninger Clinic* 20:239–262.

Wallerstein, R. S., Robbins, L. L., Sargent, H. D., and Luborsky, L. 1956. "The Psychotherapy Research Project of the Menninger Foundation: Rationale, Method, and Sample Use," *Bulletin of the Menninger Clinic* 20:221–278.

Wallerstein, R. S., and Sampson, H. 1971. "Issues in Research in the Psychoanalytic Process," *International Journal of Psychoanalysis* 52:11–50.

The Phenomena of
Depression: A Synthesis*

In view of the large number of studies of depression during the past decade, it is interesting to note that there are still wide divergencies regarding its definition, classification, cause, and treatment. Scrutiny of the recent literature indicates that the concept of depression is clouded by confusion and controversy (Beck, 1967; Mendels and Cochrane, 1968). Some writers, for instance, urge that depression is a discrete disease entity with a specific etiology, course, and prognosis. Others argue that depression is not a disease at all, but represents a reaction type or disturbance in adjustment. Still others regard depression as an aggregate of concrete diseases.

Notwithstanding the semantic difficulties, there are discrete sets of phenomena for which the discriminative label "depression" may be appropriately applied. Irrespective of their conflicting concepts of depression, investigators generally agree about the domain encompassed by these sets of behavioral phenomena. Descriptions of the correlates of depression (or "melancholia") have been surprisingly consistent for the past 2,000 years (see Grinker, Miller, Sabshin, Nunn, and Nunnally, 1961).

For purposes of definition, the phenomena of depression may be divided into five categories: (1) affective manifestations—a specific alteration in mood such as sadness, loneliness, or apathy; (2) cognitive manifestations—negative self-concept associated with a disposition to interpret experiences in a negative way and to expect negative outcome of the future; (3) motivational manifestations—desire to escape, to avoid

* The clinical investigations and the preparation of this report were supported by a grant from the Marsh Foundation and grant MH 16616-01 from the National Institute of Mental Health.

I wish to acknowledge the technical assistance of Ruth Greenberg and Mark Steinberg.

interpersonal contacts, or to commit suicide, and loss of spontaneous motivation ("paralysis of the will"); (4) vegetative manifestations—anorexia, fatigability, insomnia, loss of libido; (5) changes in motor behavior—retardation or agitation. Empirical justification for this grouping is based on a statistical study of the signs and symptoms of 1,000 depressed and nondepressed patients (Beck, 1967).

EXPLANATORY MODELS

Since the era of Hippocrates, physicians have attempted to find some unifying principle to explain the variegated and paradoxical characteristics of depression. For 2,000 years most writers postulated some biological derangement (such as an excess of black bile) to account for this disorder. The more recent influence of psychoanalysis has stimulated an exploration of psychological mechanisms in order to understand depression.

This section summarizes formulations presented by psychoanalysis (Freud), ego-psychology (Bibring, Engel, Gaylin), interpersonal psychiatry (Cohen), behavior therapy (Lazarus, Lewinsohn, Mower), existential psychiatry (Tellenbach), and cognitive theory (Lichtenberg, Beck). Though not a comprehensive review, this survey will sample the broad spectrum of psychological and behavioral theories. Biological studies of depression are not within the scope of this chapter (see Beck, 1967, for a critical review of them).

Freud (1959) spoke of depression, or melancholia, in terms of an analogy with mourning. With the exception of the melancholic's "extraordinary" loss of self-esteem, the characteristics of mourning are much the same as those of melancholia: "a profoundly painful dejection, abrogation of interest in the outside world, loss of the capacity to love, inhibition of all activity, and a lowering of the self-regarding feelings to a degree that finds utterance in self-reproaches and self-revilings, and culminates in a delusional expectation of punishment." Whereas the mourner suffers conscious loss of a loved object, a loss that painfully impoverishes the world around him, the melancholic's loss is unconscious and results in impoverishment of the ego itself. Unlike the melancholic, the mourner is eventually able to transfer his attachment to another object.

Freud reconstructed the development of melancholia in the following way: The libidinal object choice took place on a narcissistic basis, so that when the object cathexis was frustrated, there was a "regression to the primal narcissism." The free libido was

withdrawn into the ego and not directed to another object. . . . [It] served simply to establish an *identification* of the ego with the abandoned object. . . .

In this way the loss of the object became transformed into a loss in the ego, and the conflict between the ego and the loved person transformed into a cleavage between the criticizing faculty of the ego and the ego as altered by identification.

Thus, the melancholic's self-reproach is seen as •ggression against the orally incorporated object, and his inhibition as a result of preoccupation with the "work" of withdrawing the libido.[1]

According to Bibring (1953), the central feature of all depression is the subjective feeling of helplessness or powerlessness in relation to certain strongly held narcissistic goals. Bibring agreed that the orally dependent personality may be especially predisposed to depression, but he emphasized the infant's primary experience of helplessness during the oral stage. Depression stems not from a struggle between ego and id, or ego and superego, or ego and environment, but from "a tension within the ego itself." It is defined as "the emotional correlate of a partial or complete collapse of the self-esteem of the ego, since it feels unable to live up to its aspirations (ego, ideal, superego) while they are strongly maintained." Aggression is seen as irrelevant to the basic mechanism of depression.

Bibring distinguished three groups of basic aspirations which characterize the person predisposed to depression: "(1) the wish to be worthy, to be loved, to be appreciated, not to be inferior or unworthy; (2) the wish to be strong, superior, great, secure, not to be weak and insecure; and (3) the wish to be good, to be loving, not to be aggressive, hateful, and destructive." Any experience that seems to indicate that he is powerless to achieve these goals may reactivate the experience of infantile helplessness and precipitate a depression. In Bibring's view, the inhibition of activity in depressed patients is "due to the fact that certain strivings of the person become meaningless—since the ego appears incapable ever to gratify them." Finally, Bibring emphasized that for theoretical and therapeutic reasons it is important to distinguish between

(1) the basic or essential mechanism of depression (fall in self-esteem due to awareness of one's own real or imaginary, partial or total insufficiency or helplessness); (2) conditions which predispose to and help to bring about depression; (3) the attempts at restitution associated with depression; (4) conditions which complicate the basic type of depression such as aggression and orality; and (5) the secondary use which may be made—consciously or unconsciously —of an established depression (e.g. to get attention and affection or other narcissistic gratification).°

° Reprinted with permission from E. Bibring, "The Mechanism of Depression," in P. Greenacre, ed., *Affective Disorders* (New York: International Universities Press, 1953), pp. 13–48.

[1] Further contributions to the psychoanalytic literature on depression have been made by Rado (1928), Gero (1936), Klein (1948), Jacobson (1953, 1954), Abraham (1960a, 1960b, 1960c), Hammerman (1962), and Zetzel (1966). These positions have been discussed elsewhere by Mendelson (1960) and Beck (1967).

Essentially, depression is a basic reaction of the ego to frustration.

In general, one may say that everything that lowers or paralyzes the ego's self-esteem without changing the narcissistically important aims represents a condition of depression. External or internal, actual or symbolic factors may consciously or unconsciously refute the denial of weakness or defeat or danger, may dispel systems of self-deception, may destroy hope, may reveal lack of affection or respect or prove the existence in oneself of undesirable impulses or thoughts or attitudes or offer evidence that dormant or neutralized fears are actually "justified," and so forth; the subsequent results will be the same: the individual will regressively react with the feeling of powerlessness and help-lessness with regard to his loneliness, isolation, weakness, inferiority, evilness, or guilt. Whatever the external or internal objects or representations of the narcissistically important strivings may be, the mechanism of depression will be the same.

Depression will be relieved when the goals are either apparently within reach, or modified, or relinquished; when recovery mechanisms help the ego to regain its self-esteem; or when apathy, hypomania or other defenses are "directed . . . against the affect of depression as such."

Engel (1968) was similarly concerned with the feeling of helplessness in his investigation of "life-setting conducive to illness." Noting that death and disease often occur at times of personal stress—in animals and humans—Engel described a psychological pattern that seems to have a physiological effect on the ability of the organism to resist dis-ease. This "giving-up–given-up" complex is characterized by "(1) the giving-up affects of helplessness or hopelessness; (2) a depreciated image of oneself; (3) a loss of gratification from relationships or roles in life; (4) a disruption of the sense of continuity between past, present, and fu-ture; and (5) a reactivation of memories of earlier periods of giving-up."

Engel distinguished between helplessness and hopelessness: "the two affects of giving-up." A person feeling helpless believes he has been frus-trated by outside events; though he cannot help himself to overcome his difficulties, he still expects that help may come from his environment. The hopeless person feels his failures are of his own making and antici-pates no help from external sources. In either case, the patient generally feels unable to cope with change or function efficiently, to derive satis-faction from his normal roles, or to perceive any value or pleasure in the future. Like the depressed persons described by Freud and Bibring, En-gel's patients suffer from a loss of self-esteem.

In his epilogue to *The Meaning of Despair*, Gaylin (1968) traced the development of the psychoanalytic view of depression in terms of an in-creasing emphasis on the role of self-esteem. Whereas Freud believed that loss of self-esteem was a result of depression, such writers as Rado, Fenichel, and Bibring began to see the loss of a love object as merely symbolic of the essential loss of self-esteem. Gaylin asserted that these

writers are actually describing not a loss of self-esteem, which he relates
to paranoia, but a crisis in self-confidence—a fear that the ego will be
unable to cope with dangers from the environment. Depression "can be
precipitated by the loss or removal of *anything* which the ego overval-
ues in terms of its security." In his despair at losing something on which
he is dependent, the depressed person simply "gives up." Even true
symptom formation is lacking in depression as the ego makes no attempt
to maneuver itself out of the situation; there are only the "nonsymptoms
of passivity, inactivity, resignation, and despair." Finally, the depressed
person is reduced to a state of total dependency, though dependency itself may function as a restorative appeal for love and help.

In an intensive psychoanalytic study of twelve cases of manic-depressive psychosis, Cohen, Baker, Cohen, Fromm-Reichmann, and Weigert
(1954) found that all patients had a similar family background. They
tended to belong to a minority group or to a family that felt itself set
apart from society in some way. In order to compensate for these feelings, a strong mother deprecated her husband and pushed her children
toward achievement, conformity, and social advancement. Because of
his difficulties in integrating the various aspects of his mother into a unified picture, the depressed patient tends to defend himself by denying
the complexity of people, seeing them as either all good or all bad. According to these investigators, the depressed patient does not actually
suffer from feelings of guilt or regret; his persistent statements of self-reproach are an attempt to "placate authority." His hostility is due to the
annoying effect the depressed patient has on others, rather than a basic
desire to harm them.

On the basis of their clinical observation these writers concluded that
manic depressive patients showed excessive conformity, dependency, and
need for social approval. Partial support for these findings was obtained
in a loosely designed study by Gibson (1957) comparing manic depressives and schizophrenics.

More refined tests of Cohen's hypotheses within the framework of
need-achievement theory have yielded inconsistent results. Becker
(1960) found that manic depressives scored significantly higher than normal controls on measures of value achievement, authoritarian trends,
and conventional attitudes. Some doubt of the specificity of these findings was raised by a study reported by Becker, Spielberger, and Parker
(1963). These investigators found no significant difference in value
achievement or authoritarian attitude when manic depressives were
compared with neurotic depressives and schizophrenics. It was found,
however, that the psychiatric group differed significantly from normal
controls.

A study by Katkin, Sasmor, and Kann (1966) attempted to test the relative degree of conformity and achievement-related characteristics of
depressed and acute schizophrenic patients. This investigation failed to

confirm that depressed patients show high value achievement, authoritarianism, or excessive adherence to traditional family ideology. However, in an Asch-type conformity situation the depressed patients showed a greater tendency to conform to social pressure than did the schizophrenics.

Lazarus (1968) stressed the importance of stimulus-response analysis in diagnosis and therapy. Rejecting Freud's distinction between normal grief and depression, he suggested that depression might be defined in terms of a "base rate of frequent weeping, decreased food intake, frequent statements of dejection and self-reproach, psychomotor retardation, difficulties with memory and concentration, insomnia or a fitful sleep pattern, and general apathy and withdrawal." Distinguishing depression from anxiety, Lazarus stated, "Fundamentally, anxiety may be viewed as a response to noxious or threatening stimuli, and depression may be regarded as a function of inadequate or insufficient reinforcers." It may thus be precipitated by a sudden change in the environment involving the loss of an important reinforcer. In therapy, patients "require a different schedule of reinforcement and/or need to learn a way of recognizing and utilizing certain reinforcers at their disposal."

Lewinsohn, Shaffer, and Libet (1969) also assumed that depressive behaviors are elicited by a low rate of positive reinforcement, which is in turn caused by various environmental events or by personal traits such as lack of social skill. Depressive behaviors, once established, may be maintained by positive reinforcement in the form of increased attention and concern from family and friends—the secondary use of depression noted by Bibring. Also reminiscent of Bibring's formulation is the assertion that hostility is secondary to the low rate of positive reinforcement as a cause of depression; it may, however, complicate the depression by further reducing the rate of positive reinforcement from persons alienated by aggressive behaviors.

Through the use of a system for coding behavioral interaction, Lewinsohn et al. have been able to test their hypothesis that depressed patients are lacking in social skill, that is, the "ability to emit behaviors which are positively reinforced by others" and to avoid those which are negatively reinforced. Results show that depressed persons both elicit and emit fewer actions than nondepressed persons; that the messages they emit are shorter and their "timing off"; and that they tend to interact with a smaller number of people.

Mowrer (1969) developed an approach to depression based on the program at Daytop Village, Inc., and on the work of Daniel Casriel and Alfred Adler. In Mowrer's view depression stems from the emphasis our society places on productivity: The Protestant Work Ethic teaches that the love and respect of others depend on our accomplishments. When ambition and hard work are rewarded only by disillusionment, many persons may disguise as illness what is essentially a refusal to work, a retreat from reality. "According to this view of the matter, it is not that the depression is there first and produces the work inhibition, but that discouragement and muted defiance of work orders are the primal phenomena, with depression as the sequel." Casriel's strategy for the treatment of depression consists of: "(1) conscious and public (in a group) repudiation of the Work Ethic; and (2) conscious and public (in a group) affirmation of the wish and need to develop, and be loved for purely human qualities such as openness, warmth, compassion, concern, integrity."

In Mowrer's view, Casriel's approach is compatible with that of the behavior therapists, though group techniques are seen as far more powerful reinforcers than those used by the behavior therapists. Lazarus's assertion that depressed patients must be taught to use available reinforcers is seen as equivalent to Casriel's view that such persons must learn to "reach out" directly for love and personal fulfillment. Clearly, both positions recall Bibring's emphasis on the modification of narcissistic goals.

Tellenbach (1961) asserted that the premorbid personality of the depressed patient is dominated by a need for order and security and a striving to fulfill obligations, avoiding feelings of guilt. However, the patient's own conception of his obligations is so exacting, his aspirations so high, that he is forever unable to fulfill them.

According to Lichtenberg (1957), the depressed patient feels responsible for his hopeless feeling in regard to the attainment of certain goals. Three types of depression, corresponding to neurotic, agitated, and retarded depression, are distinguished on the basis of the kind of goal toward which the patient's aspirations are directed: a specific situation, a behavior style, or a generalized goal. When a person is frustrated in attaining a more differentiated goal, he will direct his expectancy to a more generalized goal.

In an attempt to present a framework to show plausible relationships among the highly diverse phenomena of depression, Beck (1963) proposed a model of depression composed of interlocking cognitive, affective, and motivational structures. This structural model assigned a primary position to the distorted concepts of the self, the outside world, and the future. The other phenomena of depression (such as the motivational, affective, and behavioral manifestations) were assigned positions as dependent variables.

Self-reports of psychotherapy patients suggested that the suicidal wishes, sadness, and loss of spontaneous motivation were the logical outcome of the patient's premise that he was locked in an insoluble position, generally as a result of his own supposed deficiencies. When the patient's belief in the insolubility of his problems was modified by specific therapeutic techniques, there was an almost immediate lifting of his mood and reduction of the intensity of his suicidal wishes, sometimes sufficient to produce a complete remission of the disorder.

DIMENSIONS OF DEPRESSION

Though most of the literature has attempted to explain depression, many investigators have felt that the isolation of subgroups, or types, of depression is a necessary first step toward revealing the essential nature and, possibly, the etiology of affective disorders.

A growing body of research has been directed to basic issues, such as whether the symptomatology currently referred to as depression is unitary in nature or whether it represents a complex structure of related but mutually exclusive syndromes. The range of behavior encompassed by popular use of the word "depression" is so large that meaningful behavioral correlates might may be obscured. Some recent research addresses itself to the problem of reducing the inherent ambiguity of the term "depression" by attempting to isolate discrete categories. The presumed dichotomies of depression (for example, endogenous vs. reactive) have been tested systematically and critically reviewed (see Beck, 1967; Kendell, 1968; Mendels and Cochrane, 1968).

This section summarizes thirteen factor analytic studies [2] designed to define specific dimensions of depression. In-patient populations were used in all studies as a source of subjects. The patient samples had either been diagnosed as depressed by a psychiatrist or had been screened as depressed by means of a cutoff criterion on one or more of the quantitative inventories described below. The number of subjects used in studies varied from 100 (Cropley and Weckowicz, 1966) to 648 (Raskin et al., 1969).

All the studies utilized instruments that permitted the quantification of data so that statistical analyses could be carried out. Instruments varied as to the context in which behavior was sampled (for example, interview, ward, hospital admission, and so on), as to the relationship of the rater to the patient (for example, ratings by psychiatrists, nurses, and self-ratings) and as to format. Except when ratings and rating categories were derived from nonstructured or semistructured interview data (Grinker et al., 1961; Michaux, Zwagg, and Kurland, 1967; Paykel, Klerman and Prusoff, 1970; Raskin et al., 1967; Raskin et al., 1969), ratings

---

[2] This survey includes the following factor analytic studies: Hamilton and White (1959); Grinker et al. (1961); Overall (1962); Cropley and Weckowicz (1966); Lorr, Sonn, and Katz (1967); Michaux, Zwagg, and Kurland (1967); Raskin, Schulterbrandt, Reatig, and Rice (1967); Weckowicz, Muir, and Cropley (1967); Zung (1967); Beck (1969); Raskin, Schulterbrandt, Reatig, and McKeon (1969); Yonge, Weckowicz, Muir, and Cropley (1969); Paykel, Klerman, and Prusoff (1970).

were taken from checklists and behavior inventories or modified forms
of those in common use (Beck Depression Inventory, Self-Rating Depression Scale, Minnesota Multiphasic Personality Inventory, Multidimensional Scale for Rating Psychiatric Patients, Ward Behavior Rating
Scale, In-Patient Multidimensional Psychiatric Scale, Brief Psychiatric
Rating Scale, Spring Grove Psychiatric Scale).

Reliability and validity of all test instruments had been determined
prior to their use as experimental tools. In many cases more than one of
these test instruments were applied to the same population (Grinker et
al., 1961, Michaux et al., 1967; Paykel et al., 1970; Raskin et al., 1967,
1969; Zung, 1967) thereby allowing intertest correlations to be performed and different descriptions of behaviors to be included in factor categories. In one study in which several different researchers rated
the same population, stereotypes derived from having the researchers
give their own ratings of typical depressive traits were run as controls
(Grinker et al., 1961).

Quantitative scores on individual items taken from the test instruments were intercorrelated and a correlation matrix prepared permitting
a factor analysis to be performed. In some studies factors were rotated
via either the promax or varimax method until factors were isolated that
accounted for some established level of variance. The factors thus isolated were in some cases tested for orthogonality (that is, independence)
and ranked according to the magnitude of factor loadings of each, with
respect to a particular test instrument. These so-called major factors
were characterized by the authors in clinical terms which gave the factors unitary labels (usually implying some kind of clinical significance)
and sometimes a short theoretical interpretation.

Despite the wide diversity in methodology, grouping criteria, and labels used by the various researchers in isolating and describing lists of
factors, the content of such lists can be said to show great consistency.
In every study cited, a factor comprising symptoms representing typical
depression was found. This factor, which sometimes appeared as a single
factor and sometimes combined with other factors, centers around the
mental and behavioral manifestations of depression (loss of self-esteem,
self-deprecation, crying, feeling sad, self-blame, and so on). A second
factor based on reduction in responsiveness was found in all studies.
This factor, most often given the name "psychomotor retardation," is
represented by apathy, emotional withdrawal, fatigue, loss of libido, and
lack of social participation. In addition a somatic factor, including sleep
disturbance, loss of appetite, and somatic preoccupation, appeared in all
factor lists.

An anxiety factor was found in eight studies. Labels that described
anxiety included anxiety, tension, agitation, and anxious depression. A
hostility factor appeared in six lists, usually under the rubrics "anger,"
"hostility," and "irritability."

Nine of the lists covered symptoms of cognitive and perceptual dys-
functions: paranoid projections, hallucinations, confusion, autistic thinking, unrealistic denial of illness, and general impairment of cognitive
functions. Not all the common factors described above appeared as separate entities but were often part of larger, more pervasive factors, such
as the general severity factor found by Paykel et al. (1970). Also these
general factors were themselves subdivided into separate factors. Hamilton and White (1959), for instance, divided anxiety into its psychic and
somatic manifestations.

Raskin et al. (1967, 1969) used the Inventory of Somatic Complaints,
the Ward Behavior Rating Scale, and a mood scale, which consisted of
items compiled by the researchers from several sources, to obtain measures. Significant correlation clusters of items that seem to be unrelated
to the criterion (depression) were derived, such as "friendliness" and
"carefree." This may indicate that instruments that sampled a broader
field of behavioral characteristics have isolated a more complete or
well-defined list of factors. This also raises the question of the degree to
which derived dimensions are limited by the instruments used or
whether test instruments can be said to inject factors into the data as an
artifact of their individual peculiarities.

In Grinker's study (Grinker et al., 1961) the experimental method was
particularly well suited to the task of extracting patterns of behavioral
and affective data. Trait checklists covering various aspects of the general diagnostic category "depressive reaction" were used. Behaviors of
in-patients used in the study were rated by psychiatrists, psychiatric residents, nurses, and members of the research team on checklists, whose
items consisted of statements corresponding to traits found in patients
commonly diagnosed as depressed and which had been taken from written clinical reports.

In the pilot study ratings were derived from written clinical reports of
semistructured interviews. Interrater reliability checks were made with
respect to each checklist, and these statistics were used to "screen" data
for consistency of judgment between raters, thereby permitting factor
analyses only where warranted by the reliability of the ratings. Also this
information was used by the researchers to determine which ratings
were to be used in the main study.

Subsequently, a full-scale study using revised versions of the trait
checklists labeled "current behavior" and "feelings and concerns" by the
researchers was done. The data from both checklists in both pilot study
and the main study were subjected to a factor analysis that yielded five
factors from the feelings and concerns list (these included three factors
found in the pilot study) and ten factors from the current behavior list.
Analysis revealed only a weak correlation between the two lists; the authors consequently described the two lists as independent. In addition, a
further analysis was performed to determine whether factors isolated

from each list could be said to go together in the sense of feelings and
their behavioral concomitants. Thus, the ten behavioral factors were
combined with the five feelings and concerns factors to yield four general patterns, each pattern consisting of a combination of the feelings
and behaviors derived from the original data. The patterns found by the
researchers were as follows:

A   Feelings   dismal, hopeless, loss of self-esteem, slight guilt feelings
    Behavior   isolated, withdrawn, apathetic, slow thought and speech,
                cognitive disturbance

B   Feelings   hopeless with low self-esteem, high guilt feelings, high
                anxiety
    Behavior   agitation, clinging demands for love

C   Feelings   abandonment, loss of love
    Behavior   agitated, demanding, hypochondriacal

D   Feelings   gloom, hopelessness, anxiety
    Behavior   demanding, angry, provocative

In both phases of the research, stereotypes taken from the researchers'
own ratings of typical depressed traits were used as controls by comparing derived stereotypes with factors isolated from the experimental data.
Stereotypes were almost identical to factor 1 isolated from the feelings
and concerns checklist (dismal, hopeless, self-castigating type of person)
in both the pilot study and the main study.

The inference was made by the researchers that factor 1 represented a
basic or underlying factor while the other factors represented defenses
causally related to factor 1.

The five affective factors were (1) person who is dismal, self-castigating; (2) person who attributes his depression to external events or persons; (3) person who feels guilty over aggression and is attempting restitution; (4) person who is anxiety-laden; (5) person who is clinging and
demanding.

The ten behavioral factors were (1) isolated and withdrawn; (2)
slowed and retarded in speech and thought; (3) disinterested and apa-
thetic; (4) angrily provocative and demanding; (5) hypochondriacal; (6)
cognitively disturbed (questionable organic brain damage); (7) agitated;
(8) rigid, immobile; (9) somatically disturbed; (10) clinging, ingratiating,
pleading for love and attention.

Grinker's study was important because it was one of the first steps
taken in the attempt to isolate basic factors in depression. It is also valuable in that it dichotomizes symptomatology into psychic and behavioral components, thereby giving a more detailed clinical picture.

HOPELESSNESS,
LOW SELF-ESTEEM,
AND LOSS

The review of the factor analytic studies suggests that the value of the purely descriptive studies of depression may have reached the point of diminishing returns. It may be appropriate at this time to take another look at the studies focused on the psychological characteristics of the depressed patient for further clues to understanding depression.

In recent years, Freud's concept of the central role of retroflected rage has been deemphasized by many writers (for example, Beck, 1963; Bibring, 1953; Gaylin, 1968; Lazarus, 1968; Lichtenberg, 1957). These writers have focused on specific aberrations of the depressed patient's ego or self-esteem. A common theme has been the patient's sense of irrevocable loss and perceived incapacity to cope with his problems.

The insoluble problem formulation of depression and suicide consists of several components: The patient regards himself as currently deprived of something that he considers essential to his happiness. This sense of loss leads to sadness even though the deprivation may not exist in reality. The types of losses reported by depressed, suicidal patients consist of categories such as the loss of gratifying interpersonal relationship, loss of a particular status in society, and loss of a positive conception of themselves (Breed, 1967).

Frequently, the patient is actively preoccupied with the idea that his future will be devoid of the essential ingredients for a reasonably satisfying existence. Alternatively, he views his future life as filled with unremitting suffering or unpleasantness. This aspect of the depressed patient's hopelessness is well described by Sarwer-Foner (1966).

It consists of the conviction that their depressed state will continue indefinitely and never end—in short, that it will last their entire lifetime and at least for some, an eternity. This deluded conviction invades and influences every aspect of their life during the depressed state. Since from the patient's viewpoint it is an integral part of his affective state and is visible in all aspects of his thought and behavior rather than mainly in his verbal productions, a clearly enunciated verbal statement to this effect is uncommon. It is thus necessary for the physician to be able to recognize this delusion since the patient himself is all too often unable to pin-point or verbally present it to the physician as psychotherapeutic "material." It rather remains material which is present in all aspects of the patient's existence during his depressive illness. Nor does it present itself as a challenge, as something to be fought, in the conscious awareness of the patient. It is rather overwhelmingly natural (i.e. ego syntonic) to his depressed state because he is unaware of any ability to cope with or change what is, for him, an accepted fact. Thus it is the task of the physician to recognize from clues in the patient's total expressive behavior, the existence of this specific psychopathological construct—a specific delusion of time consisting of the patient's conviction that he will never recover, will always be depressed, and that the depressed state is to be his permanent, ordinary lifelong state of intrapsychic being. . . .

Thus the hopelessness of such patients is directly linked to the specific delusion of the eternity of their depressed state. Since it will always continue, they have no hope of any change developing in their intense, irremediable inner loss, grief, sadness, frustrated helpless rage, and resulting depression. Along with this hopelessness, one sees very clearly their helplessness. There is nothing that they can do through their own efforts to restore their inner organs, alter their own destructiveness or a degraded worthless, empty, inner state, one so impoverished that they can do nothing well. . . .

In the more severely regressed patients, even their memory of being well is so decathected that it temporarily disappears. They literally show no evidence of any real or spontaneous ability to remember when they were not ill.[*]

The severely depressed patient generally regards himself as defective in those attributes instrumental to attaining what he considers essential gratifications, or he regards himself as blocked by external reality from attaining this essential goal. Independent support for the clinical observations underlying this formulation may be found in the systematic study of suicide notes by Bjerg (1967). This investigator reported that in 81 per cent of the notes there was a theme of "the person's seeing himself as having a desire (other than suicidal) which could not, cannot, or will not be fulfilled."

A number of other studies indicate the relationship between hopelessness, depression, and suicide. In a factor analysis of the Depression Inventory (Beck, Ward, Mendelson, Mock, and Erbaugh, 1961), Pichot and Lempérière (1964) found a factor with high loadings on just two items: pessimism and suicidal wishes. Similarly, Cropley and Weckowicz (1966) extracted a factor, the two highest loadings being pessimism (.53) and suicidal wishes (.57).

The relationship between hopelessness and suicide is also suggested by investigation by Harder (1967) who reported the case histories of eleven women who killed or tried to kill their children and then attempted suicide. In summary, he reported that these depressed women regarded their homicide-suicide plans as the only way out of what seemed to be an impossible situation. The systematic study of suicidal notes by Bjerg (1967) showed that cognitions, such as belief in personal deprivation and expectation of endless suffering, were very prominent.

A study by Wohlford (1966) indicated that compared to nondepressed patients the depressed patient sees his future as highly constricted. A similar finding in patients who had attempted suicide was reported by Farnham-Diggory (1964). Other investigators who have found a constricted time perspective in systematic studies of depression include Stuart (1962), Dilling and Rabin (1967), and Melges and Bowlby (1969).

[*] Reprinted with permission from G. J. Sarwer-Foner, "A Psychoanalytic Note on a Specific Delusion of Time in Psychotic Depression, *Canadian Psychiatric Association Journal* 11(1966): 221–228.

COGNITIVE-BEHAVIORAL STUDIES
OF DEPRESSION

Beck and his coworkers have conducted a series of correlational and experimental studies of the ideational productions and related behaviors of depressed patients. These studies led to the construction of the model of depression presented later in this chapter.

THEMATIC CONTENT OF DREAMS AND
PROJECTIVE TESTS

In a longitudinal study of twelve patients in psychotherapy Beck and Hurvich (1959) found that depressed patients showed a higher proportion of masochistic dreams than a matched control group of nondepressed patients. The ideational material consisted of a specific content: The dreamer was portrayed as a loser, in some respect; that is, he suffered deprivation of some tangible object, experienced a loss of self-esteem in an interpersonal relationship, or experienced a loss of a person to whom he was attached. Other common themes included the dreamer's being thwarted in attempting to reach a goal or being portrayed as inept, repulsive, or defective.

A second, more refined study of the most recent dreams of 218 depressed and nondepressed psychiatric patients supported the earlier findings (Beck and Ward, 1961). Significantly, more depressed than nondepressed patients reported masochistic dreams.

Beck (1961) devised a focused fantasy test consisting of a set of cards, each card containing four frames that portrayed a continuous sequence of events involving a set of identical twins. One of the twins was subjected to an unpleasant experience while the other twin avoided the unpleasant experience or had a pleasant experience. The situations were similar to those found in dreams of depressed patients; namely, the hero loses something of value, is rejected, or punished. It was found that the depressed patients identified with the loser significantly more frequently than did the nondepressed.

COGNITIVE PATTERNS IN SELF-REPORTS

In a clinical study, Beck (1963, 1964) analyzed verbatim recorded interview material of depressed and nondepressed patients in psychotherapy. He found that depressed patients tended to distort their experiences in an idiosyncratic way; they misinterpreted specific, irrelevant events in terms of personal failure, deprivation, or rejection; or they tended to greatly exaggerate or overgeneralize any event that bore any semblance of negative information about themselves. They also tended to perseverate in making indiscriminate, negative predictions of the future. Some evidence of a thinking disorder was found in their tendency to make selective abstractions, to overgeneralize, to mislabel events, and to exaggerate the significance of events.

The content of the distorted conceptualizations of the depressed patients showed a continuity with the thematic content of their dreams. This content centered around one dominant theme, namely, a negative view of the self, the outside world, and the future. This set of idiosyncratic patterns, labeled "the cognitive triad" (Beck, 1967), appeared to permeate the conscious experiences of the severely depressed patient and were present, but to a lesser extent, in the ideation of mildly depressed patients.

The initial studies of the dreams and verbal productions of depressed patients were conducted to test the hypothesis that retroflected rage is a central ingredient in depression. The results of these studies, however, suggested a more parsimonious formulation of depression: The patient experiences sadness, loss of spontaneous motivation, loss of interest, indecisiveness, and suicidal wishes because of his idiosyncratic view of the world, himself, and his future. The disturbances in affect and motivation are the logical outcome of the negative conceptualizations.

A series of correlational studies have been conducted to test these clinical observations (Vatz, Winig, and Beck, 1969). These investigators found significant correlations between the Depression Inventory and measures of pessimism ($r=0.56$) and negative self-concept ($r=0.70$). After recovery from depression, the scores on these measures showed substantial decrements. In a longitudinal study of thirty patients, the investigators found that change scores (between the time of admission and discharge) on the measures of pessimism and negative self-concept correlated 0.49 and 0.53, respectively, with the change scores on the Depression Inventory. These findings lent support to the notion that the state of depression is associated with a negative view of the self and the future.

Another line of inquiry into the phenomena of depression is the experimental manipulation of the cognitive variables and the assessment of the effects on other variables relevant to depression. According to the present hypothesis, the depressed patient is characterized by unrealistically low concepts of his capabilities. If this negative orientation can be ameliorated, then the secondary symptoms of depression, such as reduction of constructive motivation, should improve.

DIFFERENTIAL EFFECTS OF PERFORMANCE FEEDBACK

A study by Loeb, Feshbach, Beck, and Wolf (1964) indicated that patients in a superior performance group were more self-confident, rated themselves as happier, and perceived others as happier than did patients in an inferior performance group. Depressed patients were more affected than nondepressed patients by task performance in estimating how they would do in a future task. They also showed greater changes in their self-ratings on the happiness-sadness continuum. The increment or decrement in their mood, however, was short of statistical significance (0.1 level).

A later experiment (Loeb, Beck, Diggory, and Tuthill, 1967) showed that depressed patients were significantly more pessimistic about their likelihood of succeeding and made significantly lower ratings of their performance, even though they performed as well as or better than a matched control group of nondepressed patients. On a second task in this experiment, the previous experience of success or failure had different effects on the actual performance of the two groups: Success improved the performance of the depressed group, whereas failure improved the performance of the nondepressed group. The depressed patients in the success group, though positively affected by success, still showed lower probability-of-success estimates and lower self-evaluation than did the successful nondepressed group.

The success-failure experiments were designed to elicit the characteristic depressive behaviors in a controlled setting so that the specific responses could be identified, quantified, and related to one another. Though different in content from his usual life experiences, the specific experimental tasks produced the same type of reactions as might be observed in a naturalistic setting. By controlling the type of feedback regarding task performance—that is, whether positive or negative—a number of other investigators (Escalona, 1940; Rosenzweig, 1959) have also demonstrated a specific effect on responses of the depressed patient.

From a clinical standpoint, the most promising finding was that the depressed patient reacts positively to tangible evidence of successful or superior performance. The interpersonal meaning of the experimental situation, in which the subject receives positive or negative information about himself from the experimenter, may exert a particular powerful effect on the depressed patient. If this tendency to exaggerate the evaluative aspects of situations and to overgeneralize in a positive direction after success is substantiated, it offers obvious clues to the therapeutic management of depression.

A COGNITIVE-AFFECTIVE-
MOTIVATIONAL MODEL

The clinical observations and results of correlational and experimental studies may be pieced together in a logical way within a structural framework.

The principal postulate of this formulation is that in depression (as in other psychopathological conditions) a primitive or immature cognitive organization becomes dominant. This notion of a primitive organization is similar in many ways to Freud's description of the primary process and Piaget's formulation of early developmental levels of thinking. This primitive organization is composed of relatively crude concepts (schemas). In contrast to the more mature levels of organization (for example, the secondary process), these concepts are framed in absolute rather than relative terms, are dichotomous rather than graduated, and are global rather than discriminative.

Under normal conditions, the mature psychological processes are generally dominant, though more primitive levels of thinking may be detected in sporadic illogical ideas, in disproportionate or inappropriate reactions to life situations, and in dreams and waking fantasies. As depression develops, the primitive organization preempts the more mature system in areas of experience relevant to self-evaluation and expectancy. The depressed patient's negative ideation about himself and his future result from the activation of primitive schemas incorporating notions of worthlessness and pessimism. When he is engaged in problem-solving removed from these sensitive content areas, he is able to mobilize his more mature cognitive organization, that is, he can think logically, realistically, and adaptively.

The operations of the primitive organization are evident in both the content and the form of the depressed patient's ideation (in the stream of consciousness as well as in specific interpretations of external stimuli). To a large extent, the cognitive schemas dictate the meanings and even the "facts" of observation. The external stimulus configurations are screened and molded to conform to the content of these schemas. Therefore, the patient's interpretations of experiences, his expectancies, his reminiscences are congruent with the idiosyncratic schemas. The prevailing content of his ideation is consistent with the themes of the schemas: "I am bad." "I will never feel better." "I cannot cope."

The formal disturbances of thinking in depression can be attributed to the dominance of the idiosyncratic schemas, which override the demand character of the external stimulus situation. The crude, undifferentiated nature of the schemas may account for the overgeneralizations and blurring of discriminations. Extreme judgments and overgeneralizations may be explained by the properties of the primitive schemas, namely, their breadth, rigidity, and dichotomous structure. Selective abstraction and selective recall may be attributed to the extraction of data isomorphic to the dominant schemas.

THE COGNITION-AFFECT CHAIN

The intimate connection between cognition and affect may be represented metaphorically as pathways between cognitive structures and corresponding affective structures; therefore, a particular cognitive content produces an affect congruent with it. For example, ideation centering around a fantasied or symbolic loss produces the same affect as actual loss. The subjective feeling is described by adjectives such as sadness, loneliness, or disappointment. Similarly, self-critical thoughts lead to affects described by patients in terms, such as "disgust with myself."

AFFECT-COGNITION CHAIN

In order to account for the data of depression it is useful to assume that affects have stimulus properties. Irrespective of its origin, the aroused affect becomes part of the stimulus field. The affect, consequently, is subjected to cognitive processes such as monitoring, labeling, and interpretation in much the same way as are external stimuli.

The verbalized reports by severely depressed patients of their introspections suggest that they continually scan their affective field to determine how and what they are feeling. The feeling itself has a signal function. Just as anxiety is generally interpreted by the subject as a sign that danger is present, so dysphoria is assigned informational value (even though incorrect) and meaning by the patients. The patient thinks, "I'm feeling bad, so things must be bad." This axiomatic idea serves to sustain the patient's previous cognitions to the effect that his problems are insoluble. (See Sarwer-Foner's description [1966] of the delusion of hopelessness in depressives.)

THE CONTINUOUS COGNITION-AFFECT CYCLE

The downward spiraling phenomenon of depression may be explained as follows: A perceived loss (actual or fantasied) leads to negative affect (dysphoria). The dysphoria is interpreted as a sign that, "Things are not good. I'm going downhill." This interpretation is linked in an associative network to other absolute concepts, such as "My future is black. There is no possibility for happiness." This sequence leads to further feelings of sadness and loss.

The preceding paradigm may be illustrated by a brief example: A depressed patient had an assignment that he perceived as too difficult for him. This cognition led to dysphoria. The identification of the feeling of sadness led to the thought, "No sense in working, I'm feeling too bad. This feeling will go on. If I do anything, I will feel confused."

This case illustrates the sequential, or possibly simultaneous, interplay of several systems. The initial expectancy of failure leads to dysphoria, which is interpreted as indicative of weakness and incapacity to reach his goals. A by-product of this interplay of cognitive, affective, and motivational systems is the desire to escape by committing suicide.

PHYSICAL AND VEGETATIVE SYMPTOMS

The explanation of the physical and vegetative symptoms of depression in the framework of a psychological model presents certain difficulties. Mixing different conceptual levels entails the risk of confounding rather than clarifying the problem. Furthermore, whereas patients' verbal material is a rich source of information for establishing meaningful connections between the psychological variables, it provides scanty data for determining psychophysiological relationships.

With these reservations in mind, the author has attempted to relate the cognitive patterns to one of the physical correlates of depression, namely, the retardation and fatigue. Retarded patients have generally expressed attitudes of passive resignation to their supposedly terrible fate. Their attitude is expressed in such statements as "There's nothing I can do to save myself." In the most severe cases, such as the benign stupors, the patient may believe that he is already dead. In any event, the profound motor inhibition appears to be congruent with the patient's negative view of himself and loss of spontaneous motivation. When the patient's desire to do something is stimulated, retardation becomes temporarily reduced or disappears. Moreover, when the patient can entertain the idea of getting some gratification from what he is doing, there is a reduction in the subjective sense of fatigue.

The influence of psychological factors on the inertia, retardation, and fatigability has been borne out by several systematic studies. It has been noted that when depressed patients are given a concrete task, such as the digit-symbol substitution test, they mobilize sufficient motivation to perform as well as other nondepressed patients of similar severity of illness (Beck, Feshbach, and Legg, 1962). Since this test is essentially a speed test it should be particularly sensitive to psychomotor retardation. Similarly, Friedman (1964) found that depressed patients showed either no impairment or only minimal impairment when engaged in a variety of psychological tests.

The vegetative signs of depression such as loss of appetite, loss of libido, and sleep disturbance may be defined as the physiological concomitants of the particular psychological disturbance in a fashion analogous to the conceptualization of the autonomic symptoms as correlates of anxiety. To complete the description, it may be speculated that the patient's perception of his vegetative and other somatic disturbances are processed by the typical idiosyncratic cognitive schemas. Hence, the meaning attached to sleep disturbance ranges from "I am going to pieces" to "I haven't slept at all for several weeks." The perception of loss of appetite and possibly reduction of other proprioceptive sensations from the gastrointestinal system may be woven into notions such as "I am an empty shell. I have lost my internal organs."

SUMMARY

The phenomena of depression have intrigued writers since Hippocrates. Numerous theoreticians have attempted to make psychological sense out of the varied and paradoxical characteristics of this disorder. Illustrative examples of conceptual models offered by classical psychoanalytic theory, ego psychology, interpersonal psychiatry, existential psychiatry, learning theory, and cognitive theory were presented.

A different approach to illuminating the problem of depression, exemplified by the investigation by Grinker and his coworkers, has depended on the application of statistical techniques such as factor analysis to quantitative self-reports and behavioral rating scales. Factors derived from thirteen different investigations were reviewed. Certain factors were found consistently in each of these studies.

It appears that the next stage in the attempt to further understanding of the psychological dimension of depression is the systematic testing of hypotheses drawn from the theories. I presented an explanatory model of depression derived from my clinical investigations and subsequently subjected to a variety of correlational and experimental studies. This model assigns a central role to the cognitive distortions in depression. Other manifestations, such as depressed mood, loss of spontaneity, and suicidal wishes, are conceptualized as the logical outcome of the depressed patient's unrealistically negative view of himself, his world, and his future. Vegetative symptoms, such as loss of appetite and sleep disturbance, are regarded as the physiological correlates of the depressed state.

A feedback model is described to explain the continuous downward spiral. Recognition of his dysphoria, dependency, and retardation reinforce the patient's negative expectancies and debased self-image and consequently accentuate the affective, motivational, and behavioral changes. This model also offers clues for specific therapeutic interventions to alleviate or reverse the cycle of depression.

REFERENCES

Abraham, K. 1960a. "The First Pregenital Stage of the Libido" (1916). In K. Abraham, *Selected Papers on Psychoanalysis*. New York, Basic Books. Pp. 248–279.

Abraham, K. 1960b. "Notes on the Psychoanalytic Investigation and Treatment of Manic-Depressive Insanity and Allied Conditions" (1911). In K. Abraham, *Selected Papers on Psychoanalysis.* New York, Basic Books. Pp. 137–156.

Abraham, K. 1960c. "A Short Study of the Development of the Libido" (1924). In K. Abraham, *Selected Papers on Psychoanalysis.* New York, Basic Books. Pp. 418–501.

Beck, A. T. 1961. "A Systematic Investigation of Depression," *Comparative Psychiatry* 2:162–170.

Beck, A. T. 1963. "Thinking and Depression: 1. Idiosyncratic Content and Cognitive Distortions," *Archives of General Psychiatry* 9:324–335.

Beck, A. T. 1964. "Thinking and Depression: 2. Theory and Therapy," *Archives of General Psychiatry* 10:561–571.

Beck, A. T. 1967. *Depression: Clinical, Experimental, and Theoretical Aspects.* New York: Hoeber.

Beck, A. T. 1969. Factor analysis of depression inventory. Unpublished study.

Beck, A. T., Feshbach, S., and Legg, D. 1962. "The Clinical Utility of the Digit Symbol Test," *Journal of Consulting Psychology* 26:263–268.

Beck, A. T., and Hurvich, M. S. 1959. "Psychological Correlates of Depression: 1. Frequency of 'Masochistic' Dream Content in a Private Practice Sample," *Psychosomatic Medicine* 21:50–55.

Beck, A. T., and Ward, C. H. 1961. "Dreams of Depressed Patients: Characteristic Themes in Manifest Content," *Archives of General Psychiatry* 5:462–467.

Beck, A. T., Ward, C. H., Mendelson, M., Mock, J., and Erbaugh, J. 1961. "An Inventory for Measuring Depression," *Archives of General Psychiatry* 4:561–571.

Becker, J. 1960. "Achievement-Related Characteristics of Manic-Depressives," *Journal of Abnormal Social Psychology* 60:334–339.

Becker, J., Spielberger, C. D., and Parker, J. B. 1963. "Value Achievement and Authoritarian Attitudes in Psychiatric Patients," *Journal of Clinical Psychology* 19:57–61.

Bibring, E. 1953. "The Mechanism of Depression." In P. Greenacre, ed., *Affective Disorders.* New York: International Universities Press. Pp. 13–48.

Bjerg, K. 1967. "The Suicidal Life Space." In E. S. Shneidman, ed., *Essays in Self-Destruction.* New York: Science House. Pp. 475–494.

Breed, W. 1967. "Suicide and Loss in Social Interaction." In E. S. Shneidman, ed., *Essays in Self-Destruction.* New York: Science House. Pp. 188–203.

Cohen, M. B., Baker, G., Cohen, R. A., Fromm-Reichmann, F., and Weigert, E. V. 1954. "An Intensive Study of Twelve Cases of Manic-Depressive Psychosis," *Psychiatry* 17:103–137.

Cropley, A. J., and Weckowicz, T. E. 1966. "The Dimensionality of Clinical Depression," *Australian Journal of Psychology* 18:18–25.

Dilling, C. A., and Rabin, A. I. 1967. "Temporal Experience in Depressed States and Schizophrenia," *Journal of Consulting Psychology,* 31:604–608.

Engel, G. 1968. "A Life Setting Conducive to Illness: The Giving-Up—Given-Up Complex," *Bulletin of the Menninger Clinic* 32:355–365.

Escalona, S. K. 1940. "The Effects of Success and Failure upon the Level of Aspiration and Behavior in Manic-Depressive Psychosis," *University of Iowa Studies in Child Welfare* 16:199–302.

Farnham-Diggory, S. 1964. "Self-Evaluation and Subjective Life Expectancy among Suicidal and Non-Suicidal Psychotic Males," *Journal of Abnormal Social Psychology* 69:628–634.

Freud, S. 1959. "Mourning and Melancholia" (1917). In *Collected Papers,* vol. 4. New York: Basic Books. Pp. 152–170.

Friedman, A. S. 1964. "Minimal Effects of Severe Depression on Cognitive Functioning," *Journal of Abnormal Social Psychology* 69:237–243.

Gaylin, W. 1968. *The Meaning of Despair: Psychoanalytic Contributions to the Understanding of Depression.* New York: Science House.

Gero, G. 1936. "The Construction of Depression," *International Journal of Psychoanalysis* 17:423–461.

Gibson, R. W. 1957. *Comparison of the Family Background and Early Life Experience of the Manic-Depressive and Schizophrenic Patient,* Final Report on

Office of Naval Research Contract (Nonr-751(00)). Washington, D.C.: Washington School of Psychiatry.

Grinker, R. R., Sr., Miller, J., Sabshin, M., Nunn, R., and Nunnally, J. 1961. *The Phenomena of Depressions.* New York: Hoeber.

Hamilton, M., and White, J. 1959. "Clinical Syndromes in Depressive States," *Journal of Mental Science* 105:485–498.

Hammerman, S. 1962. Ego defect and depression. Paper presented to the Philadelphia Psychoanalytic Society, Philadelphia, November 7.

Harder, J. 1967. "The Psychopathology of Infanticide," *Acta Psychiatrica Scandinavica* 43:196–245.

Jacobson, E. 1953. "Contribution to the Metapsychology of Cyclothymic Depression." In P. Greenacre, ed., *Affective Disorders.* New York: International Universities Press. Pp. 49–83.

Jacobson, E. 1954. "Transference Problems in the Psychoanalytic Treatment of Severely Depressive Patients," *Journal of the American Psychoanalytic Association* 2:595–606.

Katkin, E. S., Sasmor, D. B., and Kann, R. 1966. "Conformity and Achievement: Related Characteristics of Depressed Patients," *Journal of Abnormal Psychology* 71:407–412.

Kendell, R. E. 1968. *The Classification of Depressive Illnesses.* London: Oxford University Press.

Klein, M. 1948. "A Contribution to the Psychogenesis of Manic-Depressive States (1934). In *Contributions to Psycho-Analysis, 1921–1945.* London: Hogarth Press. Pp. 282–310.

Lazarus, A. 1968. "Learning Theory and the Treatment of Depression," *Behavior Research and Therapy* 6:83–89.

Lewinsohn, P., Shaffer, M., and Libet, J. 1969. A behavioral approach to depression. Paper presented to the American Psychological Association. Miami Beach, Florida.

Lichtenberg, P. 1957. "A Definition and Analysis of Depression," *Archives of Neurology and Psychiatry* 77:516–527.

Loeb, A., Beck, A. T., Diggory, J. C., and Tuthill, R. 1967. "Expectancy, Level of Aspiration, Performance, and Self-Evaluation in Depression," *Proceedings of the Annual Convention of the American Psychological Association* 2:193–194.

Loeb, A., Feshbach, S., Beck, A. T., and Wolf, A. 1964. "Some Effects of Reward upon the Social Perception and Motivation of Psychiatric Patients Varying in Depression," *Journal of Abnormal Social Psychology* 68:609–616.

Lorr, M., Sonn, T., and Katz, M. M. 1967. "Toward a Definition of Depression," *Archives of General Psychiatry* 17:183–186.

Melges, F. T., and Bowlby, J. 1969. "Types of Hopelessness in Psychopathological Process," *Archives of General Psychiatry* 20:690–699.

Mendels, J., and Cochrane, C. 1968. "Nosology of Depression: The Endogenous-Reactive Concept," *American Journal of Psychology* 124:1–11.

Mendelson, M. 1960. *Psychoanalytic Concepts of Depression* Springfield, Illinois: Thomas.

Michaux, M. H., Zwagg, L. V., and Kurland, A. A. 1967. "Dimensions of Manifest Depression in Newly Hospitalized Patients," *British Journal of Psychiatry* 113:981–986.

Mowrer, O. H. 1969. "New Directions in the Understanding and Management of Depression," *International Psychiatry Clinics* 6:317–360.

Overall, J. E. 1962. "Dimensions of Manifest Depression," *Journal of Psychiatric Research* 1:239–245.

Paykel, E. S., Klerman, G. L., and Prusoff, B. A. 1970. "Treatment Setting and Clinical Depression," *Archives of General Psychiatry* 22:11–22.

Pichot, P., and Lempérière, T. 1964. "Analyse factorielle d'un questionnaire d'auto-evaluation des symptomes depressifs," *Revue de Psychologie Appliqué* 14:15–29.

Rado, S. 1928. "The Problem of Melancholia," *International Journal of Psychoanalysis* 9:420–438.

Raskin, A., Schulterbrandt, J., Reatig, N., and McKeon, J. 1969. "Replication of Factors of Psychopathology in Interview, Ward Behavior and Self-Report Ratings of Hospitalized Depressives," *Journal of Nervous and Mental Diseases* 148:87–97.

Raskin, A., Schulterbrandt, J., Reatig, N., and Rice, C. 1967. "Factors of Psychopathology in Interview, Ward Behavior and Self-Report Ratings of Hospitalized Depressives," *Journal of Consulting Psychology* 31:270–278.

Rosenzweig, S. 1959. The effects of failure and success on evaluation of self and others: a study of depressed patients and normals. Doctoral dissertation, Indiana University.

Sarwer-Foner, G. J. 1966. "A Psychoanalytic Note on a Specific Delusion of Time in Psychotic Depression," *Canadian Psychiatric Association Journal* 11 (suppl.): s221–s228.

Stuart, J. L. 1962. Intercorrelations of depressive tendencies, time perspective, and cognitive style variables. Doctoral dissertation, Vanderbilt University.

Tellenbach, H. 1961. *Melancholie*. West Berlin: Springer.

Vatz, K. A., Winig, H. R., and Beck, A. T. 1969. Pessimism and a sense of future time constriction as cognitive distortions in depression. Mimeographed. University of Pennsylvania.

Weckowicz, T. E., Muir, W., and Cropley, A. J. 1967. "A Factor Analysis of the Beck Depression Inventory," *Journal of Consulting Psychology* 31:23–28.

Wohlford, P. 1966. "Extension of Personal Time, Affective States, and Expectation of Personal Death. *Journal of Personality and Social Psychology* 3:559–566.

Yonge, K. A., Weckowicz, T. E., Muir, W., and Cropley, A. J. 1969. Objective therapy predictors in depression: a multivariate approach. Mimeographed.

Zetzel, E. R. 1966. "The Predisposition to Depression," *Canadian Psychiatric Association Journal* 11 (suppl): s236–s249.

Zung, W. K. 1967. "Depression in the Normal Aged," *Psychosomatics* 8:287–292.

The Treatment of
the Borderline Syndrome

Throughout history man's willing subjugation of himself to theoretical systems as an attempt to bring predictable order out of experiential chaos has been a dominant theme. The majority tends to settle for the ease a system brings at the price of ignoring experiential contradictions. It is tempting when but a small leap of armchair theorizing can land one safely in one camp or another. The unusual man struggles to comprehend these contradictions in a wider synthesis, and may experience the loneliness of being out in front. Leonardo da Vinci, for instance, has been criticized for the incompleteness of his systems, in painting, anatomy, invention, and psychology. These complaints can best be met not by demonstrating that this type of man has been a better systemizer than he is credited to be, but by indicating that such goals were foreign to him. A real explorer has to stay close to the observational data. This brand of conservatism is essential to voyaging in the real world where constructing theory is useful only if it means increasing our capacity to learn from experience, and testing preexisting theory is essential to getting rid of useless baggage.

In *The Borderline Syndrome*, Grinker, Werble, and Drye (1968) fashioned a sophisticated operational tool out of psychoanalytic ego psychology and used it with the help of computer technology to tease out a valid category of aberrant behavior which is distinguishable from schizophrenia, neurosis, and character disorder. This use of statistical analysis to see what traits cluster together most reliably is as exciting as a detective story. It eliminates the question of do they belong together theoretically and addresses itself to the question of which behavioral events appear together regularly. Intracluster relationships can then be studied from a variety of theoretical points of view.

We have had our Kraepelinian period of using nosology for the purpose of projecting the course of a hypothesized disease process. This tendency was heavily corrected for by the psychotherapists who found it necessary to put the patient back into an interpersonal context in order to treat him. Therapeutic optimism swung very close to diagnostic nihilism when psychotherapy was conceived of as a single set of techniques of broad applicability. Swinging back half-circle with Dr. Grinker we agree that mental events are not divorced from their behavioral manifestations. If we avoid the pitfall of simply looking at the topography of behavior rather than its functional relationship with the environment, we can begin to see the lawfulness that exists between inner and outer events. Just as the artist by his vision creates his world for us to see, so every man creates by his own behavior the physiognomy of the world that he sees during each minute of his life. The psychotherapists who reacted against Kraepelinian nosology had two valid complaints: One was their distaste for the prediction of an immutable fate for the patient from the form of his behavior. The other was the practical matter that you cannot treat a category or even a cluster of traits; you must treat the individual. The formal aspects of a patient's disorder, however, do dictate treatment strategies and even tactical priorities in many other diagnostic categories, so why not this one?

SUMMARY OF THE BORDERLINE SYNDROME AS DESCRIBED BY DR. GRINKER

Complex statistic analyses designed for the data gathered in this investigation revealed four distinct subgroups. When translated into clinical syndromes, the subgroups coincided with clinical experience. Group 1 was closest to the psychotic border. Members of this group gave up attempts at relationships but at the same time overtly, in behavior and affect, displayed anger toward the world. Their ego integrations were endangered by this strong affect.

Clinically, group 2 was seen as representing the core process of the borderline. Persons in group 2 were inconsistent. They would move toward others for relations, and then with acted-out repulsion initiate a return to a lonely and depressed state of isolation. Thus, shifts from anger to depression would accompany their continual back and forth movements.

Group 3 was the most adaptive, compliant, and lacking in identity. These patients seemed to have given up their search for identity. Their world was an empty place. They did not have the angry reactions characteristic of group 1. Instead they passively awaited cues from others and behaved in complementarity—"as if."

Subjects in group 4 were the closest to the neurotic border. They were searching for a lost and unattainable symbiotic relation with a mother figure. They revealed what might be called an anaclitic depression, weeping and feeling neglected and sorry for themselves. These patients had sometimes been mistakenly categorized as suffering from depression.

A GLIMPSE OF SOME
PERTINENT LITERATURE

I wish Dr. Grinker did not feel quite so strongly the necessity to eliminate any association with the term "schizophrenia," however. I would be quite happy to add the borderline group to Dr. Beck's series of schizophrenias designated S1 to S6 by giving them a higher number. Goodness knows it is a loose term itself, but the Rosenthal and Kety (1968) data from Denmark strongly suggest that the borderline group belongs with the chronic schizophrenics on the basis of inheritance. Curiously enough, the acute schizophrenics belong with neither in Kety's data. Similarly, Heston's (1970) data show that the borderline syndrome as part of "schizoidia" may fit into his genetic theory too.

In the Ekstein and Wallerstein (1954) work, cited as evidence of the existence of the borderline syndrome in childhood, emphasis is placed on the sensitivity of the borderline child's ego to stimuli both from within and from without. In their case material it is well documented that a relatively slight cue from the therapist leads to a wide excursion of behavioral response. These data, from a behavioral point of view, suggest that the child's earlier experiences had shaped this inordinate response as an appropriate one, in some sense, within the family.

Margaret Mahler's (1949) benign cases of childhood psychosis, which she attributes to constitutional or genetic factors, strike me as having many features in common with the less clinically disturbed children who showed a short reality span, described by Geleerd (1946, 1958), and by Weil (1953a, 1953b). Dr. Weil makes the point that these children were observed to be "different" from early infancy and that they showed conspicuous delay in ego development (but not in motor and intellectual development). This delay was described as "a marked deficiency in the development of object relationship with all its consequences (giving up of omnipotence, of magical thinking, acceptance of the reality principle), in reality testing, in the development of the synthetic function, and in the proper use of age-adequate defenses."

From the point of view of an inheritable characteristic this set of observations could be equally well related to Paul Meehl's (1962) fantasy of schizotaxy, schizotypy, and schizophrenia (the inherent source trait, the schizoid development, and the schizophrenic development) based on a common denominator of failure of inhibition in cortical data-processing. He proffers several different neurophysiological mechanisms, but this one seems to tie into such a broad set of other observations that it appeals to me the most. John Frosch's (1964) description of the psychotic character as having the problem of "closeness of the ego and id" with easy interpenetration of primary process into the stream of consciousness and Searles's (1965) comments on sensory deprivation phenomena in schizophrenics in which sensory input very easily becomes "noise" in the system seem to me to support this hypothesis of failure of inhibition.

Furthermore, Rothenberg's (1970) description of the Janusian, or creative, character gives us an excellent description of what may be the most highly adapted end of this spectrum. These creative thinkers have mastered conventional symbolic activity while still retaining the capacity for novel integration of paradoxical or incongruous thoughts. He points out the affective acceleration that accompanies the simultaneous awareness of disparate and contradictory aspects of experience. Failure of inhibition could certainly account for the chaotic affect produced by paradoxical ideation. Is it any wonder then that there are heroic efforts made at affective inhibition by the schizophrenic, and less total but comparable efforts made by the borderline whose affect storms elude his grasp a fair amount of the time? It was also interesting to note that when Ekstein and Wallerstein's schizophrenic children improved in treatment they were then described as borderline. This would fit the pattern of movement on a continuum.

Stephen Pittel of the Haight-Ashbury Research Project (1970) reported at a recent conference on the drug subculture that many of those youngsters who were the most frequent casualties of the drug scene were "fantasy-poor" before they started with drugs. In describing their current fantasy life Kempner (1970) reports that they now exhibit perceptual distortion and regression, but without secondary elaboration. In other words, their fantasy life is still impoverished. If this is the case then we might take at face value their wish to "open the doors of perception." Unfortunately, for those who need it the doors open onto a barren landscape. It takes more cerebration to live in our society today than in a more traditional one. As David Rioch (1970) says, "To live in our society you have to start and stop quickly . . . change directions smoothly without losing track of where you are going." This may well help account for the rising incidence of the diagnosis of borderline syndrome.

DISCUSSION OF THE BOOK AND
RELATED LITERATURE

As it emerges, the borderline syndrome with its four clinical subgroups is a useful concept. By useful, I mean that it permits us to recognize a configuration of traits that could evoke the response of doing some technical things differently in therapy than if the patient were seen as schizophrenic or neurotic. All four groups are seen in the book as suffering from an arrested development of ego functions rather than regressing in response to internal or external stress. This would then place their defects in the conflict-free sphere of ego development.

In this business of seeing or of recognizing patterns we get into many fascinating problems. Looking at a puzzle picture, once we see the face in the tree our perception is changed even though the drawing has not changed. This recognition is made in an associative context having to do with tendencies to respond. An example would be the patient in treatment who in spite of our thoughtful exploration of a number of relevant early and current interpersonal problems grows steadily worse. One day we recognize that her solitary drinking, of which we were peripherally aware, is at this point the major problem. Suddenly there is a rush of recall of supporting data to fill in the picture of an alcoholic and, we hope, some change in the treatment program.

The illustrative case material from Grinker's and other people's work demonstrates more than anything else the remarkably short sequences of interactive behavior these patients are capable of maintaining. Their needs are not necessarily different from anyone else's, but it is obvious that they have a harder time getting their needs met than most. The sequences fail to complete in many instances and thus deprive the patient of even the satisfaction of closure. Certainly failure of inhibition of "noise" in the system is apt to lead to cognitive slippage, to anhedonia, to anger. At the behavioral level the case material shows that type 1 had the shortest and most abruptly broken sequences with consequent rage. On the other hand, even though types 3 and 4 have longer behavior sequences, they appear to peter out in apathy or depression rather than achieving satisfaction. Thus it would seem that in many instances the distress of the borderline patient is more a failure of holding a focus than a lack of exceptional effort.

So often it seems that to entertain the notion of a biological flaw is tantamount to therapeutic despair. In the case of the borderline syndrome, however, pessimism is common among therapists without regard to their theories of etiology. I suspect that this sense that "nothing can be done about it," which the book conveys, may be a reaction to the patient's "helplessness of the ego" as described by Geleerd (1958). If we think of myth or symbol as the cultural carriers (cultural genes as it were), the family must be the major transmitter in view of our long period of dependency in childhood. If the family is as described: inconsistent, affect poor rather than malevolent, and isolated rather than part of an extended family, it probably is a poor transmitter of cultural genes. Particularly so if the child has trouble tracking an input because of noise in the system.

The adolescent seen in psychotherapy shows many of the features described as borderline (A. Freud, 1966; Jacobson, 1964): for example, lots of anger, vacillation between approach and withdrawal, "as if" behavior in testing out new roles, and anaclitic depression in brief but often dramatic bursts. Could we consider this relatively common phenomenon as a normal borderline traverse from the family system of symbols and myths to the broader community? Then considering the borderline patients to have a difficult time at best in holding onto and manipulating reality, need we be surprised when they seem to have gotten to adolescence and then curiously found themselves unable to move either forward or backward? This phenomenon could be related both to a genetic handicap, making it difficult to integrate myth and symbol, as well as to distinctive features of the particular family of origin. At least it would appear that successful adolescence is a period in which the myths and symbols established in childhood are modulated to conform to the standards of the emerging adult peer group.

In this connection I was struck by the fact that in their admittedly cursory family studies, Grinker and his group found none of the communication problems reported to abound in the families of schizophrenics. Is it possible that the genetic defect failed to flower into chronic schizophrenia because of a slightly more favorable family environment? I have had the notion for some time that many of my borderline patients have been reared in environments with special characteristics, not malevolent or neglectful but special and not broadly generalizable to other environments, which may account in part for Dr. Grinker's patients' reluctance to separate from their families of origin. Certainly, the family background is intriguing enough to demand further investigation.

What stood out in the book, second only to the syndrome itself, was the significant lack of change for better or for worse in the follow-up study. Many of these people had what were described as "psychiatric contacts" in the follow-up period. What does contact mean? Like the expression "being followed by a psychiatrist" it leaves one wondering if it is significantly different from being watched by one of the original observers from the project. Robert and Mabel Cohen (1961) have described psychotherapy as an undefined technique applied to unspecified problems with unpredictable outcome. Maybe this is what these subjects had. I jest, but could we not be a bit more specific than is our custom about what works and what does not? Thanks to Dr. Grinker and his

group, we have now some specifications for the relative ego incompetencies of this group of patients put into operational terms, but as Skinner
(1961) says, "knowledge is not to be identified with how things look to
us, but with what we do about them." How do we as therapists "capture" this behavior to make it sensitive to intervention?

TREATMENT CONSIDERATIONS

If anything is impressive in the literature on the treatment of borderline
syndrome, it is that nothing seems to work very well, or for very long.
Some of the most strenuous and long-lasting efforts appear eventually to
wash away like sand castles before the eyes of the tired therapist. Perhaps the main reason the term "borderline" has not caught on with psychiatrists is the paucity of techniques to deal specifically with such a
syndrome. The psychoanalysts who have written about it at some length
have found no easy answers. There is general agreement among Knight
(1954), Frosch (1964), Kernberg (1960, 1970), Searles (1965), and Spotnitz (1969) that you have to be yourself rather than anonymous, work
with the negative affect, and reinforce the relationship with reality and
the sense of reality. The paradox that emerges is that to do this work
you have to be yourself, your very best self, and technique makes it possible. You become it and that is why it works and why technique is so
important. Perhaps you can be anonymous and technical with a neurotic
patient, though I doubt it. With the borderline schizophrenic group it is
impossible to be anonymous but all the more reason to be technically
relevant. That we have kept at it may simply be evidence that we have
accepted the responsibility to work intensively with patients for whom
simpler measures do not work.

It seems quite likely that we have made the task unnecessarily difficult for ourselves by our reluctance to use psychotropic drugs to cushion
the impact of drives against inadequate impulse control. With children
it appears that using a mild tranquilizer, such as Benadryl, the stimulus
barrier can be raised, or the distractibility reduced to the point that interactions become more capable of providing satisfaction for the child.
The borderline adults I have seen are a more complex problem. In the
first place, a larger number of them are self-medicators, not in the sense
of the recreational use of drugs necessarily, but their irregular and often
excessive use of alcohol or other drugs is tied to efforts at adjusting an
intolerable feeling state. Euphoriants appeal to them, though the response may be unreliable and often paradoxical. Phenothiazines (which
should theoretically help with our postulated lack of central inhibition)
have in my experience yielded no improved affect; possibly they
lengthen the periods of unpleasant affect, and in larger doses result in

depressed behavior. As Eisenstein (1952) put it, "They are plagued by negative feelings . . . sensitive to both warmth and aloofness." In spite of their being porcupines it seems we are thrown back on having to treat them on an interpersonal basis. In this regard we may have tended to focus too much on the capacity for dyadic intimacy rather than on reliable intragroup behavior, which is more dependent on role identification than on the dissipation of approach anxiety. As John Frosch (1964) points out in his paper on the psychotic character, the major problem is their performance in relation to reality and their deficient sense of reality, rather than a problem of reality-testing. I would say, then, ease and reliability before intimacy applies to the borderline group as well as to the schizophrenic. It would seem logical that for the borderline patient in particular, the "heavy" types of group therapy such as are being used for addicts (Synanon, Gateway, "The Family" at Mendocino State Hospital, and so on) would be a treatment to consider for improving both their performance in relation to reality as well as their sense of reality.

My suggestion that heavy group therapy would be an interesting treatment to try does not mean that I am suggesting that we abandon our efforts at the individual treatment of the borderline syndrome. My question is rather: If we set ourselves the task of treating specific ego incompetencies, is psychoanalysis the technique of choice? Not if we think of its task as primarily the removal of resistances and repressions so that impulse and fantasy can make their way into consciousness. The answer is very likely affirmative if we think of interpretation as a way of helping the patient to develop sharper discrimination of both inner and outer stimuli, and the emotional interaction between analyst and patient as a way of shaping his behavior with direct consequences. When we consider the ego to be primarily a control device concerned with the appropriate staging and phasing of behaviors, we are remaining in the realm of observable data. It is my position that psychoanalysis is a relevant theoretical system for treating such patients, since it relies heavily on the assumption that even seemingly erratic behavior is in fact consequential, often at a level below awareness, and that the elucidation of its consequences is the major vehicle for treatment. That is, making the unconscious conscious means making explicit not only the reasons for a patient's maladaptive behavior but the details of those behavior patterns and their alternatives as well. I believe this expanded view is essential for one to develop an effective therapeutic approach to the borderline patient.

Psychoanalysts perhaps more than any other workers in the field have had the opportunity for the intensive study of the individual case. It is thus incumbent on us to exhibit an even greater rigor in the validation of our findings than we require of the worker who spends much less time with his patient.

In order to allay the uneasy feeling that to be really scientific we must use large numbers, may I remind the reader that Chassan (1960) demonstrated that intensive vs. extensive design can be shown to have the greater reliability when the background and other characteristics of the single case are specified and used as the ground against which the specified interaction can be studied. Thus, we have no necessity for any national statistical pool of psychoanalytic results in order to come to a scientific conclusion. The individual analyst with his patient can generate perfectly satisfactory research if he chooses to do so.

Ezriel (1956) at the Tavistock clinic has made a beginning in describing the psychoanalytic session as an experimental situation and demonstrated how testable data can be generated from such a plan. His definition of the psychoanalytic situation as an ahistorical dynamic and not a genetic method, though based on an analysis of the transference, is curiously enough described by him with the same three adjectives used in Skinner's description of the operant behavior paradigm. I say curiously, because much has been made of the imagined polar differences that separate behavior research from psychoanalysis. Yet, as we become more explicit we find that we stand on common ground. Ezriel's "law" of behavior is open to test replication by any of us.

If we set up a field by putting two persons together, the one of whom is a patient in need of treatment and the other a therapist able to gratify this need, and if the therapist assumes a passive, nondirective role, then the patient will display in his words and actions a manifest form of behavior in which a conscious and an unconscious impulse can be distinguished. If then the therapist gives an interpretation, i.e. points out the unconscious impulse and the phantasy-determined reason for its rejection then the subsequent material will contain the hitherto unconscious impulse in a less repressed form.

To paraphrase only slightly we could arrive at another "law" of behavior congenial to a behaviorist: "If we set up a field by putting two persons together, the one of whom is a patient in need of treatment and the other a therapist able to gratify this need, and if the therapist assumes a passive, nondirective role, then the patient will display in his words and actions a manifest form of behavior in which a variety of themes can be distinguished. If then the therapist gives an interpretation, that is, points out a theme and approves its expression, then the subsequent material will contain the approved theme or related material in greater abundance."

Granted that we have all observed the working of Ezriel's law on the verbal productions of many patients. We know that interpretation does influence behavior. Is this enough, or do we wish to examine more closely the remarkable property of some language used within the psychoanalytic session to influence behavior not only in the session but beyond it in time and space? It seems clear that conceptualizations of the purpose and significance of one's own behavior and the situation in which it occurs exercise powerful directive effects on overt behavior

much as strategies determine tactics (Hunt and Dyrud, 1968). It is my
view that what we can most profitably study is not the question of results but the question of effectiveness. That is, equating effectiveness
with having an effect, what can we say objectively about the effects we
produce? What is characteristic of a mutative interpretation? One characteristic we have been taught is that it includes three elements: (1)
consideration of the transference, (2) consideration of the current life situation, and (3) consideration of the patient's past experience. The other
characteristic is that when timely it works. When it works we can only
state that there is a higher degree of congruence between the patient's
witting and unwitting behaviors and that the move has been toward a
higher degree of appropriateness in the here and now.

The interpretation itself may well be dependent on a careful scrutiny
of the patient's statements about past behavior and experiences with
others to define the nuance of the behavior currently under study. Many
authors express a critical concern with the historical truth of such interpretation (see Wisdom, 1967). This correctness or truth of the interpretation is, however, usually validated not by extraanalytic attempts at historical research but by its consequences in modifying the patient's
behavior—either interrupting ongoing behavior or instating new behavior. What is often lost sight of is that the nature of preliminary interpretations and working through is such that whatever truth is established is
tautological truth; thus the question of validity is more properly addressed to the behavioral change rather than the ultimate historical
truth of the concept used in bringing it about. The concept is important
in another sense in that it not only specifies what new behavior may be
engaged in but in what behaviors are prohibited. This adaptational aspect of interpretation is often ignored because our theory suggests that
new, appropriate behaviors simply appear once the unconscious conflict
has been brought into awareness. With the borderline patient in particular this pattern-making must be explicit within awareness because it
does not occur automatically as in the neurotic.

Granted that we do not need as much interpretation as we do clarification, we can provide organizing concepts or explanations that help
lengthen foresight and stem premature discharge. Cohen and Cohen
(1961) have written of the value of psychoanalytic theory and interpretation in organizing the information overload the patient is experiencing
in treatment. It may be that the experience of overload occurs only
when one's behavior is unreliable in producing satisfaction. One of the
unique features of the psychoanalytic situation with its rigid constraints
on adventitious or unearned gratification and apparent lack of direction
may well be that it provides a setting in which one's behavior is least
likely to produce satisfaction. The response to such deprivation is, in
most experimental animals, a rapid increase in both rate and intensity of
effort. In a real sense we experimentally provide a situation in which in-

formation overload does occur so that our responses can become critical in influencing the patient's subsequent behavior. Our apparent lack of direction can become tragic in its consequences if we indeed have no direction. The heightening of the effect of our interventions by restraint leads to a moment when our single interpretation can restructure the patient's responses not only in the treatment hour but in his life outside, his historical view of himself, and his prospect for the future.

Whenever a figure of speech is multivalent, as in the case of the mutative interpretation described above, it is poetic language. That is, instead of sharply restricting the language to a single implication, we exploit the metaphorical quality of an expression that has multiple and simultaneous levels of meaning. Obviously there is great economy of expression in such usage. Very likely there is maximum impact, much as in an effective poem. But why does it work?

The technical maxim "When the patient speaks of the present refer to the past, and vice versa" can be viewed as an invitation to help lift the patient from a literal locked-in view to a metaphorical view of himself. Interpreting present behavior in terms of early experience can usefully highlight a metaphor hidden in a context of conventional usage, thus rendering that response more recognizable as only one out of the several that might be in varying degrees appropriate to the situation.

Freud also emphasized the complexity in the language of our patients when he wrote that treating an obsessional patient by talking is like trying to help him dry himself with the towel he was holding when he fell out of the boat. This semantic paradox holds not only with obsessional patients but with any subject of psychotherapy. The patient enters treatment with faulty interpersonal language, and yet the vehicle for treatment is language. This is not as it may seem on first glance, a perverse compounding of the difficulty. The choice of psychotherapy over other approaches to the problem of behavior modification has been forced on us by the recognition that our behavior is a manifestation of symbolic activity that, if it is to become nonstereotyped and appropriate, must be freed from the literalness of representation that limits the person's response alternatives. It is not the "innerness" of these mediational behaviors that presents a problem to us, but the identification of their linkages with the overt behavior they control. Thus, it may be said that the goal of treatment is a richness of response alternatives rather than simply a different set of behaviors.

In attempting any form of psychotherapy it is mandatory to have some systematic notions of how to order one's clinical observations. The possession of a system of related concepts covering the major areas of human interaction is of value in the sense of having a systematic basis for scanning the patient's response potential. Psychoanalytic theory is undoubtedly the most complete system available. The unique value of the psychoanalytic system is that it is metaphor expressed in body language, which lies closest to our feeling experience. The risk is that it is so good that it can easily be taken literally and thus lose its poetic power to elicit the optimum response. The importance of this body aspect of the strength of the psychoanalytic figures of speech is illustrated by the observation that, in practice, metapsychological concepts have had less acceptance over the years the further they are removed from body feeling. There is no denying the genital shrinking sensation we associate with castration anxiety.

I do not mean to suggest that the analyst use psychoanalytic language in the consulting room. What I do mean to suggest is that, with the borderline patient in particular, there is a specific need for labeling of previously unverbalized and inchoate experiences by an analogical approach that broadens the range of expressive behavior he holds in common with his fellows. There are some metaphors we use purely for the therapist's convenience in organizing his own mind, such as Otto Kernberg's (1967) notion of ego-splitting or Hyman Spotnitz's (1969) notion of the obliteration of the object field. These artifices are designed to help the therapist conceive of the other person as he reveals himself. On the other hand, there are some we use to help the patient organize his mind, such as Spotnitz's stress on murder and suicide. I do wonder if murder and suicide always lie at the heart of it though, or if they might not be ways of talking about frustration. By this I mean simply that verbalizing the negative affects is essential but inexhaustible unless something is done about the uncompleted acts that maintain the frustration.

The language of the consulting room has been slow to come under scrutiny in our publications for good and sufficient reasons. It is difficult to see the value of describing the one apt way we found to interpret a particular patient's separation anxiety at a particular point in the analysis out of the myriad ways we might have presented it. Yet to me, this aptness is the intriguing point of the story. We know the generality, but what made for the success of the particular way it was put? Indeed, what are our criteria for thinking that that particular remark among the many verbal and nonverbal transactions between therapist and patient effected the change?

This question may sound nihilistic, but it is not. I think we do have criteria or could establish them where we do not have them. The lack of explicitness of many of our criteria is not due to the extreme difficulty of defining them, but rather to our convention of relying on intuition and peripheral vision not only in the process of treatment but carrying it over into our attempts to evaluate the data. Reluctance to tape record one's session, for example, is understandable when we try it and begin to glimpse the staggering amount of time it takes to make profitable use of it. Yet this step is only a small one away from the sweeping generalizations and retrospective lapses of note-taking either during or after the session. Reporting to a supervisor a few days up to a week after the notes were made tends to require further abstraction and generalization when four or five hours are being accounted for in one.

This economy falls in so smoothly with our innately human myth-making or pattern-making propensity that its departure from the data, the events themselves, may pass unnoticed.

Several years ago Howard Hunt and I (1968) wrote several paragraphs on human ethology that seem particularly pertinent here.

To give a speculative example of our own, man is a "pattern-maker"; he not only uses symbols, but he organizes them into some sort of a scheme which, in a sense, provides answers to questions that appear to grow implicitly out of his experience. These "patterns" or answers feel like and are known to us as beliefs, theories, hunches, guesses, suspicions, hypotheses, and the like. They also appear to serve some sort of a strategic guiding function (discriminative?) in behavior. We invent them, indefatigably, ubiquitously, and out of the most marginal data. Once grown they resist change until their asynchrony with experience becomes gross and overwhelming. And often they are bolstered, in defense against other patterns or disconfirmatory events, by subsidiary patterns developed for that purpose, or by force and violence. Such patterns appear to play a uniquely important function in our behavioral economy: for example, again, almost everybody has some sort of a religion or metaphysic, however primitive and simple, and feels "lost" without it. To tinker with such a pattern, however ridiculous it may be, is blasphemous and heavily sanctioned. Attacks against one pattern usually emerge from the vantage point of a competing one, and many of us seem to have almost a compulsion to have our particular patterns adopted as universals, unless we have had special training. Such training—to be able to say "I don't know" and to be comfortably uncommitted—is given to scientists who may practice the agnostic arts in their area of special expertise, but usually scientists are as pattern-bound as anyone else otherwise.

These tendencies appear in the fantasies and fairy stories of children, their ideas about where they came from, where they go when they die or go to sleep, and the like. In adults, they appear in, among other things, the best scientific theories and intellectual inventions, conspiratorial views of politics and history, delusional systems and compulsive rituals, and in the rationalizations that pass for explanations and understandings in everyday life. In most cases, they appear somehow to purvey a comfort, and kind of security, that makes them hard to relinquish. Perhaps this tendency achieves its most exquisite expression in the widely and closely held (but usually unanalyzed) notion that in some way or another a proposition, inference, or idea can be "true" in some referential sense, that an experience or an idea can and must represent "something" with some sort of fidelity.°

I speak of these difficulties in staying sensitively close to the data because I believe they lie at the heart of the problem. Freud (1959) set out to make a scientific study of human behavior using the best models of scientific thought that were available to him. In his encyclopedia article under the heading "Psychoanalysis an Empirical Science" he stated:

° Reprinted with permission from Howard Hunt and J. Dyrud, "Review and Outlook of Behavior Analysis," in *Research in Psychotherapy* (Washington, D.C.: American Psychological Association, 1968).

Psychoanalysis is not, like philosophies, a system starting out from a few sharply defined basic concepts, seeking to grasp the whole universe with the help of these and, once it is completed, having no room for fresh discoveries or better understanding. On the contrary, it keeps close to the facts in its field of study, seeks to solve the immediate problems of observation, gropes its way forward by the help of experience, is always incomplete and always ready to correct or modify its theories. There is no incongruity (any more than in the case of physics or chemistry) if its most general concepts lack clarity and if its postulates are provisional; it leaves their more precise definition to the results of future work.

Over the years the view of psychoanalysis as an art form has been in the ascendancy for the very reason that we have had difficulty in establishing an objective methodology for studying our only source of data—the therapeutic session. Our emphasis on psychic reality as opposed to external reality seems to have had the curious side effect of deprecating the systematic collection of objective data as if this effort in some way would detract from comprehending the patient's symbolic activity, even though our only access to the patient's psychic reality is by way of his manifest behavior in the psychotherapeutic session. The data of unconsciousness can only remain a heuristic muddle until we specify the manifest data from which our inferences about unconscious processes are drawn.

Dr. Grinker's work on the borderline syndrome has helped us to focus on the borderline between intuition and rational planning where we as therapists must work. Perhaps more essentially he has given us a model of maturity, specifically his ability to keep questions open until all the relevant data are in.

REFERENCES

Beck, S. J. 1959. "Schizophrenia Without Psychosis," *Archives of Neurology and Psychiatry* 81:85.

Beck, S. J. 1965. "Psychological Processes in the Schizophrenic Adaptation." New York: Grune & Stratton.

Chassan, J. B. 1960. "Statistical Inference and the Single Case," *Psychiatry* 23:173–184.

Cohen, R., and Cohen, M. 1961. "Research in Psychotherapy: A Preliminary Report," *Psychiatry* 24 (2, suppl.): 46–61.

Ekstein, R., and Wallerstein, J. 1954. "Observations on the Psychology of Borderline and Psychotic Children." In *Psychoanalytic Study of the Child*. New York: International Universities Press. Vol. 9, 344–369.

Eisenstein, W. W. 1952. "Differential Psychotherapy of Borderline States." In G. Bychowski and J. L. Despert, eds., *Specialized Techniques in Psychotherapy*. New York: Basic Books. Pp. 303–323.

Ezriel, H. 1956. "Experimentation within the Psychoanalytic Session," *British Journal for the Philosophy of Science* 7:29–48.

Freud, A. 1966. *Normality and Pathology in Childhood.* New York: International Universities Press.

Freud, S. 1959. "Psychoanalysis" (1922). C. P. Standard Edition, 18:253–254.

Frosch, J. 1964. "The Psychotic Character: Clinical Psychiatric Considerations," *Psychiatric Quarterly* 38:81–96.

Frosch, J. 1970. "Psychoanalytic Considerations of the Psychotic Character," *Journal of the American Psychoanalytic Association* 18:24–49.

Geleerd, E. 1946. "A Contribution to the Problem of Psychoses in Childhood." In *Psychoanalytic Study of the Child.* New York: International Universities Press. Vol. 2, pp. 271–291.

Geleerd, E. 1958. "Borderline States in Childhood and Adolescence." In *Psychoanalytic Study of the Child.* New York: International Universities Press. Vol. 13, pp. 279–295.

Grinker, R. R., Sr., Werble, B., and Drye, R. 1968. *The Borderline Syndrome.* New York: Basic Books.

Haight-Ashbury Research Clinic. 1970. Personal communication.

Heston, L. L. 1970. "Schizoidia and Schizophrenia All Associated with Biochemical Defects Transmitted by a Single Mode of Inheritance," *Science* January 16.

Hunt, H., and Dyrud, J. 1968. "Review and Outlook of Behavior Analysis." In *Research in Psychotherapy,* Vol. 3. American Psychological Association. Washington: Pp. 140–152.

Jacobson, E. 1969. *The Self and the Object World.* New York: International Universities Press.

Kempner, P. 1970. Personal communication.

Kernberg, O. 1960. "The Treatment of Patients with Borderline Personality Organization," *International Journal of Psychoanalysis* 49:600–619.

Kernberg, O. 1967. "Borderline Personality Organization," *Journal of the American Psychoanalytic Association* 15:641–685.

Kernberg, O. 1970. "Treatment of Narcissistic Personalities," *Journal of the American Psychoanalytic Association* 18:51–85.

Knight, R. P. 1954. "Borderline States." In R. P. Knight *Psychoanalytic Psychiatry and Psychology.* New York: International Universities Press. Pp. 97–109.

Mahler, M. 1949. "Clinical Studies in Benign and Malignant Cases of Childhood Psychoses," *American Journal of Orthopsychiatry* Vol. 19.

Meehl, P. 1962. "Schizotaxy, Schizotypy and Schizophrenia," *American Psychologist.* 17:827–838.

Rioch, D. McK. 1970. Personal communication.

Rosenthal, D., and Kety, S., eds. 1968. *The Transmission of Schizophrenia.* New York: Pergamon Press.

Rothenberg, A. 1971. "The Process of Janusian Thinking and Creativity," *Archives of General Psychiatry* 24:195–205.

Searles, H. 1965. *Collected Papers on Schizophrenia and Related Subjects.* New York: International Universities Press.

Skinner, B. F. 1953. *Science and Human Behavior.* New York: Macmillan.

Skinner, B. F. 1961. *Cumulative Record.* New York: Macmillan.

Spotnitz, H. 1969. *Modern Psychoanalysis of the Schizophrenic Patient.* New York: Grune & Stratton.

Weil, A. 1953a. "Certain Severe Disturbances of Ego Development in Childhood." In *Psychoanalytic Study of the Child.* New York: International Universities Press. Vol. 8, pp. 271–287.

Weil, A. 1953b. "Clinical Data and Dynamic Considerations in Certain Cases of Childhood Schizophrenia," *American Journal of Orthopsychiatry* 23:518–529.

Wisdom, J. O. 1967. "Testing an Interpretation Within a Session," *International Journal of Psychoanalysis* 48:44–52.

The Worcester State Hospital Research on Schizophrenia

An excellent history of the Worcester State Hospital from 1830 to 1920, *The State and the Mentally Ill*, by Gerald N. Grob (1966), has recently appeared. This volume covers two of the three golden ages—those of Samuel B. Woodward and Adolf Meyer—that Worcester passed through during the approximate century and a half since its founding. Grob's account does not, however, reach into its third golden age—that of William A. Bryan—which began with Bryan's taking over the superintendency at Worcester in 1921 and extended some half-dozen years beyond his departure in 1940, years in which his influence continued to assert itself vigorously.

A volume that supplements Grob, covering the varied facets of fruitfulness of this hospital's activities for the period beyond 1920, cries for the writing. I am most hopeful that some psychohistorian will undertake the task. My intention in the present essay is to contribute to this goal by describing one major aspect of this history, that relating to the Research Service. For the two decades between 1927 and 1946, the research operation at Worcester occupied itself intensively with the study of schizophrenic conditions. The fruitfulness of this program, its impact on the field of psychiatry and related areas, its high standards, comprehensiveness, and productivity in data, hypotheses, and research workers, who subsequently played important roles elsewhere, all justify detailed consideration.

I am in the relatively unique position of being able to view the Worcester program from both the vantage and disadvantage points of a participant-observer who was there for all but the first year of the period mentioned. During this time I was deeply involved in the general planning of the multidisciplinary research as a whole, and in the intensive planning of all of the psychology and some part of the psychiatry program. Though my account will therefore have the inestimable benefit of intimate personal knowledge, it may at the same time suffer from some narcissistic bias. To protect myself against this danger, I have sought out critical reviews from colleagues who spent at least some time at Worcester during this period.

Actually, I was at the Worcester State Hospital for two periods. The first, as assistant in psychology to the then head of the Psychology Department, Dr. Grace H. Kent, extended from July 1924 (following the receipt of my bachelor's degree) to September 1925, when I went back for graduate study. In the second period I returned to Worcester to serve as Chief Psychologist and Director of Psychological Research from August 1928 through July 1946.

The hospital had two divisions during this time. The old, or Summer Street division, in downtown Worcester, had been established in 1833. The other, the so-called main hospital, on the eastern border of Worcester overlooking Lake Quinsigamond, had been established in 1877. The 2,400 patients on the hospital books were equally divided as to sex with 400 out "on visit." The main hospital housed 1,800 patients and Summer Street 600, the most chronic residing at the latter. Of the approximately 530 new yearly admissions, about 20 per cent were schizophrenic; however, about 50 per cent of the resident population were so diagnosed.

The Worcester State Hospital was, for several decades at a time during different periods of its history, the outstanding state hospital in the nation. When first established, it was the seventh state hospital in the country and the first in New England, both Horace Mann and Dorothea Lynde Dix having been involved in its founding.

Dr. Samuel B. Woodward, the first superintendent, who served from 1833 to 1846, was responsible for its initial golden age. His regime is particularly noted for its emphasis on moral treatment. He helped to organize the Association of Medical Superintendents of American Institutions for the Insane in 1844 (later the American Psychiatric Association) and was its first president.

The second golden age came half a century later under the aegis of Adolf Meyer, who served as Director of Clinics and pathologist from 1896 to 1902. He developed the clinical, research, and teaching activities of the hospital to a very high level. Throughout his period of leadership he emphasized complete participation by the medical staff not only in the full range of medical activities but in full-time devotion to the general goals of the hospital. He contributed greatly to building up a substantial, and remarkably complete, specialized library in psychiatry, neurology, psychology, medical research, and related fields. In addition to establishing an important pathological institute, he initiated close relationships with Clark University (then under the presidency of its founder, G. Stanley Hall), where he served as docent. Among his students and staff at Worcester were persons of the caliber of Barrett, Bassoe, Coriat, Diefendorf, Dunlap, Kirby, Kline, who later became outstanding leaders in psychiatry.

The third golden age was that of William A. Bryan, who served as superintendent from 1921 to 1940. Most of the credit for the eminence achieved by the Worcester State Hospital in the third to fifth decades of this century belongs to him. His innovations were endless. He established a variety of advisory councils, a research service, a medical service, a chaplaincy service under Anton Boisen (Guiles, 1947) (which was the original source of training and inspiration for the development of the nation-wide program of pastoral training for theological students), the first cafeteria for patients, and some of the earliest group therapy and industrial (as a major complement to occupational) therapy programs. His *Administrative Psychiatry* (1936), the first book ever published on that subject, describes many of the modernizations he introduced at Worcester.

RESEARCH SERVICE: HISTORY

Against this sketchy general background let us consider the Research Service set up at Worcester to study schizophrenia. In this essay, only the structure and general accomplishments of the service can be considered. I shall make no attempt to deal with its substantive findings, which must be left for another occasion. This program actually had its beginnings in the period between 1925 and 1927 when Dr. Lewis B. Hill (the assistant superintendent), with the aid of Dr. Francis Sleeper (the senior assistant physician), undertook some limited research in schizophrenia. This work, carried out in the context of the day-by-day clinical activities of the institution, was supported from the regular hospital appropriation.

The major and systematic development of the research in schizophrenia, however, began in 1927 with the coming of Dr. Roy G. Hoskins. Several years prior to 1927, Mrs. Katherine Dexter McCormick had become interested in establishing a research program on schizophrenia. In 1926 Dr. Walter B. Cannon, from whom Mrs. McCormick had sought advice, asked Dr. Hoskins, an old student of his and at that time a professor of physiology at Ohio State University, to submit a plan for such an investigation. Mrs. McCormick wished the program to embody a physiological, mainly endocrinological, approach to the problem. The proposal was to finance the research from the estate of her husband, Stanley McCormick, the son of the inventor of the reaper, who was suffering from schizophrenia. After considerable discussion with representatives of the estate, a general plan was agreed on which included the establishment of the Foundation for Neuro-Endocrine Research. Dr. Hoskins, whose specialty was endocrinology, was engaged to supervise its execution, entering on active duty early in 1927.

Because the Commonwealth of Massachusetts had already attained a recognized position of leadership in the care of mental disorders, it seemed desirable to institute the research in this state. Boston was the obvious choice for location of the central office. Professor Cannon, who had displayed a keen interest in the project from the start, graciously proffered office and laboratory quarters at the Harvard Medical School.

For the contemplated clinical work several choices among institutions were available. Dr. Cannon intimated that the operation would be welcome at the McLean Hospital, an institution associated with the Harvard Medical School. Through the kind offices of Dr. Allan Winter Rowe (head of the Evans Memorial, Massachusetts Memorial Hospitals, Boston University School of Medicine), contact was also made with the commonwealth's Department of Mental Diseases. At a dinner meeting given by Dr. Rowe, Dr. Hoskins discovered that Dr. George M. Kline, Commissioner of Mental Diseases, and the superintendents of three of the state hospitals were deeply interested in the proposed research. On behalf of the department, the commissioner offered entree to any of the state hospitals that Dr. Hoskins might select. Preliminary consideration narrowed the choice of institutions to two; but subsequent study indicated that the Worcester State Hospital seemed to hold out the greatest advantages and was selected. This decision was dictated by several facts. The superintendent, Dr. William A. Bryan, and other administrators of that hospital were enthusiastically interested in research and had already taken initial steps toward organizing a project for the investigation of schizophrenia. Since the Worcester plan and that of the Foundation were found to be in substantial accord, joint prosecution of the study was readily agreed on. Dr. F. H. Sleeper, by that time clinical director, represented the hospital as resident director, while Dr. Hoskins represented the Foundation as director. Dr. Sleeper, on Dr. Hill's departure, was subsequently made assistant superintendent of the hospital but continued his relationship with the research. An excellent spirit of cooperation between the two agencies quickly developed and was effectively maintained throughout the duration of the partnership.

It was decided originally that the work should be approached through applied physiology (emphasizing biochemistry as well as physiology). It seemed expedient, therefore, to attack the problems of schizophrenia both directly, utilizing patients as subjects, and indirectly, through animal studies of those physiological functions that seemed most likely to be involved in the psychosis. Dr. Milton O. Lee, who had previously been associated with Dr. Hoskins in the Department of Physiology at Ohio State University, was employed to devote his attention primarily to the animal experimentation.

PERIODS OF RESEARCH OPERATION

The schizophrenia research operation, which extended over a period of twenty years, divides itself into three major periods. The first two, including a period of establishment and a period of reorganization, together lasting for three to four years, were relatively brief. The third period, with the service working at full force, involved a succession of many separate programs and extended over approximately sixteen years.

1. INITIAL ENDOCRINE-METABOLIC (1927–1929)

Several factors, only in part influenced by Mrs. McCormick's views, governed the decision to initiate the patient work at Worcester with a diagnostic endocrine survey. Among these considerations were the notion, accepted by many in the field at that time, of the intrinsic probability that endocrine factors were significant in determining the psychosis; the spontaneous interest of the Worcester group in that subject; and the familiarity of Dr. Hoskins, the director, with endocrine theory and practice. (From 1917 to 1940 Dr. Hoskins was the editor of the official organ of the Society for the Study of Internal Secretions, *Endocrinology*.) The outstanding difficulty lay in the fact that techniques for endocrine diagnosis were relatively inadequate. The most promising methodology seemed to be that which had been developed by Dr. Rowe and his colleagues at the Evans Memorial. Dr. Rowe's material was, at that time, still unpublished, but Dr. Hoskins was extended the privilege of making a first-hand study of Rowe's methods at the Evans, and these were considered for their possible adaptability to schizophrenic patients.

Whether, in fact, a diagnostic system designed for, and based on studies of, nonpsychotic patients could be taken over bodily for use in schizophrenic patients was a question that only experience could answer. Using the patients available on hospital wards, the investigators were able, however, to make an immediate test of the inner consistency of the system. It was found that diagnoses of conditions in individual patients, when made independently by Dr. Sleeper and by Dr. Hoskins, practically always coincided. In a considerable number of instances, Dr. Rowe also made independent diagnoses, and these, too, commonly agreed with those made by the Worcester group.

By the diagnostic criteria employed, considerably more than half the patients studied during this early period showed endocrine deficiencies, principally thyroid and pituitary, but a considerable number appeared to be suffering from "endocrine deficiency, unclassified." The diagnoses having been made, endocrine treatment was administered in accordance with the deficiencies diagnosed. The early results were, like early therapeutic results generally, notably encouraging. Many patients with dubious or definitely bad prognoses showed improvement, and a substantial number recovered sufficiently to return home. The growing enthusiasm of the group was kept within bounds, however, by a realization that schizophrenic patients in surprisingly large numbers show temporary improvement under almost any type of therapeutic attention. Furthermore, the group knew that a considerable number of patients with this psychosis can be counted on to improve irrespective of treatment. Nevertheless, after making allowance for these various difficulties, there seemed some basis for the belief that the group was making genuine progress in the treatment of schizophrenia.

2. RECOGNITION OF COMPLEXITIES; STOCKTAKING; PLANS FOR REORGANIZATION AND EXPANSION (1930–1931)

By approximately the third year of the program, various changes had gradually taken place in the Research Service, with more to follow in the succeeding years. This initial period of reorganization involved deepening recognition of the complexities of the problem and much planning for reorganization and expansion. Because of the importance of this period for subsequent developments in the Research Service, I shall consider its high points in some detail.

The special wards set up for the service had had the opportunity to become established, and the staff had been considerably enlarged. A growing sophistication had begun to permeate the group, and some changes in its goals and enthusiasms were emerging.

Let us begin by examining the many factors underlying these changes. There was to start with, the original staff's increasing experience with schizophrenics; and then there was the broadening of both the number and range of activities of the investigators. In 1928, I, a psychologist, joined the group, and during 1930, Dr. Harry Freeman, a general internist, Drs. Milton H. Erickson and James R. Linton, psychiatrists, and Paul E. Huston, an experimental psychologist, were recruited. The psychologists and psychiatrists were psychologically, rather than organically, oriented, and the internist had a broader, or at least a somewhat different, range of physiological interests than that represented in the original group. This program expansion continued further in 1931 with the recruitment of additional staff, the most prominent of whom were E. Morton Jellinek, who came to head biometry, Dr. Joseph C. Looney, to head biochemistry, and Drs. Hugh T. Carmichael and Joseph C. Rheingold, who were added to the psychiatric group.

From the constant interactions of the now quite diversified staff developed an awareness of the high degree of ignorance existing about schizo-

phrenia. The group also became increasingly persuaded that the literature, particularly the research literature in the various areas, both
descriptive and therapeutic, was quite undependable and unsatisfactory.
The conviction grew that an adequate definition of schizophrenia based
on accurate clinical and experimental descriptions was a primary need.
We also felt that as far as practicable, these descriptions should be
stated in quantitative terms; or, when not, that they at least be expressed in terms of publicly derivable systematic procedures, such as
rating scales.

In this atmosphere of increased experience, wider ranging approaches,
and growing sophistication, the earlier, relatively simplistic approach
centering around endocrine therapy gradually changed. By the third
year of the study, too, certain disturbing facts about this therapy had
become clearly visible. In some cases the apparent good effects of the
endocrine medication began to wane, and the patients in the end
seemed as badly off as, or sometimes even worse off than, before. The
patients who had been sent home began to return to the hospital in re-
lapse with distressing frequency. But most disconcerting of all was the
increasingly manifest fact that the metabolic picture in a given patient
was likely to undergo a spontaneous change. Thus a man diagnosed as
suffering from thyroid deficiency at one time might appear, three months
later, to show pituitary deficiency. Genuine endocrine disorders were
known not to be so labile.

Certain underlying problems troubled us even more than the above.
The most fundamental concerned diagnosis. It had become obvious that
the routine diagnoses made by either the house service or by the Boston
Psychopathic Hospital, from whom we received a considerable number
of our patients, did not always agree with our own. A clearer definition
of schizophrenia was essential if we were to achieve consistency and
maintain confidence in our research.

Since I have presented fairly detailed consideration of this topic elsewhere (Shakow, 1966, 1969), I shall here merely present some highlights.
Before going into nosology, we set up criteria for excluding from our
sample patients contaminated by irrelevant conflicting factors. The criteria for inclusion were determined in three organized and systematic
stages: an initial mental status examination based on a five-point rating
scale of symptoms and traits, the process of establishment of syndromes,
and a group diagnostic process, a part of which was devoted to determining the suitability of the patient for the research program.

If deemed suitable for study, the patient was interviewed frequently
(ordinarily daily, or at least several times a week) by the psychiatrist
throughout the periods of active investigation, and notes, summarized
into weekly records, were made of his mental state and behavior. (The
rating scale system previously mentioned was used at prescribed intervals.) When necessary, special conferences were called to make changes

in the original diagnosis and subtype category of any patient. In addition to the psychiatrist's observations, daily notes were entered in the patient's ward record by the nursing staff and by two specially trained ward observers charged with reporting selected aspects of a patient's behavior. Less frequent notations about the patient were also made during any "rest" periods (the periods between active study).

When we put the stringent selective standards just described into active use, a serious problem developed. The hitherto conspicuous supply of schizophrenics was reduced, and those suitable for study diminished even more strikingly. The three sources on which we had previously depended for patients—the resident hospital population, the schizophrenics among the new admission cases coming directly to the hospital, and the schizophrenics among the patients comprising our share of the routine transfers from the Boston Psychopathic Hospital to individual state hospitals—were far from sufficient. We therefore approached the Boston Psychopathic Hospital to send us more than our usual share of schizophrenic patients, and they complied. We also arranged with the other state hospitals in our geographic area to transfer potentially suitable patients to our service, and we received their cooperation as well. Even with these improvements, we were barely able to fulfill the subject requirements of our research.

The period of stocktaking, which took place in the late 1920's and the early 1930's, resulted in a number of immediate changes in the program and laid the ground for still others in the succeeding years. These modifications included the extension of the disciplines represented on the high-quality staff, broadening of the program beyond the endocrine, more careful definition and selection of subjects, intensification of the search for suitable subjects, increased dependence on our own studies of both schizophrenic and normal subjects, special attention to the problem of variability in the schizophrenic group, and generally increased vigilance about the manifold problems involved in achieving satisfactory experimental and statistical control. These advances took place in a context of both greater and wider financial support, improved physical facilities in wards, laboratories and library, widened range of groups studied, and improvement of procedures for carrying through projects from initiation to final publication.

3. RESEARCH PROGRAM AFTER 1931

During the course of the approximate decade and a half of activity of the research service, which in part paralleled, but mainly followed, the period of rethinking and reorganization just described, a constant stream of different kinds of organized schedules of experiments and tests occupied us. They varied considerably in the breadth of involvement of different investigators and disciplines. A rough categorization of these indicates four levels of compass: (1) comprehensive multidisciplinary; (2) moderate multidisciplinary; (3) limited, bi-, or intradisciplinary; (4) exploratory. This last group, by far the most restricted in scope, both in nature and extent, were, with few exceptions, carried out for relatively short periods of time by one, or at most two, investigators.

Level 1 Comprehensive Multidisciplinary

In the period spanned, this class of most extensive schedules included only three programs. Each involved practically all the investigators of the various disciplines then represented on the service. (During the same periods, however, many individual investigators also carried out concurrent independent projects.) Though not planned with such an object in mind, it turned out that the three programs served in many ways to complement one another; they provided us with a more rounded picture of schizophrenia because of the different degrees of patient chronicity and stress situations involved in the separate programs.

Seven-months Study (1931–1933): Since a contemporary paper (Hoskins, Sleeper, Shakow, Jellinek, Looney, and Erickson, 1933) describes in detail the earliest of these, the so-called seven-months study, I shall here merely point to some of its prominent features. Though the active period of this, the most elaborate investigation we ever conducted, lasted for less than two years, 1931–1933, it provided enough material for many further years of analysis and report.

The major goal of the study was to obtain baseline data on schizophrenics and to determine both their intraindividual and interindividual variability, the finding that had so impressed us in the earliest studies. Since we sought as careful a description of the patient in his states of "spontaneous variability" as possible, we tried to keep him as free from manipulation as we could. We therefore avoided deliberately organized therapy of any sort and attempted to keep the ward and other living conditions as equable and as routine as possible. At the same time, we recognized how unavoidable was the unintended and uncontrollable stimulation, therapy, and trauma that any human, particularly hospital, environment provides. The former program thus took care of the experimental controls; we provided statistical controls for the latter as far as we could by recording and attempting to partial out any such events that we were able to observe. That these were incomplete goes without saying.

During the course of the active period of the study we worked with approximately 100 patients. We intended for each to go through the multidisciplinary schedule (Hoskins et al., 1933) three times at intervals of three months, thus yielding data from a period of seven months of study. In the end we came up with data for only ninety-five patients— sixty-three who participated in the three (some even in four) periods, five in two periods, and twenty-seven in one period.

The patients were predominantly chronic, with a mean age of not quite thirty-two years and a mean hospitalization period of somewhat more than five years. The mean hospitalization period for Massachusetts patients during approximately the same period was eleven years, a statistic indicating that ours was not a markedly chronic group. As controls, we used whatever satisfactory data on normal subjects we could find in the literature. Most often, however, we were compelled to resort to groups of normals whom we selected and studied ourselves. For comparison purposes, we also used data available in the literature on schizophrenic patients, and on a number of occasions, we also studied non-schizophrenic patients.

Our studies fell mainly into the biological, psychological, and psychiatric spheres. Special efforts were expended in the last area to obtain quantitative data in addition to the conventional clinical data. Each group of investigators following the schedule indicated, carried out its own selected studies. The details of the specific studies are provided in Hoskins et al. (1933). As test, experiment, or examination data were collected by the different investigators, they were reported to the biometrics department for tabulation and analysis.

The amount of data generated by the seven-months study was of such volume (our Biometrics Department estimated that during the course of the study some 500,000 quantitative observations had actually been recorded) that it soon became evident that we could not both analyze these and write up our results, while simultaneously pursuing further empirical studies. We therefore declared a moratorium during which the group devoted itself practically full time to the analysis and preparation of manuscripts on this material. This period, depending on the particular project, lasted for six to twelve months. Though a heavy flow of publications resulted in the immediately following years, papers stemming from the study continued to appear for many years subsequently, some as late as a decade and a half after the study was completed.

The yield of the seven-months study was of three kinds: (1) the publications resulting from fairly rigorous pursuit of the outlined program; (2) publications based on studies carried out at approximately the same time but that did not follow the exacting time schedule or other criteria prescribed for the central project; (3) general impact on the program.

The first group encompasses papers from work in the various fields: general (Hoskins, Sleeper, Shakow, Jellinek, Looney, and Erickson, 1933); biological (Freeman, 1933a, 1933b; Hoskins, 1933; Hoskins and Sleeper, 1933; Looney and Childs, 1933; Carmichael and Linder, 1934; Freeman, 1934a; Freeman and Looney, 1934a, 1934b; Jellinek, 1936a, 1936b); psychological (Huston, 1932; Shakow, 1932; Huston, 1934; Shakow and Huston, 1936; Huston, Shakow, and Riggs, 1937; Huston and Shakow, 1946, 1948, 1949); and combinations of disciplines (Hoskins and Jellinek, 1933).

Aside from the specific products, the general impact of the seven-months study was considerable. Because of the practice involved in the exercise of multidisciplinary, cooperative techniques, it helped us to become a smoothly functioning research group, laid the ground for many subsequent studies, and developed in us high standards of performance. Besides reimpressing us with the striking degree of variability inherent in the group with whom we were working, it provided us, above all, with a sounder basis for further theorizing. The number of significant methodological studies resulting from the program was another of its positive adjuncts (see, especially, those in the psychiatric [Erickson and Hoskins, 1931], biological [Looney and Childs, 1934; Looney and Jellinek, 1935; Jellinek and Fertig, 1936], and psychological [Huston and Hayes, 1934; Rosenzweig, 1935] realms).

Insulin Study (1936–1937; 1937–1938): In March 1936, under the major impetus of D. Ewen Cameron (Worcester State Hospital, 1936), the first insulin therapy studies in the United States were initiated at the Worcester State Hospital. Sakel had begun his insulin work on isolated addicts and psychotic patients in 1927 in Berlin. But it was only in Poetzl's clinic in Vienna in 1933 (Sakel, 1958) that he was able, with the aid of such associates as Dussik, to extend his study to larger groups of schizophrenic patients. Various other European psychiatrists, such as Müller, Benedek, Berglas, and Sušić, followed with studies in the immediately succeeding years. Our Worcester group considered Cameron's work during this first of two successive periods of work with insulin as either exploratory or as falling into our level 3 studies.

The background for Worcester's heavier involvement in insulin research during the latter phase was a good deal more complex than during the first. For one, natural underlying guilt feelings gnawed at some members of the group for having devoted themselves so exclusively to either careful descriptive studies or to pure research, while neglecting efforts to provide active and immediate help to patients. This uneasiness was aggravated by the glowing reports coming from the European scene, of Glueck (1936) and others, about the therapeutic effectiveness of the new technique. The fact that it was a device with some semblance of rationale, that it was organic, but especially that it was endocrine and therefore fitted the history of the Worcester attack on schizophrenia so appropriately, only underscored this neglect. Most important, perhaps, for the majority of our group was, however, the opportunity the drug seemed to afford for the study of induced changes in the organism— changes that would enable us to observe the same patient in both schizophrenic and normalized states. Such interest had been long-standing among us; in fact, a similar proposal had been advanced explicitly quite early in our research, in the paper describing the seven-months study (Hoskins et al., 1933).

The return of Ewen Cameron as an actual member of the service also influenced the decision to instigate the study at that time. (Cameron had spent three months as a visiting researcher on our service in 1935, and came back in February 1936 as a permanent member of the research psychiatry staff.) Though his return to Worcester was related to his continuing interest in our schizophrenia research, he could not, of course, have so deep a commitment to the previous program as did the established staff. Thus, it was easier for him to adopt this new approach and endeavor to introduce the investigation of insulin into the program of a group that he considered to be among those in the forefront of research in schizophrenia.

During the first stage of insulin study, from March 1936 through August 1937, Cameron worked with Hoskins mainly along therapeutic lines, to attempt to clarify the technical procedures of insulin administration. At the same time he was associated with a few other members of the staff in trying to define some of the aspects of treatment as they related to other endocrine functions, lipoid metabolism, tissue oxygenation, and the electroencephalographic response.

During the summer of 1937, however, when different groups of investigators were encouraged to prepare various frame programs that might serve as overall group programs, Cameron, in association with Jellinek, presented a comprehensive program on insulin. Theirs was accepted as the major prospectus for the group and ran for about a year. Some of the other frame programs were also adopted and conducted in parallel, but not on so extensive a scale.

Some members of the group initially opposed elevating the insulin study from a moderate multidisciplinary to a comprehensive level program because of its apparently heavy therapeutic emphasis. These objections were stilled, however, when the program was clearly organized around basic biological and psychological studies directed at understanding such processes in the changing patient.

The specific goals of this second major investigation were primarily the delineation of the patient's characteristics, measured by the multiplicity of devices which the group had been refining over the years, to follow his change from the schizophrenic to the recovered (or at least improved) state. In addition, many of us (perhaps all, though some may have been reluctant to admit openly such a contamination of the scientist's role) were interested in the process of cure. The more psychologically oriented among us were intrigued, as well, by what the psychodynamic implications might be of the profound shock to the schizophrenic organism induced by insulin coma.

As I have already indicated, besides the procedures dealing with the conditions of the patients and a description of the techniques used in inducing insulin coma (Cameron, 1938; Cameron and Hoskins, 1937a, 1937b), the studies in the earlier phase of the insulin research centered mainly on the electroencephalographic response (Hoagland, Cameron, and Rubin, 1937a, 1937b; Hoagland, Rubin, and Cameron, 1937a, 1937b) and to some extent on glucose tolerance (Looney and Cameron, 1937). The findings were, in general, reported in the context of the total group, without regard to condition or outcome.

In the second phase, it seemed more appropriate to consider the findings in relation to good or poor therapeutic outcome. Aside from a general report on the methodology and description of the group, both with respect to selection and behavior (Jellinek, 1939), studies were carried out in a variety of other areas. The biological parameters included pulse rate and blood pressure (Cameron and Jellinek, 1939), serum lipids (Randall, Cameron and Looney, 1938; Randall and Jellinek, 1939a; Randall, 1940a, 1940b), cholinesterase activity of the blood serum (Randall and Jellinek, 1939b; Randall, 1940a), and blood minerals (Looney, Jellinek, and Dyer, 1939). The psychological studies provided data on various parameters of the Stanford-Binet, the Kent-Rosanoff Association Test, and the Aspiration Procedure (Schnack, Shakow, and Lively, 1945a, 1945b). In addition, one psychiatric study of behavior (Wall, 1940) and one animal study on the effects of insulin on the lipid composition of rabbit tissues (Randall, 1940b) were carried out.

With respect to the results of our studies, I can only permit myself to remark that the series of papers, both in the earlier and later phases, generally expressed considerably greater skepticism regarding the therapeutic efficacy of insulin than did the European reports. This skepticism seems to have been justified by subsequent events.

Military Study (1942–1944): After the comprehensive studies on insulin, which in actuality lasted through 1939 because of the need for analyzing and reporting the collected data, a period of concentration on individual studies followed. But by the beginning of 1940 a military smell was already pervading the research service air. With the passage of the Selective Service Act in September 1940, some members of the group had actually become involved in working on problems of general adjustment. Their planning was focused around the broad notion of a person's ability to cope with greater than ordinary demands made on the organism. They hoped to establish predictive tests of the ability to deal with situations involving frustration and stress. The organization of a National Youth Administration Camp at the nearby town of Spencer afforded the group an unusual opportunity to conduct such a study. Their expectation was that it might contribute to standards for the selection of recruits for the armed services (Rodnick, Rubin, and Freeman, 1943).

Following the Japanese attack on Pearl Harbor in December 1941 and our entry into the war, the atmosphere and tempo of the Research Service changed radically. Though exerting great pressure on itself for immediately completing ongoing projects and papers, the research group simultaneously devoted much thought to ways in which it could be of optimum service in the national emergency. The thinking naturally revolved around making the contribution through the most appropriate use of the research organization so laboriously developed over the preceding decade and a half.

The actual objectives of the program as outlined at the time (Research Service, 1942) were even more far-reaching than those I have just stated. We believed that our studies would result in contributions not only to military psychiatry but to furthering the general understanding of schizophrenia.

More specifically, with respect to military psychiatry, we expected to provide suggestions for selective criteria for the possible elimination of potential schizophrenics, the appraisal of military conditions that might precipitate schizophrenia, the evaluation of the most effective therapeutic methods, and the determination of differential criteria for final disposition of the soldier (such as return to armed forces or to civil life, or further hospitalization).

In relation to the understanding of schizophrenia in general, we believed that the factors giving rise to schizophrenia in the military situation would presumably be easier to determine than those in civilian life because of the several peculiar conditions that surrounded the former. These included more acute onset, faster course, and the generally good prognosis reported in many of the cases; the relatively homogeneous nature of the group and relatively uniform conditions under which the psychosis developed, that is, the military situation; and, finally, the relatively high degree of control over the consequences of the development of the psychosis, that is, removal of the patient from the military situation.

A varied investigative and therapeutic program was organized, the investigative techniques being, as in earlier studies, multidisciplinary. They tended, however, to emphasize sociopsychiatric factors much more than had previous studies, as well as therapeutic and follow-up studies.

Since the above program of research required a fairly large number of patients in an early acute condition, we deemed it important to get them to our own hospital as soon as possible after onset, which it was possible to do with the help of both the Department of Mental Health and the military authorities. Altogether about 175 military patients came through this program, approximately half of whom were diagnosed by us as schizophrenic. The 40 per cent of the total group who were Massachusetts residents were usually kept in the hospital until they could be sent out on visit or discharged. The 60 per cent of non-Massachusetts residents remained at the hospital for varying periods of up to sixty days, the average stay being about three weeks.

When the data from the various initial studies on a patient had been accumulated and analyzed, a staff conference, attended by all the investigators, was held. At this meeting a detailed presentation of the gathered material was presented, an interview with the patient (when deemed advisable) was held, and after a discussion of the dynamics of both the underlying and precipitating factors of the psychosis, an evaluation of prognosis and recommendations as to the course of therapy to be followed were made.

In the case of Massachusetts residents, we attempted to carry out the recommended therapeutic programs, following up the patients during the posthospital period. Since out-of-state patients ordinarily had to be returned to their home states (via the army hospital) we could go no further than to recommend the therapeutic program agreed upon.

The products of the military program were of three kinds: the papers actually published or presented at meetings, the concrete changes resulting in the nature of the operation of the research service, and its general impact.

The published papers fell into four categories: (1) sociopsychiatric (Malamud and Malamud, 1943; Shakow, 1943; Malamud and Stephenson, 1944; Kant, 1945; Malamud and Malamud, 1945), (2) psychological (Shakow, Rodnick, and Lebeaux, 1945), (3) biological (Freeman, 1946), (4) and combined disciplinary (Rodnick et al., 1943; Freeman, Rodnick, Shakow, and Lebeaux, 1944).

Among the number of changes in the subsequent operations of the research service was the institution of more intensive staff conferences giving much greater attention to detailed social histories, psychodynamics, prognosis, and plans for therapy and rehabilitation. Deeper consideration of socioenvironmental factors was emphasized particularly. For the first time, too, intensive efforts at posthospital rehabilitation under the supervision of social workers, as well as definite programs of varied therapies, were effected.

The general impact of the program was to awaken in the participants a broader appreciation of the complexities of schizophrenia. The built-in emphasis on the effects of stress encouraged concentration, in the immediately succeeding years, on studying the effects of emotional/environmental stresses on such biological measures as 17-ketosteroids and lymphocytes. Many of these studies were carried out in association with the Worcester Foundation for Experimental Biology (Freeman and Elmadjian, 1946; Hoagland, Elmadjian, and Pincus, 1946; Pincus, 1947; Pincus, Hoagland, Freeman, and Elmadjian, 1949).

Level 2 Schedules

As I have already indicated, the level 2 schedules, though also involving several disciplines, were considerably less extensive than those in level 1. They were carried out at different periods, from the early 1930's through the mid-1940's and may be thought of as being either generally descriptive in character or as directed at creating changes in the patient in order to study various functions during both unstimulated and stimulated conditions.

In the descriptive category may be found such studies as the comparative studies, the old schizophrenia studies, and the orientation studies. Those involving change included studies using drugs, both endocrine and nonendocrine, shock, or stress. I shall make no attempt in this section to be all-inclusive, merely reporting a sampling of these studies with a few descriptive sentences giving the compass of each, and, when the study was reported, indicate the publications resulting.

Of the general descriptive studies at this level, three were most prominent. The so-called comparative study, conducted during 1934–1935, compared approximately thirty patients and thirty hospitalized normal controls on a series of tests (Jellinek, 1936; Freeman, 1938; Looney and Freeman, 1938). A group of aged schizophrenics, mostly long-hospitalized, were studied to determine whether aging per se had any additional effect on the subjects beyond the psychosis (Freeman, Pincus, Elmadjian, and Romanoff, 1955a, 1955b). Beginning in 1942, we instituted an orientation schedule that put all new patients coming onto the service through an intensive preliminary investigation.

In the other category of studies at level 2, we were concerned mainly with the effects of inducing changes in the patients by one or another form of deliberate stimulation to compare these with the baseline data.

Among these were several endocrine studies. During 1939–1941 and 1940–1941, two series of studies were carried out using sex hormones (Looney and Romanoff, 1940; Randall, 1940c; Rosenzweig and Freeman, 1942). Among the nonendocrine studies, a first investigation of the effects of dinitrophenol took place in 1933 (Freeman, 1934b; Looney and Hoskins, 1934, 1935), and a second in 1939 (Hoagland, Rubin, and Cameron, 1939). Though many different shock therapy studies were carried out, the major ones were on Metrazol, one begun in 1938 and a series continued for some time thereafter (Cohen, 1938a, 1938b, 1939a, 1939b, 1939c, 1939d; Rodnick, 1942; Schnack et al., 1945a, 1945b; Simon and Holt, 1946).

Beginning in 1943, studies of stress in relation to biochemical and behavioral changes began. The earliest of these were concerned with the natural diurnal cycle of 17-ketosteroids in normals and schizophrenics (Elmadjian and Pincus, 1946); the later ones, with the effects of special stress situations on 17-ketosteroids and lymphocytes (Hoagland et al., 1946; Pincus and Elmadjian, 1946). At one stage in these studies the experiments were organized around a hypothesis which assumed the existence of four levels of stress: (1) natural stress—the stress that appeared to result from the ordinary diurnal cycle, for example, the effect of waking up in the morning; (2) environmental stress, such as changes due to unusual cold or heat in the temperature surroundings, while the subject remained passive and not subject to demands for any voluntary response; (3) impersonal task stress, for example, where the subject was actively involved in a pursuitmeter task while being subjected to impersonal stresses (such as bright lights or loud noises) from the environment, for which he had no responsibility; (4) personal stress, for example, where the subject was led to recall his own failures in life adjustment, either directly by interview confrontation, or indirectly by the use of projective devices, such as the Thematic Apperception Test. Not all aspects of the data derived from this experiment were fully analyzed, though parts of the study were reported and some even published (Pincus, 1947; Pincus et al., 1949).

Level 3 Studies

As I indicated earlier, the level 3 studies were even more limited in character than the level 2 studies, though they might well have been equally systematic. The projects would ordinarily be confined to a single investigator, or to relatively few investigators representing either one, or at most two, disciplines.

At this level, as well, the studies fell into two major categories: descriptive and manipulative. In the former, patients were studied in a baseline state; in the latter, they were studied under stimulated conditions (either by drug administration or environmental stress) and then compared with their functioning under nonstimulated conditions. Over the years, about 100 papers resulted from the fifty or so studies which belonged at this level.

Level 4 Studies

This group consisted, as earlier indicated, of exploratory studies. With rare exceptions these were carried out by a single investigator to test some tentative hypothesis. If the exploration showed promise, a project was set up, usually at level 3; if the idea did not appear well founded it was dropped. The great majority of these preliminary studies fell into the latter category. I shall not discuss any of these projects, despite their great psychological importance, for from the point of view of the research operation as a whole their major significance lay in the welcomed opportunity such studies afforded the investigators for free and unrestricted action.

EVALUATION

In making a judgment about the success of the Worcester State Hospital schizophrenia research operation with the utmost criticalness I can muster, I would rate it highly. It seems to me to deserve this evaluation for several reasons: in part, for the substantive contribution it made by its systematic gropings toward understanding the complex problems of schizophrenia; in part, for being so early an example of successful multidisciplinary research; and in part, for having accomplished so much in the initially unpromising setting of a state hospital. Aside from the care taken in the definition of schizophrenia, the achievement of rigorous standards and control for the research itself, and skillful handling of the complexities and perplexities inherent to a group endeavor, our enterprise involved, as well, the need for training a personnel entirely foreign to research participation. Despite these and other handicaps, much was accomplished in attaining reasonably high standards and research productivity. Difficulties, of course, always beset us, but with the years, at least, they tended to be substantially reduced.

The Worcester research, because of its early multidisciplinary character, had few models. It therefore had to find its own way, not only through the tortuosities of the research process itself, but also through the entanglements of the interpersonal relations that so delicate a process introduces (Luszki, 1958). As individual investigators, we managed first to wend our way with reasonable success from simultaneous, concurrent studies to collaborative, intradisciplinary ones. We could then go on to concurrent and collaborative multidisciplinary studies, and with time, were actually able to carry out a number of truly integral interdisciplinary studies. Despite the different foci of interest and conceptual systems represented in our group—a posture we encouraged even within individual disciplines—we managed to achieve a surprising degree of harmony. Not that we did not have our lively arguments, our marked disagreements, and much mutual criticism—about methods, hypotheses, and goals. But, with rare exceptions, these were not ad hominem, and generally took place against a background of considerable accord.

The harmony had a paradoxical base that stemmed mainly, I believe, from an underlying discontent, truly a "divine discontent." However, it went beyond being merely "the germ of the first upgrowth of virtue," as Kingsley had it. The discontent served as the perennial source of sustenance for whatever virtue we did have. We were discontented with the state of knowledge and standards in our separate disciplines, with our methods, with what we were accomplishing, with our research facilities. We were discontented with what others claimed to know about schizophrenia and, it goes without saying, with what we ourselves knew. Though our aspirations always remained above our achievement, they were not inordinately high. And only as our achievement slowly rose did our aspirations rise concomitantly. Above all, we showed unconscious wisdom in not placing the blame for the discontent elsewhere. In line with that well-tested principle of good prognosis, we were ready to accept the blame for the situation; if we were dissatisfied with the state of the universe, it was up to us to improve conditions.

The fact that there was much to be contented with helped. The relatively isolated setting of the Worcester hospital, where we all resided on the grounds, made for a compact social, as well as professional, community. Other factors contributing to this cohesiveness were the twenty-four-hour nature of our jobs, the intimate contact with patients that ensued and, above all, our involvement in full-time investigation. The last was indeed a rare privilege during that era. It provided the opportunity to devote oneself almost exclusively to research without having to squeeze such activity into the crevices of full-time preoccupation with clinical work, teaching, administration, or various combinations of these —the usual fate of researchers elsewhere. Altogether, this advantage, along with the general feeling we had of being innovators, breaking new ground on many frontiers, helped to impress us with our exceptional fortune.

Outsiders might very well have assumed that our environment was a regimented one. As evidence, they could have cited our limitation to the study of schizophrenia, our being required to follow certain procedures for the initiation and execution of projects, as well as our obligation to participate in a group approach to the problem. Actually, the situation was quite different; we enjoyed surprising freedom. For one, in our initial decision to join the research service, we had voluntarily accepted the delimitation of central area. The procedures we followed relating to the execution of projects were the result of general agreement by the senior staff members on the best way of protecting both the individual investigator's and the group's interest in the maintenance of high standards for work emanating from Worcester. In regard to group participation, we were not only free to choose the particular tasks (whether test or experiment) that we contributed to the multidisciplinary effort, but each investigator was also granted considerable opportunity to carry out individual studies on schizophrenics or other groups, exploratory studies, or scholarly work.

The pervading tone of the hospital—its democratic organization, general preoccupation with innovative activities, and extensive teaching programs—provided additional ground for our positive attitudes. (In a paper dealing with the problems of working in government institutions [Shakow, 1968], I have described such factors in greater detail.) What helped above all was the mutual respect that we, with a few exceptions, had for each other as persons, a level of respect that allowed for the harshest mutual criticism of manuscripts that I have experienced in my professional life. From this developed a corresponding respect for the different disciplines represented among us. The resulting atmosphere of tolerance and understanding laid a solid base for our joint attack on schizophrenia.

Having indicated the areas of contentment, and merely stated the areas of discontent in the broadest of terms, let me be somewhat more specific about the latter. We were discontented about the physical conditions under which we worked, the wards, the biochemical laboratory, the psychological laboratory, the library, and we did something about them. We were troubled about the shortage of funds for expensive equipment, and we went out and got them. These realistic difficulties seemed to us minor compared with the bureaucratic annoyances which came from being part of a fairly tight-knit state system. But in one way or another we managed to deal with these as well. Another serious problem was the procurement and training of nursing personnel, but after much effort we were also able to deal reasonably well with this problem, achieving a reasonably high level of quality and stability in this area.

These difficulties notwithstanding, the principal source of discontent was ourselves. This self-dissatisfaction, both as investigators and as theorists, led to our persistently examining and reexamining the techniques, methods, goals, and substance of our studies. Not only did we have our own individual superegos to contend with, but, because of the harmonious relationships existing among ourselves as colleagues, we were able to overcome any natural reluctance to serve as altersuperegos for each other. The most potent influence of all, however, toward hewing to a straight line was the group's unanimously accepted supersuperego—the head of our Biometrics Department, E. Morton Jellinek.

As to goals, we differed as to whether to use "pink pill" (simplified medicinal therapies) or practical approaches rather than basic approaches; we constantly faced problems about accuracy or relevancy with respect to technique and method; and, with regard to substantive findings, we always had to be ready to face our colleagues' questionings about the import of our findings, to answer their what-of-it's.

Another issue creating discontent was the contribution that others had made or were making to the field of schizophrenia. When we embarked on our program we had expected to be able to take off from the shoulders of our predecessors. We were disappointed, however, to find that instead we would have to start from the ground up. This dissatisfaction stemmed neither from a false sense of superiority nor from superficial examination of the literature. On the contrary, we were far from smug, and before undertaking a study we made exhaustive studies of the relevant literature. However, repeated perusals of both the literature on schizophrenia, in the specific area of our concern, and the relevant normal literature left us dissatisfied. We so frequently found marked inconsistencies in results, based either on inadequate samples, techniques or analysis, or on several or all these, both for schizophrenic and normal subjects, that we were forced to carry out our own parallel studies of both groups. (Not that others did not raise similar questions about our studies.)

A major source of concern was the unavoidable one of nosology, the perennial problem of psychopathology. We were also troubled about the state of the demographic data in the field generally, since these were integrally related to problems of sampling.

Though we later shifted to other goals, our initial period of systematic cooperative study had to be devoted mainly to providing reasonably accurate baseline descriptive data, both for our own purposes and for the literature. It was directed at answering the complex question: What is a schizophrenic really like? This called for a detailed examination of the total research process, from initiatory questioning and observation to final theorizing. We had to concern ourselves with problems of selection and nosology, techniques and methods of study, conditions of study, timing of studies, appropriate controls, relevant kinds of statistics, areas of study, hypotheses, and theories that developed. We dealt with these problems in many ways, some with undoubtedly greater adequacy and efficiency than others. As will be seen from an examination of our published papers, these aspects were sometimes handled implicitly, but most often explicitly. I myself have discussed some of the general methodological aspects in detail (Shakow, 1969), and several members of our group have at one time or another considered both the general (Hoskins et al., 1933; Jellinek, 1937) and specific aspects of such problems (Jellinek, 1936; Jellinek and Looney, 1936; Lengyel and Freeman, 1938; Shakow, 1966).

To achieve these descriptive goals, as well as the goals we later pursued, which went beyond description, we employed a variety of devices: the exploratory project, the protocoled individual project, the group project, research council discussions, ad hoc committees for the evaluation of projects, memoranda from the biometrics group analyzing the findings, group conferences to react to early reports evaluating the findings, and editorial committees to evaluate original and final write-ups.

Of this process let me single out what was probably the unique aspect that characterized our research operation for a long time, the acceptance of the Biometrics Department as the central core of our evaluation process.

This department, organized in May 1931 with the coming of Jellinek, played an important role in the research operation almost from the beginning. The department's major responsibilities were two: compiling the data of the research operation into analyzable form and providing analyzed material that might serve as a basis for publication. The latter entailed the provision of statistical analyses that included not only the establishment of the constants by both large and small sample statistics, as required by the nature of the data, but also the correlations and other appropriate statistical analyses. The complete analyses were provided to the individual investigators in the form of detailed tables and scatter diagrams to complement the detailed memorandum on the project which discussed the statistical and logical implications of the data. Such memoranda were usually prepared by Jellinek himself. Not only were they masterpieces of statistical and logical insight, but frequently were suffused with his marvelous sense of irony and humor. For the preparation of his paper, the individual investigator was then left to provide his own specialized disciplinary and personal background to complement and integrate the statistical material.

After some eight years of operation, there was widespread feeling among the members of the Research Service that the time for a major stocktaking had arrived. The directors, reacting favorably to this sentiment, appointed a task force (the Committee on Coordination of Research) in the fall of 1934, which consisted of four senior members of the service from different disciplines under the chairmanship of Jellinek. The committee was instructed to make a survey of the state of the Research Service, its adequacies and inadequacies, but particularly to point out our shortcomings and to make recommendations for improvement. After approximately a year of frequent meetings, the committee, having considered all aspects of the research program from their own points of view, reviewed memoranda provided by other members of the service, consulted with both intramural and extramural experts in relevant areas, and submitted a report to the directors on October 11, 1935. This report was immediately made available to the senior researchers for consideration and comment. The body of the report was 176 pages long (double-spaced) and had 98 pages of appendixes. It was an excellent hard-hitting document, which, though it listed many of the Research Service's accomplishments, did not spare the failings of any of the researchers, from the director down. Though many of the substantive and methodological findings from our researches were listed in the appropriate sections, the thrust was directed most at the consideration of areas where we had been remiss with respect to method, substance, or goals.

One problem that had troubled us greatly over the years—the role of psychiatry in our program—was singled out for detailed consideration. Though our psychiatric group had played an essential role, it had on the whole been a minor one. Though they carried out their routine responsibilities in the diagnosis and interviewing of patients in excellent fashion, they appeared intimidated about pushing for psychiatric projects, per se. Instead they tended to associate themselves with either organic or, more rarely, strictly psychological projects.

Many of the subsequent Research Council meetings in 1935 and 1936 were spent considering the recommendations made in the report, resulting in many specific improvements in the service, including the achievement of a more important relative role for psychiatry. Perhaps the greatest impact of the report was the indirect effect it had—reemphasizing to us subtly, but at the same time with great explicitness, the underlying values which presumably were motivating our research.

What was the effect on our own group of the program as a whole and of the manner in which it was conducted? With few exceptions, I believe it to have been positive.

It broadened our vision as participants converging from varied disciplines who had often had little, if any, previous contact with some of the other disciplines represented. It fostered in us not only an understanding of the goals and methods of other disciplines, but actually led to some welcomed encroachments on the boundaries of one another's areas.

It provided the individual investigators with many opportunities for carrying out cooperative activities leading not only to many joint projects within a discipline but also to much multidisciplinary and cross-disciplinary activity. It led, as well, to considerable mutual understanding of the research needs of other disciplines.

The heavy preoccupation with collaborative and cooperative research led to another kind of sensitivity, namely, sensitivity to the needs of the investigators. Recognition of the importance of the release that comes from some kinds of independent activity led to our providing adequate opportunity to carry out individual studies. These might be either exploratory or systematic, conducted individually or in association with another investigator, but they had above all to be personal, not tainted by any suggestion of total group involvement.

A substantial by-product of the research operation was the training ground it provided for young researchers. Many of the young investigators who trained at Worcester in those days frequently attested years later to the lasting value of this early experience.

Of course, we had our share of dissent. Sometimes these were *fach* conflicts, most frequently revolving around theory and interpretation. Less frequently they concerned methodology. Such clashes were not only worked through, but appeared to serve a useful, often implicit, maturing and broadening function. At times the disagreements were about administrative decisions. And there were occasions when friction appeared to stem from personality differences, from the personal difficulties that are bound to arise among colleagues working in such close association. But in one way or another these conflicts were resolved.

The effects of the Worcester research on the field as a whole are much more difficult to evaluate. They may in part be judged by the status of the papers coming from the Worcester project in the schizophrenia literature generally; the respect with which the then Worcester staff were greeted at meetings and conferences during the period of active research (and even are being met with up to this day); the positions occupied by Worcester representatives in professional organizations; the important roles, both administrative and substantive, that Worcester staff members have subsequently played in the fields and disciplines in which they have worked; and the fact that Worcester studies are still being quoted today. Further evidence of the impact is to be found in the offshoots of the Worcester program which I shall consider shortly. Only a study dealing with this problem directly can with any adequacy determine the extent of Worcester influence, a task much beyond the scope of this essay.

Any attempt to appraise the specific contributions of the Worcester program must consider its neglects and failures as well as its heeds and successes. Let us first attend to the inadequacies.

Though we carried out an immense number of studies dealing with single or even multiple functions, we did not ever truly succeed in achieving unity from this diversity, in reintegrating the individual from the multiplicity of aspects on which we obtained data about him. Perhaps the nearest we came to this goal was during the later period of the research program, during the early 1940's, when we were occupied with the military cases, and the immediately following years until the program's termination.

Another limitation of our studies was the tendency to treat the patient as an independent entity, as if he had no relevant context. We seemed oblivious—particularly during the early days of the research—to the significant role played by familial and social group factors. This failing, too, was partly remedied in our military study, where, because of its much improved social work and greater emphasis on rehabilitation, the group context assumed a more important role. The coming of Dr. Otto Kant to the Research Service, among whose particular interests was prognosis (Kant, 1940, 1941a, 1941b, 1942a, 1942b, 1944), also contributed to our attending more to familial factors.

Associated with the neglect of familial factors was the program's relative neglect of genetic factors in the etiology of the psychosis. Though some thought had been given to the need for work in this area during the early days of our research little was actually done in this respect. Here, again, the military study with its improved family histories led to greater attention to the topic. Dr. Kant, through his studies of family inheritance based on the careful examination of the social histories of patients (Kant, 1942a, 1942b), helped to advance our thinking in this area as well.

How about that other facet of the research process, its pattern? Though we performed an enormous amount of multidisciplinary research, to my mind, at least, a disproportionate amount of it was concurrent-collaborative, rather than integral, in nature. It is true that as the research program progressed more integral research, characterized by cross-disciplinary design and theorizing, became increasingly prominent, but in toto we were probably remiss in this respect.

Another problem presented by our program, and one more difficult to evaluate, was the burden that our frequently heavy research schedules imposed on the patients. Did this impair, improve, or have little effect on a patient's functioning? We were never able to decide. My guess is, that like so many situations in psychopathology, and indeed, in psychological studies generally, the differential effects were great: Some patients being affected favorably, others unfavorably, and some not at all.

Let us turn now to the adequacies of the program. I cannot, of course, attempt to summarize here the two decades of the program's specific substantive contributions, an extensive task which I must reserve for another occasion. But in a cursory way I can at least list the broad areas in which the Worcester program added to our knowledge of schizophrenic and normal functioning. I shall, besides, describe in somewhat more detail its contributions both to methodology in schizophrenia and psychopathology research, and to general standards of research in mental disorder. I shall also indicate some of Worcester's offshoots and consider some general lessons that may be drawn from our experience with the group research process.

Despite much of the backing and filling that resulted from our persistent concern with adequacy of techniques, the Worcester operation added much in the way of specific data about substantive functioning in schizophrenics. Constants in the form of means, measures of variation, and measures of relationship were obtained on patients under both non-stimulated (basal or ordinary) and stimulated conditions. The latter type of findings stemmed generally from either a manipulative context or a therapeutic context. These data covered almost the full range of biological and psychological functions. Though the studies concentrated on schizophrenia, a not inconsiderable number were devoted to the functioning of normal subjects.

The Worcester operation added much valuable material to schizophrenia methodology, particularly in defining and refining the techniques of nosology and in suggesting ways for dealing with selection processes in research.

The Worcester program also contributed greatly to standards of research in psychopathology. Here I feel more confident in asserting my doubts as to whether there had ever before been devoted such careful consideration to high standards of statistical and experimental control in schizophrenia research. (This judgment must be recognized as relative, rather than absolute; our program was far from having achieved the true excellence we strove for.) It is in relation to these standards of research that I believe the Worcester program to have been preeminent at the time as well. The elaborate and sophisticated statistics provided by our Biometrics Department were unmatched. But sophistication in itself is, of course, impertinent if it is not both relevant and applied to high-quality data. With respect to the former we had utmost faith in Jellinek; with respect to the latter, we were constantly being pressed, by both our own and the group's superegos, to achieve such quality. Our concern with instrumental control, though perhaps not reaching quite so high a level as in the statistical area, in part corroborated such search for standards.

Our concern with person controls, though far from meeting ideal standards, was still considerably beyond the general standards of the day. We were, I believe, one of the first, if not the very first, groups in psychopathological research to bring normal controls to live on the same wards as the patients in order to reduce the effects of differential environment, activity, and nutrition (Jellinek, 1936a; Freeman, 1938; Looney and Freeman, 1938).

Another contribution to methodology in research relates to the already described procedures developed for carrying out the various stages of the research process, both individual and group, from the idea or hypothesis stage through the publication stage.

The prevailing atmosphere of the hospital and the Research Service encouraged the writing of papers of a broader and more general scope than those deriving directly from the central research projects as other immediate responsibilities of the staff. Thus, a considerable number of literary, scholarly, and encyclopedia articles, monographs, and books appeared during the years under discussion.

I shall have occasion to discuss the viability of elaborate research programs shortly, but let me consider one aspect of viability, which relates to the numerous offspring of the Worcester research. These include the programs of the Worcester Foundation for Experimental Biology, Huston, at Iowa, Gottlieb, first at Iowa, but mainly in his subsequent work at Wayne, Cameron's work at McGill, Harry Freeman's work at Worcester and then at Medfield, the Duke program under Rodnick and Garmezy, Phillips's projects at Worcester and then at Boston College, and my own studies at Illinois and those with which I was associated at the National Institute of Mental Health.

It is inevitable, especially in an expository context, for one to oversimplify complex processes. Such a propensity arises in dealing with a complicated process such as the one I am attempting to describe—the Worcester group research program. One tends to confine oneself rather to what may merely amount to no more than a description of the original intention or organized plan, thus neglecting many of the actual and considerably more elaborate steps entailed in the execution of the program. In reality, carrying out such a project ordinarily involves an almost limitless number of elements forming an intricate pattern that, though detectable with effort, leaves one with the immediate impression of lack of design. Perhaps the most apt analogy is that of a loom fabricating cloth for a coat of many colors. The pattern of cloth produced by the loom changes not only in color composition but in the speed with which it produces the cloth. At times the loom weaves with a predominating color, interrupted by relatively few intrusions of different colored strands; at other times, the pattern is composed of many equally represented, interwoven strands of variegated colors and shades, some of which have even appeared seemingly on their own. At times the loom bowls along at a fast clip, at others it barely meanders. A large research operation, such as the Worcester one, exhibits the full range of such characteristics and is therefore most difficult to portray simply. The various levels of research programs I have described, the simultaneous interweaving of levels, the general moratoria established to provide the different participants with time for writing up their constituent projects, as well as the individual moratoria following on independent projects— these were all intertwined in such fused fashion.

Speaking less metaphorically, a sizable and long-lived group research program may have many intended trends, but, if democratically, rather than autocratically, administered, it also has many spontaneous and re-active, even unintended, trends. It proceeds according to a rhythm of its own, affected by many cyclic activities, actions and counteractions, bal-ancings, and counterbalancings. Thus individual projects balance group projects, exploratory projects balance defined projects, basic projects balance practical projects, immediately oriented projects balance re-motely oriented projects, therapeutic projects balance descriptive proj-ects, pink pill projects balance pure projects, theoretical projects bal-ance empirical projects, casual projects balance precise projects, projects on acute patients balance those on chronic patients, and projects highly susceptible to the current Zeitgeist are balanced by obstinately self-re-liant projects strongly resistant to any pressures from the outside. It is this diversified, dynamic, constantly changing, impressionistic motile portrait that depicts what actually happened at Worcester over the years so much more accurately than does the most detailed naturalistic pictorialization.

A point, having even wider implications, also needs making about the program. My impression is that an enterprise such as ours has a limited period of viability. A generation—about two decades—appears to be its natural term of life. This span encompasses the phase of birth, continu-ing growth, the achievement of a high plateau and that of slow decline. Even though losses of personnel—even of important personnel—may occur in the early phases, recruiting results in their replacement by new and equally effective personnel. The operation proceeds with essentially the same general goals, though specific goals may change frequently. This pattern appears to me to have been true of the Worcester opera-tion. In the last phase, however, that of decline, such renewal does not seem to occur.

Still another generalization relates to the changes taking place in indi-vidual participants with respect to the theoretical positions they adopt. I have already mentioned changes that occur in the general maturing of viewpoints. I can most readily illustrate this in myself, though I ob-served shifts of this kind in a number of my colleagues. During the early days of my participation in the group, the marked organic flavor then current had a natural balancing effect in strengthening my already es-tablished psychogenic predilections. With the passage of time, however, my stand became less rigid, especially when the viewpoints represented in the group were considerably broadened. I became more sympathetic to organic views, and more understanding of genetic arguments.

A further generalization applicable to a research program such as ours relates to the progress with time of the group as a whole in specific sophistication and independence. During the early days of our program we depended quite heavily on what our predecessors had done, to the point of accepting existing literature as gospel. In the course of time, however, we recognized how fallible these sources were. Again, in the early days we tended to be naïve pink pillers, expecting miraculous cures. But as time passed, we became aware of how involved the schizophrenia problem really was. We thus moved from a stage of naïve enthusiasms to one of being impressed by the complexities of the group we were studying, from simplistic organic views to an appreciation of the intricate psychobiological processes we were dealing with.

A long-lived multidisciplinary program has other rewards worth mentioning. Isolated, short-term projects, of course, make their contributions. But multifaceted projects that last over many years add dimensions and have values that myriads of projects of the former type cannot possibly approach because of the cumulative, serendipitous quality resulting from such operations. Many unplanned gains accrue to long-lasting programs and to permanent members of the group, from operations that inevitably entail working at different times with varieties of patients, with different hypotheses and techniques, and with changing experimenters. This progression of changes results in findings that cannot be achieved through any degree of multiplication of short-term projects.

But probably the most important (perhaps obvious?) lesson I learned from the Worcester operation, a lesson corroborated subsequently by repeated experiences, was that it is not the institutional structures set up to facilitate a program that are essential for success, but rather the particular persons who control these structures. Individuals, and not machinery, determine the outcome of any operation; the mechanisms certainly help to advance one toward the achievement of wished-for goals, but they are never the vital element. In this fundamental respect Worcester was fortunate, and that is why its participants were so satisfied. Because of the difficulties we faced and the necessity for surmounting them, because of our youth, because of the opportunity for growth, because of our readiness to accept a common goal in order to deal with a most complex problem, and, above all, because of the persons who participated, we developed into a highly motivated group. From this combination arose at least a trace of the kind of spirit that must have animated early kibbutz groups.

SUMMARY

In this essay I have attempted to portray the history and nature of the Worcester State Hospital schizophrenia research program in its general aspects, and indicate some of the lessons pointed up by this two decades of experience. I have made no effort, however, to deal with the substantive findings of the program. Despite its admittedly nostalgic air, I trust that my account represents more than an oldster's hankering for his youth or merely a vision rosed by the aging of his glasses.

Against the background history and description of the Worcester State Hospital itself, I dealt with the establishment of the Research Service, and the three periods through which it went. I described briefly the two earlier periods: the endocrine-metabolic period of 1927–1930, with its relatively limited personnel and range of attack on schizophrenia, and the period of stocktaking which followed in 1930 and 1931, when the complexity of the problem of schizophrenia was increasingly recognized and the group and the total operation were greatly expanded. This involved a much broadened approach to schizophrenia, both in relation to the disciplines represented and the areas covered, and considerable growth in its physical and financial structure. In the context of such developments, the program entered into its third and major period, one lasting for about one and one-half decades—from 1931 to 1946.

In considering this period, I first described and defined the four types of schedules which occupied us during the time: the comprehensive multidisciplinary, the moderate multidisciplinary, the limited bi- or unidisciplinary, and the exploratory. I went into some detail in the description of the three dominant multidisciplinary projects of level 1—the seven-months study, the insulin study, and the military study. In much less detail, I considered the various studies which were of level 2 scope, barely described a selection of examples from the level 3 group, and only considered the general nature of the level 4 studies.

Approximately half of the essay is devoted to an overall evaluation of the Worcester schizophrenia program—its failings and its strengths, both as multidisciplinary project and as research program in psychopathology and schizophrenia. I dealt in general terms with our "divine discontents" and the factors resulting in the harmony it led to, as well as with the broader effects on the participants. The more specific section that followed detailed the inadequacies of the program as well as the adequacies represented in a variety of formal contributions to knowledge about schizophrenia, to methodology in schizophrenia research, and to standards in this area of research.

I also devoted a section to some rather general lessons highlighted by the experience with this long-lasting group program—about the nature of such research, its cyclic character, its natural history and decline, and its growth-producing potential, both for the individual and the group.

The place of the Worcester State Hospital research in the history of research on the problem of schizophrenia may be hard to evaluate adequately. But that it did have a place, and an important place, is difficult to deny. For me the Worcester period was my most profitable professional experience. If I have to some extent conveyed to the reader the warm feelings that the period at Worcester still arouses in me, then I have accomplished at least part of what I intended.

REFERENCES

Bryan, W. A. 1936. *Administrative Psychiatry*. New York: Norton.

Cameron, D. E. 1938. "Further Experiences in the Insulin-Hypoglycemia Treatment of Schizophrenia," *Journal of Nervous and Mental Disease* 87:14–25.

Cameron, D. E., and Hoskins, R. G. 1937a. "Experiences in the Insulin-Hypoglycemia Treatment of Schizophrenia," *Journal of the American Medical Association* 109:1246–1249.

Cameron, D. E., and Hoskins, R. G. 1937b. "Some Observations on Sakel's Insulin-Hypoglycemia Treatment of Schizophrenia," *Archives Suisses de Neurologie et Psychiatrie* 39:180–182.

Cameron, D. E., and Jellinek, E. M. 1939. "Physiological Studies in Insulin Treatment of Acute Schizophrenia: II. Pulse Rate and Blood Pressure," *Endocrinology* 25:100–104.

Carmichael, H. T., and Linder, F. E. 1934. "The Relation between Oral and Rectal Temperatures in Normal and Schizophrenic Subjects," *American Journal of the Medical Sciences* 188:68–75.

Cohen, L. H. 1938a. "The Early Effects of Metrazol Therapy in Chronic Psychotic Over-activity," *American Journal of Psychiatry* 95:327–333.

Cohen, L. H. 1938b. "Observations on the Convulsant Treatment of Schizophrenia with Metrazol: A Report of Seven Cases," *New England Journal of Medicine* 218: 1002–1007.

Cohen, L. H. 1939a. "Factors Involved in the Stability of the Therapeutic Effect in the Metrazol Treatment of Schizophrenia: Report of 146 Cases," *New England Journal of Medicine* 220:780–783.

Cohen, L. H. 1939b. "The Pharmacologic Antagonism of Metrazol and Sodium Amytal as Seen in Human Individuals (Schizophrenic Patients)," *Journal of Laboratory and Clinical Medicine* 24:681–684.

Cohen, L. H. 1939c. "Return of Cognitive Conscious Functions after Convulsions Induced with Metrazol," *Archives of Neurology and Psychiatry* 41:489–494.

Cohen, L. H. 1939d. "The Therapeutic Significance of Fear in the Metrazol Treatment of Schizophrenia," *American Journal of Psychiatry* 95:1349–1357.

Elmadjian, F., and Pincus, G. 1946. "A Study of the Diurnal Variations in Circulating Lymphocytes in Normal and Psychotic Subjects," *Journal of Clinical Endocrinology* 6:287–294.

Erickson, M. H., and Hoskins, R. G. 1931. "Grading of Patients in Mental Hospitals as a Therapeutic Measure," *American Journal of Psychiatry* 11:103–109.

Freeman, H. 1933a. "Effects of 'Habituation' on Blood Pressure in Schizophrenia," *Archives of Neurology and Psychiatry* 29:139–147.

Freeman, H. 1933b. "Sedimentation Rate of the Blood in Schizophrenia," *Archives of Neurology and Psychiatry* 30:1298–1308.

Freeman, H. 1934a. "Arm-to-Carotid Circulation Time in Normal and Schizophrenia Subjects," *Psychiatric Quarterly* 8:290–299.

Freeman, H. 1934b. "The Effect of Dinitrophenol upon Circulation Time," *Journal of Pharmacology and Experimental Therapeutics* 51:477–481.

Freeman, H. 1938. "Variability of Circulation Time in Normal and in Schizophrenic Subjects," *Archives of Neurology and Psychiatry* 39:488–493.

Freeman, H. 1946. "Resistance to Insulin in Mentally Disturbed Soldiers," *Archives of Neurology and Psychiatry* 56:74–78.

Freeman, H., and Elmadjian, F. 1946. "The Relationship between Blood Sugar and Lymphocyte Levels in Normal Individuals," *Journal of Clinical Endocrinology* 6: 668–674.

Freeman, H., Pincus, G., Elmadjian, F., and Romanoff, L. P. 1955a. "Adrenal Responsivity in Aged Psychotic Patients," *Geriatrics* 10:72–77.

Freeman, H., Pincus, G., Elmadjian, F., and Romanoff, L. P. 1955b. "Adrenocortical Reactivity in Aged Schizophrenic Patients," *Ciba Foundation Colloquium on Aging* 1:219–236.

Freeman, H., Rodnick, E. H., Shakow, D., and Lebeaux, T. 1944. "The Carbohydrate Tolerance of Mentally Disturbed Soldiers," *Psychosomatic Medicine* 6: 311–317.

Freeman, W., and Looney, J. M. 1934a. "Phytotoxic Index: I. Results of Studies with 68 Male Schizophrenic Patients," *Archives of Neurology and Psychiatry* 32: 554–559.

Freeman, W., and Looney, J. M. 1934b. "Studies on the Phytotoxic Index: II. Menstrual Toxin ('menotoxin'), *Journal of Pharmacology and Experimental Therapeutics* 52:179–183.

Glueck, B. 1936. "The Hypoglycemic State in the Treatment of Schizophrenia," *Journal of the American Medical Association* 107:1029–1031.

Grob, G. N. 1966. *The State and the Mentally Ill.* Chapel Hill: University of North Carolina Press.

Guiles, A. P. 1947. "The Beginnings of Clinical Training in New England," *Andover Newton Theological School Bulletin* 40:1–20.

Hoagland, H., Cameron, D. E., and Rubin, M. A. 1937a. "The 'Delta Index' of the Electrencephalogram in Relation to Insulin Treatments of Schizophrenia," *Psychological Record* 1:196–202.

Hoagland, H., Cameron, D. E., and Rubin, M. A. 1937b. "The Electrencephalogram of Schizophrenics during Insulin Treatments: The 'Delta Index' as a Clinical Measure," *American Journal of Psychiatry* 94:183-208.

Hoagland, H., Elmadjian, F., and Pincus, G. 1946. "Stressful Psychomotor Performance and Adrenal Cortical Function as Indicated by the Lymphocyte Response," *Journal of Clinical Endocrinology* 6:301–311.

Hoagland, H., Rubin, M. A., and Cameron, D. E. 1937a. "The Electrencephalogram of Schizophrenics during Insulin Hypoglycemia and Recovery," *American Journal of Physiology* 120:559–570.

Hoagland, H., Rubin, M. A., and Cameron, D. E. 1937b. "Electrical Brain Waves in Schizophrenics during Insulin Treatments," *Journal of Psychology* 3:513–519.

Hoagland, H., Rubin, M. A., and Cameron, D. E. 1939. "Brain Wave Frequencies and Cellular Metabolism Effects of Dinitrophenol," *Journal of Neurophysiology* 2: 170–172.

Hoskins, R. G., 1933. "Schizophrenia from the Physiological Point of View," *Annals of Internal Medicine* 7:445–456.

Hoskins, R. G., and Jellinek, E. M. 1933. "The Schizophrenic Personality with Special Regard to Psychologic and Organic Concomitants," *Proceedings of the Association for Research in Nervous and Mental Disease* 14:211–233.

Hoskins, R. G., and Sleeper, F. H. 1933. "Organic functions in Schizophrenia," *Archives of Neurology and Psychiatry* 30:123–140.

Hoskins, R. G., Sleeper, F. H., Shakow, D., Jellinek, E. M., Looney, J. M., and Erickson, M. H. 1933. "A Cooperative Research in Schizophrenia," *Archives of Neurology and Psychiatry* 30:388–401.

Huston, P. E. 1932. "Eye-Hand Coordination in Schizophrenic Patients and Normals as Measured by the Pursuit Meter," *Psychological Bulletin* 29:662

Huston, P. E. 1934. "Sensory Threshold to Direct Current Stimulation in Schizophrenic and in Normal Subjects," *Archives of Neurology and Psychiatry* 31: 590–596.

Huston, P. E., and Hayes, J. G. 1934. "Apparatus for the Study of Continuous Reaction," *Journal of Experimental Psychology* 17:885–891.

Huston, P. E., and Shakow, D. 1946. "Studies of Motor Function in Schizophrenia: III. Steadiness," *Journal of General Psychology* 34:119–126.

Huston, P. E., and Shakow, D. 1948. "Learning in Schizophrenia: I. Pursuit Learning," *Journal of Personality* 17:52–74.

Huston, P. E., and Shakow, D. 1949. "Learning Capacity in Schizophrenia: With Special Reference to the Concept of Deterioration," *American Journal of Psychiatry* 105:881–888.

Huston, P. E., Shakow, D., and Riggs, L. A. 1937. "Studies of Motor Function in Schizophrenia: II. Reaction Time," *Journal of General Psychology* 16:39–82.

Jellinek, E. M. 1936a. "Estimates of Intra-Individual and Inter-Individual Variation of the Erythrocyte and Leukocyte Counts in Man," *Human Biology* 8:581–591.

Jellinek, E. M. 1936b. "Measurements of the Consistency of Fasting Oxygen Consumption Rates in Schizophrenic Patients and Normal Controls. *Biometric Bulletin* 1 (no. 1):15–43.

Jellinek, E. M. 1937. "Some Uses and Abuses of Statistical Method in Psychiatry. *Biometric Bulletin* 1 (no. 3):97–108.

Jellinek, E. M. 1939. "Physiological Studies in Insulin Treatment of Acute Schizophrenia: I. Methods," *Endocrinology* 25:96–99.

Jellinek, E. M., and Fertig, J. W. 1936. "A Method for the Estimation of Average Heart Rates from Cardiochronographic Records," *Journal of Psychology* 1: 193–199.

Jellinek, E. M., and Looney, J. M. 1936. "Studies in Seasonal Variation of Physiological Functions: I. The Seasonal Variation of Blood Cholesterol," *Biometric Bulletin* 1 (no. 2):83–95.

Kant, O. 1940. "Types and Analyses of the Clinical Pictures of Recovered Schizophrenics," *Psychiatric Quarterly* 14:676–700.

Kant, O. 1941a. A Comparative Study of Recovered and Deteriorated Schizophrenic Patients," *Journal of Nervous and Mental Disease* 93:616–624.

Kant, O. 1941b. "Study of a Group of Recovered Schizophrenic Patients," *Psychiatric Quarterly* 15:262–283.

Kant, O. 1942a. "The Incidence of Psychoses and Other Mental Abnormalities in the Families of Recovered and Deteriorated Schizophrenic Patients. *Psychiatric Quarterly* 16:176–186.

Kant, O. 1942b. "The Problem of Psychogenic Precipitation in Schizophrenia," *Psychiatric Quarterly* 16:341–350.

Kant, O. 1944. "The Evaluation of Prognostic Criteria in Schizophrenia," *Journal of Nervous and Mental Disease* 100:598–605.

Kant, O. 1945. "Types of Psychiatric Casualty in the Armed Forces," *Mental Hygiene* 29:656–665.

Lengyel, B. A., and Freeman, H. 1938. "Analysis of the Effects of High Humidity on Skin Temperature," *Biometric Bulletin* 1 (no. 4):139–149.

Looney, J. M., and Cameron, D. E. 1937. "Effect of Prolonged Insulin Therapy on Glucose Tolerance in Schizophrenic Patients." *Proceedings of the Society for Experimental Biology and Medicine* 37:253–257.

Looney, J. M., and Childs, H. M. 1933. "The Blood Cholesterol in Schizophrenia," *Archives of Neurology and Psychiatry* 30:567–579.

Looney, J. M., and Childs, H. M. 1934. "A Comparison of the Methods for the Collection of Blood to Be Used in the Determination of Gases," *Journal of Biological Chemistry* 104:53–58.

Looney, J. M., and Freeman, H. 1938. "Oxygen and Carbon Dioxide Contents of Arterial and Venous Blood of Schizophrenic Patients. *Archives of Neurology and Psychiatry* 39:276–283.

Looney, J. M., and Hoskins, R. G. 1934. "The Effect of Dinitrophenol on the Metabolism as Seen in Schizophrenic Patients," *New England Journal of Medicine* 210: 1206–1213.

Looney, J. M., and Hoskins, R. G. 1935. "The Therapeutic Use of Dinitrophenol and 3:5 Dinitro-ortho-cresol in Schizophrenia," *American Journal of Psychiatry* 91: 1009–1017.

Looney, J. M., and Jellinek, E. M. 1935. "Galactose Tolerance as Measured by the Folin Micro and Macro Blood Sugar Methods," *Proceedings of the American Society for Biological Chemists* April 10–13.

Looney, J. M., Jellinek, E. M., and Dyer, C. G. 1939. "Physiological Studies in Insulin Treatment of Acute Schizophrenia: V. The Blood Minerals," *Endorcinology* 25:282–285.

Looney, J. M., and Romanoff, E. B. 1940. "The Effect of Testosterone on the Serum Lipids of Normal Subjects," *Journal of Biological Chemistry* 136:479–481.

Luszki, M. B. 1958. *Interdisciplinary Team Research: Methods and Problems.* New York: New York University Press.

Malamud, I. T. and Stephenson, R. B. 1944. "A Study of the Rehabilitation of Neuro-Psychiatric Casualties Occurring in the Armed Forces," *Applied Anthropology* 3:1–16.

Malamud, W., and Malamud, I. T. 1943. "A Socio-Psychiatric Investigation of Schizophrenia Occurring in the Armed Forces," *Psychosomatic Medicine* 5:364–375.

Malamud, W., and Malamud, I. T. 1945. "Socio-Psychiatric Problems in Rehabilitation," *Diseases of the Nervous System* 6:134–142.

Pincus, G. 1947. "Studies of the Role of the Adrenal Cortex in the Stress of Human Subjects." In *Recent Progress in Hormone Research.* New York: Academic Press. Vol. 1, pp. 123–145.

Pincus, G., and Elmadjian, F. 1946. "The Lymphocyte Response to Heat Stress in Normal and Psychotic Subjects," *Journal of Clinical Endocrinology* 6:295–300.

Pincus, G., Hoagland, H., Freeman, H., and Elmadjian, F. 1949. "Adrenal Function in Mental Disease," *Recent Progress in Hormone Research,* Proceedings of the Laurentian Hormone Conference 4:291–322.

Randall, L. O. 1940a. "The Effects of Insulin on Serum Lipids and Choline Esterase in Schizophrenia," *Journal of Laboratory and Clinical Medicine* 25:1025–1028.

Randall, L. O. 1940b. "Effect of Repeated Insulin Hypoglycemia on the Lipid Composition of Rabbit Tissues," *Journal of Biological Chemistry* 133:129–136.

Randall, L. O. 1940c. "Effect of Testosterone on Serum Lipids in Schizophrenia," *Journal of Biological Chemistry* 133:137–140.

Randall, L. O., Cameron, D. E., and Looney, J. M. 1938. "Changes in Blood Lipids during Insulin Treatment of Schizophrenia," *American Journal of the Medical Sciences* 195:802–809.

Randall, L. O., and Cohen, L. H.. 1939. "The Serum Lipids in Schizophrenia," *Psychiatric Quarterly* 13:441–448.

Randall, L. O., and Jellinek, E. M. 1939a. "Physiological Studies in Insulin Treatment of Acute Schizophrenia: III. The Serum Lipids," *Endocrinology* 25:105–110.

Randall, L. O., and Jellinek, E. M., 1939b. "Physiological Studies in Insulin Treatment of Acute Schizophrenia: IV. The Choline Esterase Activity of the Blood Serum," *Endocrinology* 25:278–281.

Research Service. 1942. Memorandum, June 24.

Rodnick, E. H. 1942. "The Effect of Metrazol Shock upon Habit Systems," *Journal of Abnormal Social Psychology* 37:560–565.

Rodnick, E. H., Rubin, M. A., and Freeman, H. 1943. "Related Studies on Adjustment: Reactions to Experimentally Induced Stresses," *American Journal of Psychiatry* 99:872–880.

Rosenzweig, S. 1935. "Outline of a Cooperative Project for Validating the Rorschach Test," *American Journal of Orthopsychiatry* 5:121–123.

Rosenzweig, S., and Freeman, H. 1942. "A 'Blind Test' of Sex-Hormone Potency in Schizophrenic Patients," *Psychosomatic Medicine* 4:159–165.

Sakel, M. 1958. *Schizophrenia.* New York: Philosophical Library.

Schnack, G. F., Shakow, D., and Lively, M. L. 1945a. "Studies in Insulin and Metrazol Therapy: I. The Differential Prognostic Value of Some Psychological Tests." *Journal of Personality* 14:106–124.

Schnack, G. F., Shakow, D., and Lively, M. L. 1945b. "Studies in Insulin and Metrazol Therapy: II. Differential Effects on Some Psychological Functions," *Journal of Personality* 14:125–149.

Shakow, D. 1932. "A Study of Certain Aspects of Motor Coordination in Schizophrenia with the Prod Meter," *Psychological Bulletin* 29:661.

Shakow, D. 1943. "Some Aspects of the Problem of Unsuccessful Personality Screening," *Transactions of the N.Y. Academy of Sciences* 5:199.

Shakow, D. 1966. "The Role of Classification in the Development of the Science of Psychopathology with Particular Reference to Research." In M. M. Katz, J. O. Cole, and W. E. Barton, eds., *The Role and Methodology of Classification in Psychiatry and Psychopathology*. Washington, D.C.: U.S. Government Printing Office. Pp. 116–142.

Shakow, D. 1968. "On the Rewards (and, Alas, Frustrations) of Public Service," *American Psychologist* 23:87–96.

Shakow, D. 1969. "On Doing Research in Schizophrenia," *Archives of General Psychiatry* 20:618–642.

Shakow, D., and Huston, P. E. 1936. "Studies of Motor Function in Schizophrenia: I. Speed of Tapping," *Journal of General Psychology* 15:63–106.

Shakow, D., Rodnick, E. H., and Lebeaux, T. 1945. "A Psychological Study of a Schizophrenic: Exemplification of a Method," *Journal of Abnormal and Social Psychology* 40:154–174.

Simon, B., and Holt, W. L., Jr. 1946. "The Relief of Specific Psychiatric Symptoms by Convulsive Shock Therapy," *Diseases of the Nervous System* 7:241–244.

Wall, C. 1940. "Observations on the Behavior of Schizophrenic Patients Undergoing Insulin Shock Therapy," *Journal of Nervous and Mental Disease* 91:1–8.

Worcester State Hospital. 1936. Annual report.

13]
DANIEL OFFER,
DANIEL X. FREEDMAN,
JUDITH L. OFFER

The Psychiatrist as Researcher

A man's professional life as a researcher in psychiatry is obviously far more than the sum total of his published documents. In the pursuit of his curiosity generated in clinical encounters, or about an issue relevant to the understanding and control of the regulations governing behavior, the researcher's biography in psychiatry (or the psychiatrist's biography in research) may be traced through a variety of trajectories. No one of these patterns need be prototypical. There is a growing literature addressed to psychiatric research and researchers. We will review this literature as it bears on concern for generating research personnel and values in contemporary psychiatry.

In the aggregate, it is the thrust of the scientific mode of inquiry and the scientific adjudication of data that encompasses and gives a crucial stamp to an array of activities we call psychiatric. At some apt time of reckoning, the record of this thrust and its impact will show what the discipline has accomplished and, more importantly, what it required of itself to do. In such an assessment, one would expect to find a few lonely but signal occasions of discovery, if not generative ideas, perhaps several distinctions of clarity, and certainly an occasional therapeutic consequence, as well as the familiar and expected fads, cults, postures, presumptions, derelictions, and philosophies. The dimensional coordinates on which one can trace and construct evolving knowledge of both mental events and behavior are intrinsically difficult to conceive and stipulate. Yet, the study of the subjective as well as the objective in behavior has indeed advanced since Plato, as it has since Freud. Apart from the application of a range of specific methods for treatment and diagnosis, we tend now to see behavior disorder not only as not alien, but also, where not yet comprehendible, then to be comprehended. There are several generations behind us that, though not solving many fundamental mysteries, have made it less mysterious to probe the organization and development of the human psyche, its experiential and social forms and variations, and their place in the broad scheme of biological events (Freedman, 1968).

Our focus is on research activity as a part of a whole. This has been suggested candidly by Grinker's life as researcher, teacher, editor, and leading force in policy-making decisions on a national and local level, as well as a chairman of a department. Whatever his particular contributions to new knowledge, he has continually encouraged psychiatrists undergoing training to care well for patients and to care better than most by asking questions. He has encouraged many to search and to research. This is the essential activity that nurtures not only the life of ideas but also the approach to action. The wise application of knowledge so derived on behalf of the patient is the essence and definition of the best that modern medicine can deliver.

A SYSTEMS VIEW OF
PSYCHIATRIC RESEARCH

At the heart of the matter is the fact that discovery in a vacuum is fruitless. A field is required in which transactions among the generations of teachers, colleagues, students, and patients render ideas viable in the entire process of articulated training, service, and research. These transactions—whatever the gaps in time and fashion—bring research into play within the field and in so doing, if the research can generate interest, change the field. Broadly, the act of inquiry and the valuing of it are of defining importance both to the history of individuals and the profession they claim.

The facts are that communications do actually occur in a community. The community is comprised of those investigators in the biobehavioral sciences relevant to psychiatry, teachers educating the next generations of students, groups and individuals delivering services to the mentally ill, and planners formulating mental health policies and programs. An absorption either in health delivery or in a concern for the resources derived from science tap different personal preferences and evoke varying emphases and values. Yet, both interests comprise a part of a total fabric. The intrafield communications occur in the many informal colleges and colloquia of interested parties and against a background of ideas and practices.

The rules of communication are formally defined by the scientific method and its obligation for orderly reflection and critique, which can follow wild invention. But the rules are informally sustained by a range of as yet poorly examined practices, consisting of peer reviews, debates in societies, laboratory gossip, preparation and publication of articles, selections of boards and committees, and meetings of all kinds; both unconscious and group processes and social aspirations affect each of these. All these social practices have the consequences—and none the obligation—of adjudicating information and certifying significances as they evolve in inquiry and debate. We could add that such socially valued characteristics as some care for knowledge, the love of learning, and some intolerance of foolishness are relevant to efforts to conserve, transmit, and acquire knowledge of human behavior.

Delivering services to the mentally ill is, of course, central. It reflects the nature of the psychiatric obligation to deliver care and to do so in the absence of certain knowledge. However, there is also an obligation to be receptive to knowledge, one that is not always fulfilled. Incorporated within the current retreat from complexity, intellectual challenge, and history is a misrepresentation of research activities as occurring in isolation from modern medical needs or practices. Research is neither alien to nor necessarily competitive with concerns for health delivery. Linkages between research and service occur commonly in medicine—and finally are beginning to occur in psychiatry—as research findings are more widely appreciated and occasionally applied. Ironically, when findings are applied—such as penicillin for syphilis, or crisis intervention on the battlefield to prevent shell shock—the origins of such steps in prior inquiry and knowledge are generally promptly forgotten. Nevertheless, there always is a need for a commonality and community of practices and ideas out of which investigative activities are activated and from which they gain both meaning and relevance. Thus, for the life of inquiry, both researchers and an audience of participating students and practitioners are necessary.

It is precisely the advancement of this pattern of practice, reflection, investigation, and communication in psychiatry that is really meant when we focus on the practical issues of advancing research in departments of psychiatry. Though the literature reviewed might well be applicable to the behavioral sciences generally, and though we are vitally interested in the relevance of related biobehavioral sciences, our explicit emphasis here is on psychiatrists themselves working in research and the importance of this for sound growth of the discipline. The fact is that it is the scholarly or research-minded psychiatrist who must introduce his clinical colleagues to developments occurring outside the confines of the discipline. We need psychiatrists who acquire the special competencies necessary for rendering such translations less corrupted and banal than they often have been. This bridging work should also make findings from other areas more readily and soundly applicable to the core interests of psychiatry. But if psychiatrists are to develop special skills and

competencies, it will take work to foster the understanding for the required academic developments. Such an understanding is needed if we
are to have sane national policies and priorities, guidelines for programs, as well as guidance for individuals in their development.

In the topic area—the problems of fostering psychiatric research—it
is our impression and confession that there is much more in the way of
informal discussion and partially thought-through issues than can yet
be brought into systematic form. A complete agenda has not really been
formulated either by ourselves or others. Of those who have publicly articulated viewpoints, often in response to specific tasks, six major areas
of interest appear to be central.

1. The need for research. The literature shows a concern with motivating the psychiatrist to devote time to research. Some of these articles
are based on the assumption that the necessity has not been self-explan-
atory to the discipline. This, sadly, has been true, and the failure of
many in psychiatry to distinguish special clinical skills or training from
explicit concern with the scientific status of information, its contingency,
or the procedures of gaining new knowledge is no doubt accountable for
this abiding theme. Some have held that it is demeaning to "sell the un-
washed," but in view of the powerful trends toward anti-intellectualism,
it may be necessary to consider a number of explicit steps in educating
not only the public but perhaps many researchers, as well as clinicians,
about their heritage.

2. An identification of the psychiatric researcher. We review some of
the parameters that have been used to identify the researcher in psychiatry. This diagnostic question becomes an issue of importance when
grants, goals, identity, academic positions, and self-esteem are the
stakes.

3. Types of research in psychiatry. Integral to the question of identification of the researcher is the question of what constitutes research in
psychiatry. We review disparate positions, illustrating the existence of
the ever-continuing debates about the philosophical understructure of
scientific procedure.

4. Qualifications of the researcher. We meet with a category of questions relating to the capacities of the man in research and the personality and training qualifications required for pursuing research activities.
In part, it was with the notion that the researcher's qualifications will
determine the quality of his research that the age and maturity of the researcher became an issue. The theme focuses on the vulnerability of career development and research output to the vicissitudes of the researcher's personality growth and development.

5. Research atmospheres. Roles of chairmen and supervisors and the
nature of residency programs and of rewards and recognition are considered.

6. Attitudes within psychiatry toward research. Divergent attitudes on the respectability of a career as a researcher in psychiatry are considered. The psychiatrist who does research, by even the most liberal of interpretations, is not a typical member of his profession. We discuss the researcher as a member of his profession and the beginning emergence of styles of clinical investigations—with all their problems—as a meaningful link to the relevant systems which, in fact, comprise the profession.

ON THE NEED FOR RESEARCH

In terms of shared interests and topical concerns and common points of origin, a generation in scientific research generally can be defined by intervals of approximately seven years. Most disciplines can point to a recurring crop of investigators who mutually share concerns around an exciting scientific issue. But in terms of a firm history of sustained inquiry within a university and with regard for transmitting special skills and methods, psychiatry has a history of less than twenty-five years in the production of explicitly research-oriented psychiatrists. Accordingly, there are deficiencies in the numbers and gaps in the deployment of such individuals. The question of whether research in psychiatry is a luxury can be answered only in the negative. Not only can it not be forfeited for additional service units to psychiatry, in all probability—on close examination—it absolutely need not. Federal expenditures required for research training and for core research projects specifically in psychiatry (and not including demonstration projects) probably total $100 million, a sum that—if donated in total to the State of Illinois's annual mental health care budget—would only cover one-third of the expenditure.

It is essential to finance basic research if we are to be able to continuously supplement, challenge, and change our fund of knowledge and support the centers that conserve and generate knowledge. At times research findings will give rise to exceptionally pragmatic solutions to serious problems. Lustman (1969) recalls the history of polio control and the role of Enders and the many basic scientists who made tissue culture possible. When the vaccine was developed by others, the problems of therapy and service to polio victims almost solved themselves. From the arcane probings into the chemistry of the nervous system (which, under psychiatric auspices, accelerated during the late 1950's), there have been challenges to biologists' conceptions of brain function and the role of chemistry in it. But there was also the spin-off of a useful therapy for Parkinsonism in L-dopa and a new investigative thrust in the affective disorders. From the clinic—after ignored true starts and a number of false starts—has emerged an astonishing effect of lithium in preventing recurrence of manic psychoses. Perhaps we are astonished because an ion (affecting amine metabolism and membrane function) not only may prevent recurrences of a mental disorder but does so without any grounds for the charge of producing a zombie, that is, being without subjective awareness of its effects when used in maintenance dosages.

Yet, the point is not that control of the chemistry of the nervous system can in itself bring the effective management and treatment of all and any behavior disorders, but rather that fundamental questions about the role of chemistry in either the causes or repair of some behavioral disorganizations are advanced. Neither the specific sources nor the directions from which new advances will be found seem predictable. How the pieces of the various puzzles involving the morphological, chemical, functional, and electrical aspects of the nervous system may be put together, how new sorting of behavioral observations will influence diagnosis, treatment, and prevention, remains a matter for discovery and very likely surprise. But given the application of method, a continuing engagement with the issues and receptivity to information, discovery will undoubtedly come and be recognized.

Psychiatry in the United States has been dominated for a number of years by an absorption in and adaptation to the promises and consequences of psychodynamic theories, most of which are central to or derivative of psychoanalysis. Psychodynamic inquiry has helped develop our comprehension of human behavior, but this hardly indicates whether the place of psychodynamic theory will be of greater or lesser importance in the explication of specific multitransactional systems of predictable variables that occur over time. The problem is to locate where and when in the sequential analysis of behavioral systems and in the mastery of their consequences the thought and practice of the depth psychologies may be relevant. The analogy between the role of inflammation in medicine and psychodynamics in psychiatry might be applicable.

The natural course of inflammation is of general relevance in pathology and medicine. It can be experimentally produced. The cellular processes and reactions we call inflammation are undeniable attributes to the organism. We can map with clarity a sequence of organic events which every medical student studies in pathology. These chartings, however, do not have specific or total explanatory powers. For example, the tissue inflammatory response and general course of a streptococcal pharyngeal infection are easy to describe. The inflammatory response is limited to the throat, produces a predictable reaction in the host, and is over in a few days. The more serious pathological reactions are much more difficult to explain or predict. These depend on a complex of factors in the bacteria and the host. For example, variables such as the bacteria's virulence, its metabolites, its adaptability, and its reproductive capacities interact with variables in the host, such as level of resistance or immunity and susceptibility of vulnerable organs. Many of these factors—in their specificity and detail—were unknown and unpredicted by informed scientists during the 1950's. Yet, these are essential factors for predicting, understanding, and controlling diseases such as rheumatic fever or glomerulonephritis. Further, the consequences of such organ damage to the organism are explicable, not on the molecular level, but in terms of the physiology of organ systems. None of these facts invalidates the fundamental beauty or veracity of the inflammation model.

Similarly, we would expect that, from psychodynamic theory, we will find paradigmatic value, but not specific explanatory powers for dealing with a large variety of psychiatric disorders. The same would hold true where we are speaking of psychoanalytic theory (in its dynamic, economic, genetic, topographic, and structural viewpoints). The three basic axes of behavior—biological, psychological, and social—require a variety of explanatory and analytical systems that may be relevant. We have moved beyond the age when motivational models alone are sufficient; the capacity of the organism also becomes a factor, whether we call this ego strength or refer to discretely analyzable biobehavioral variables. If psychogenetics describes the how and what of certain aspects of personality development, it need not (and does not) sufficiently explain the whys.

More information is required both to expand the psychodynamic models and to develop new approaches with a potential toward formulating theoretical bases for understanding specific psychopathological developments. That theory may relate apparently unrelated phenomena and events is clear, and this fact represents the risk any theoretician takes. There is room not only for such creative conjecture within the framework of learning or psychoanalytic theories, but also for special specialists as well as general specialists—individuals who seek through a sustained and profound study, for example, of depression, narcissism, or character structure, a firmer grasp of issues in order to pose critical questions and test them. We need scientists who spend their time probing and investigating because our knowledge is clearly not at the point where any kind of closure can be considered. We emphasize contingent knowledge and the obligation of the researcher to find in each answer an irritating doubt and an exciting new question. Nevertheless, it is the obligation of both researchers and clinicians to place in reasonable perspective—even to defend—the sources and degree of certainty with which certain specific gains have been made, the confidence with which many practices are implemented, lest our impatience with our ignorance and imperfections be exploited by the ever-present forces of know-nothingism.

THE IDENTIFICATION OF THE
PSYCHIATRIC RESEARCHER

The investigator's professional qualifications, the auspices under which research is conducted, the subject studied, and the audience to whom the research is communicated are all factors in the general identification of researchers in psychiatry. Further, the choice of methodology affects the investigator's acceptability as a psychiatric researcher and determines his status within the various subgroupings of the psychiatric community. Indeed there now tends to be a careeristic view of the researcher that requires that he engage in skilled systematic inquiry, distinguishing him sharply from the gentleman scholar model, the clinician who is inquisitive and enjoys sharing his thoughts and perceptions. To illustrate, we cite the identification of the research psychiatrist which appeared in the Group for the Advancement of Psychiatry (GAP) report, *The Recruitment and Training of the Research Psychiatrist* (1967):

In this report a research psychiatrist will be identified as a medical scientist, trained in clinical psychiatry, who devotes a major part of his energy to investigations bearing on the etiology, diagnosis, treatment, or prevention of mental illness. He may work as a scholar with scientific literature, be concerned with theory and methodology, depend mainly upon clinical observation, or work in a laboratory; and he may use the techniques of any scientific discipline to study human or infrahuman behavior.

The definition provides a baseline for a generalized qualification of the psychiatric researcher. The danger lies in the application. When the choice—"Is he or is he not a researcher?"—must be made for administrative, bureaucratic, or other purposes, adjudication is not easy. Nor is the issue readily understood, especially by those concerned with total health delivery (or delivery of primary clinical care) or those from strictly academic disciplines. Thus, at this juncture, the judge must add his specific position paper. Is a case reporter whose paper refines psychoanalytic theory, ever increasing its beauty of structure and power of explication, to be considered a researcher? What about the professional identities of the psychiatrists who drug rats or interview normal populations? And should this work clarify methods of coping important for rehabilitation, or define mechanisms of drug action relevant to new therapies? Is such work, then, properly psychiatric research? The application of the GAP definition stretches with an unpredictable elasticity.

Further complexities arise. At the time of publication of the GAP report (1967), the phrase "medical scientist, trained in clinical psychiatry" meant a psychiatrist with an internship. Training of psychiatrists now need no longer include an internship, thus very probably deemphasizing the need for clinical-medical skills.

The point is that our definition changes from group to group and according to the needs of the role under consideration or the position to be filled. Perhaps the most comprehensive criterion identifying the researcher is the pattern of sustained activities with investigative intent that leads to a genuine and communicable grasp of a topic area. To treat one patient is not to do research, though ideas generated from the single case or, more frequently, a series of cases, and synthesized by the therapist can instigate or influence psychiatric research.

What is crucial is a recognition of specific competencies, whatever the label we give to the worker in the field. This is true because problem-posing derived from theoretical insights is reciprocally linked to hypothesis-testing and to the hard (and frequently insulting) problems posed as a result of empirical inquiry. The establishment of validity, relevance, and reliability taps various competencies and frequently those of various individuals.

TYPES OF RESEARCH IN PSYCHIATRY

What constitutes research within the field of psychiatry? Ultimately, if not immediately, the question leads to "What is scientific methodology?" A third question, depending on the concept of scientific methodology accepted, is either "How can psychiatric research be conducted within the framework of scientific methodology?" or "Why need we be scientific?" And this properly means: Why worry about certain methods? We noted that choice of investigative method is relevant to the classification of the psychiatrist's activities as research and that the word "method" becomes elusive as we observe its many interpretations.

During the past two decades there have been numerous publications dealing with the scientific nature of the behavioral and social sciences in general and psychology and psychoanalysis in particular. The publications range from the purely philosophical (Feigl and Scriven, 1956; Feigl, Scriven, and Maxwell, 1958; Kaplan, 1964; Sherwood, 1969; and Hook, 1958) to the how-to-do-it books (Jahoda, Deutsch, and Cook, 1951; Levitt, 1961; Chassen, 1967). In the middle are publications that address themselves to a variety of topics. We find comprehensive reports, such as the GAP (1967a, 1967b) reports or Koch's (1959–1963) *Psychology: A Study of a Science*, describing the theory and practice of social and behavioral scientists. The three-volume series, *Research in Psychotherapy*, edited by Rubinstein and Parloff (1959), Strupp and Luborsky (1962), and Schlien (1968), addresses itself specifically to psychotherapy research, its nature, weakness, and potential promise. Rapaport (1959) and Pumpian-Mindlin (1952) discuss the scientific state of the theory in psychoanalysis, attempting to differentiate the experimentally documented from speculation. Grinker (1953) has done the same for the field of psychosomatics. In addition, symposia have been organized primarily in order to discuss the nature of psychiatric research (Brosin, 1959; Heseltine, 1969). The objectivity of research data has been questioned further through studies devoted to the examination of the impact that the investigator has on the nature of the data he collects in the behavioral and social sciences (Rosenthal, 1966; Friedman, 1967).

A concern for the correct method can be a strength as well as a hindrance. That pattern of overconcern called "scientism" can surely produce mediocre results. Bakan (1967, p. xiv) remarked: "Good research into the unknown cannot be well designed. . . . One has to allow the investigator to be guided by the experiences of the investigation." Methodology, statistics, computers, and the perfectly designed experiment are meaningful when relevant questions are asked. The perennial question in our field—How much of psychiatry is art and how much is science? —often leads to a head-on collision between the humanist and the experimenter-scientist.

The personal preferences of the researcher are significant determinants of the methodological posture that will be adopted (Freedman, 1968). The research project grows from genetic and environmental conditioning of the researcher—his mental capacity, his education and reference groups, and his value orientation, if not his phobias. The volume of collected GAP reports, *Clinical Psychiatry* (1967a), includes an extensive discussion of the effect of value orientations (see below) on the formulation and mode of exploration of research hypotheses. Elegance, confidence, and utility become operant values (Sargent, 1956). A researcher may decide, for example, that to be valid, conclusions must be quantitative and replicable; confidence, here, is his first priority. Another investigator, giving priority to the meaning and explanatory power of a theory (elegance), will discard psychological data that have been translated into statistical formulas, a process he views as utilizing part observations that are meaningless when separated from the individual's total patterns. We must add that bridging kinds of research, for example, Grinker's on anxiety, are also in evidence; these rest on the belief that one type of methodology should act as complementary to another when utilized within the context of the same investigation. The risk lies in the compromises that must be made for the sake of the merger.

Divisions of this nature exist in every field. Dissertations on the philosophy of science may remind the "hard" scientist of the subjectivity of his data or of unproven assumptions underlying his experimental hypotheses (and we confess a preference that Whitehead, if not Popper, should be a part of the training of any such scientists). The nature of the experiment and its deficiencies can usually be recognized by the researcher or his critic. What is important for each is the acknowledgment that each has made a deliberate (or overdetermined) choice and that evaluations of procedures can be based on informed critiques, as well as value judgments. What is troublesome is the investigator who disregards data derived from methods with which he is unfamiliar, because of ignorance or fear of them, or one who is unwilling to accept the intrinsic limits to his preferred approach, systematic or unsystematic. Again, some good practices now lacking—such as a respect for scholarship and for the history of a problem—might be helpful.

QUALIFICATIONS OF THE RESEARCHER

The interrelationship between the researcher as a psychological being, his values, and his actual research has been the springboard for the request to develop researchers free of restrictive biases. From this perspective, the mental health of the investigator as well as his level of knowledge of human behavior must be secured and protected for the sake of the field of psychiatry.

The character and the ability of the investigator himself as the prime mover of the research is a significant factor in funding every project and program or recruiting for positions. For many years the emotional stability of the investigator has been assessed informally within the closed sessions of the grant-givers or the university departmental meetings. Yet the criteria for the evaluation of the applicant's emotional health and its relationship to his capacity to work have been ill defined. They remain ill defined, but today we do have literature presenting a rationale for utilizing emotional characteristics as criteria for potential research productivity. If it were possible to obtain some clear guidelines for measuring a researcher's potential qualitative and quantitative productivity, we could utilize them not only as selection criteria but also in developing training and research programs.

Within clinical psychiatry the accuracy of the investigator's perceptions clearly involves emotional factors. The question is not whether it does (the psychology of observer bias dates back to the early nineteenth century) but when, where, and how the perception is distorted. However, the trend in the literature concerned with the investigator's psychological constitution seems to imply not only that there are agreed-on criteria for mental health, but also that sound mental health is the prime predictor of the reliability and value of the man's research project. We shall consider the argument and present our evaluation.

STRESSES OF THE CAREER

The dialogue on the emotional life of the researcher has been opened by Kubie (1953a, 1953b, 1954, 1959). His assertations have led to a focusing on those issues that were never before recognized as issues within literature concerning research careers. He describes the stresses of being a career researcher in psychiatry. These stresses fall into two categories: (1) those induced by the nature of the work; (2) the additional stresses created by a host of deficiencies in our present methods of orienting, funding, and scheduling the researcher for his work.

Engaging in research taxes one's psychological stability. According to Kubie (1959), "In subtle ways scientific research makes demands for a flexibility and for the ability to make quick shifts of roles which are greater than the demands which are made by any other field of human activity. . . ." Kubie (Brosin, 1959, p. 216) describes the complex series of role changes the scientific investigator must make as the investigation proceeds from one state to the next:

Scientific investigation proceeds by a series of steps, but not in a fixed or rigid sequence. Each step casts the scientist in a different role which makes different demands on him. In one he must be an observer, who attempts to see clearly, and not through eyes which are colored by his own biases. In another he constructs theoretical hypotheses concerning the interactions among the items he observes. In still a third, he makes experiments in which he takes some fragment of what he has observed in nature and reproduces simultaneous variables. Thereupon he must observe this creature of his own making with an objectivity which approximates the objectivity with which he made his original observations. He must record these observations without bias, make fresh theoretical hypotheses concerning them, and retest them by fresh experiments. This new set of experiments again becomes the facsimile of a fragment of nature to be observed, theorized about and subjected to a new order of tests. Thereupon he must once again become an observer, once again climbing out of his own skin, divesting himself of his pride of authorship, looking upon his mental progeny as though they were somebody else's children to be observed and tested dispassionately.

The investigator is called on to be objective about data in which he has vested interests. Kubie (Brosin, 1959, p. 217) states that at different times the investigator must act as "receiver, recorder and responder." "In many ways the investigative process calls upon qualities of human nature which run the gamut of all psychopathological mechanisms." A personal maturity comprised of an understanding of oneself and others is thus held to be a special necessity for the investigator in the behavioral sciences. For him these mental gymnastics and multiple roles become still more complex, and perceptions become still more vulnerable to distortion. The dire alternatives include such behavior as the scientist hiding behind the rules of science while being pushed by his neurotic problems.

One of Kubie's pleas and an effect of his own communications is to inform the potential researcher of both the psychological and financial complications for himself and for his family that will result from his career choice. Generally the researcher's income will be lower than that of his colleague in private practice. Longer hours at his desk, bench, or clinic do not necessarily mean a higher income. He is dependent on grants with little in the way of long-term security.

Problems of scheduling and phasing therapeutic, research, and teaching activities further complicate the picture. The researcher's services are in demand from many quarters: Anthony (1969) writes of dividing his time into four-thirds! The psychiatric investigator must treat patients in order to learn about human behavior, but for research purposes his objectivity about data from his own patients is poor.[1] Even teaching can be too heady wine, according to Kubie, as the teacher begins to believe what he parrots and goes unchallenged by students who know less of his subject than he does.

TRAINING RECOMMENDATIONS

Kubie presents dollars and cents reasons for subsidizing lengthy training periods and subsequent research activities with funds adequate to the researcher's emotional and physical health. How should the training program be designed? Kubie answers (Brosin, 1959, p. 222):

There should be a minimal age of consent before which it becomes statutory rape to force anyone to teach psychiatry, or to do research in psychiatry. Perhaps about 45 or 50 years would be a proper age to set for this. Before that the young psychiatrist should be generously subsidized: first for the minimum of five years of half-time analytic training, followed second by ten years of

---

[1] We recognize the problem of bringing clinical investigations to the same level of energy and expertise that characterizes other investigative areas in research psychiatry. This topic is being informally discussed among the leaders in the field of psychiatry. There are many complex issues for which no systematic agenda has yet been detailed. For example, there is unresolved debate on the question of the compatibility of doing one's own research while functioning as a therapist. The benefits of the therapeutic alliance for collection of data are lauded by Stein (1970). In *The Psychological World of the Teenager* (Offer, 1969), the research alliance between subject and psychiatrist is described in order to draw attention to the fine line existing between data collection and therapy even with nonpatient populations. The subject who is not a patient may implicitly expect help from an interviewing psychiatrist; similarly, the psychiatrist is tempted to slip into a therapeutic stance. A research alliance is evolved to attain maximum cooperation from the subject and to keep an explicit awareness of the effects of the relationship between the psychiatrist and the subject on the data collected. Maslow (1966), a psychologist long involved in clinical investigations, argues for the special advantages for research of caring objectivity. This can be obtained by the scientist who does not need to distort his perceptions to accommodate personal neuroses. Maslow contends that for certain purposes this type of knowing is superior to so-called detached observations. Kubie appears to argue that therapeutic experience is necessary for psychiatric research and for the researcher, but that the role of therapist does not readily allow for the level of objectivity required for investigations.

half-time analytic practice. The other half-time of these 15 years would be spent in laboratories of experimental psychology, neurophysiology, neuro-biochemistry, mathematics and modern electronic neurophysiology. This will give us a new generation of mature clinicians and mature experimentalists.

The trained researcher should be given frequent periods of a year or two when he can draw away from both therapy and teaching in order to regain objectivity. Men over sixty ought to be awarded fellowships for elder statesmen; their maturity in psychiatric research must be valued to the highest degree.

COMMENTARY

Research and certainly research in psychiatry is vulnerable to distortions from restrictive perception, but at least three interrelated questions remain: (1) What constitutes maturity for engaging in research? (2) When should the psychiatrist assume responsibility for engaging in research? (3) How do we operate in a situation that will be financially less than ideal?

Maturity

Maturity, like normality and health, is a value-laden concept. In *Webster's New Twentieth Century Dictionary* (1956), maturity is defined as (1) "being fully developed"; and (2) "being perfect, complete or ready." To say, therefore, that a researcher should be mature could mean that he should be "fully developed and ready for the tasks—a perfect specimen." In *Normality* (Offer and Sabshin, 1966), four different perspectives of normality are described: Normality as health; as utopia; as average; and as process. Each perspective connotes a different value orientation. We suspect that the concept of maturity can be similarly treated.

To state that a researcher should be flexible, free of neurotic conflicts, imbued with self-knowledge, and trained in clinical experiences and theoretical knowledge is not enough. First we must identify our perspective on maturity and then, stipulate how each criterion comprising this emotional and intellectual maturity will be defined, achieved, or rated.

We do not know even how relevant the characteristics selected by Kubie are to quality or quantity of research output. As Kubie also states, there have been too few studies on researchers and research productivity to isolate particular recurrent characteristics appearing in their work histories. We proceed on intuitions. Nor have we integrated our own experiences with those of nonpsychiatric colleagues, in areas where research departments are more firmly developed than in psychiatry (see also Hamburg, 1970). It is extraordinarily difficult to program for the researcher's emotional and educational maturity when we are neither sure of what is best for the researcher's work or what are the best ways of getting there.

Age of Onset of Research Responsibility

We suspect from our pragmatic experience that a man should begin to develop the habit of learning and using research methods when he has the energy and momentum of youth. The topic area in which the psychiatrist does his research constitutes an important aspect of this issue. For areas other than the neurotic process (for example, certain aspects of psychosis, psychopharmacology, social and community psychiatry, or group process) the field requires resource persons; for the execution of their work the order of maturity required of an analyst is not directly relevant. If such research takes place in the kind of community we have talked about, then in the design and the execution of the research, the collaboration and critique of mature scholars of human development should be available and utilized by the young investigator. This kind of environment in which the researcher would be taught, guided, and encouraged is what the very best of academic settings are seeking to provide. We doubt that Kubie really would impose abstention and rather believe that his major concerns are with the social as well as psychological pressures that can distort the potentially broad development he envisages for the bright young man.

Redlich's (1961) words constitute a persuasive dissent on the issue of lengthy training periods:

If residents engage in actual research, they should be encouraged to start whenever they are ready for the particular problem. Admittedly, there are problems that should be left for old age, but there are a great many which can be tackled by the young investigator before "maturity" is reached, either alone or in collaboration with an experienced investigator. The notion of clinical maturity has intimidated many young and fresh minds. This has been particularly true in clinical psychiatry and most of all in psychoanalytic research. The complex subject matter and long training, with a powerful tendency to indoctrinate and even to infantilize the student, have had quite a devastating effect in sterilizing originality and productivity. Sigmund Freud pointed to the brilliance and originality of the child's mind which is steadily and progressively blunted as he grows older. . . . Mastery and insight in our field come later than in the physical sciences. Yet in my opinion this is not a sufficient reason to bar young minds from research. I feel certain that the middle-aged men who made important discoveries were not discouraged from thinking and experimenting in their young days. Even in psychiatry, experience is not a satisfactory substitute for originality and creativity. I am really quite concerned that our system of indoctrination is apt to blunt and destroy these traits in our gifted trainees.

Ignorance and neuroses may not be the single most important enemies of significant results. Nor may ignorance and neuroses be cured most effectively by longer training periods. Redlich has cited indoctrination and a blunting of creativity as deleterious effects of long training peri-

ods. And even with neuroses-guided projects, the results are not necessarily scientifically contaminated. Often the original goals of the
research project recede into the background as an unlooked-for by-
product becomes the most important and lasting aspect of the work.
Accordingly, to understand the influence of the investigator's personality
on the value of his research will involve far more probing and empirical
research than has been done to date.

Finances

The question of economic feasibility ignores the conceptual purity of
a training situation designated as ideal. Glaser (1969, p. 1120) summarized the situation for medical research financing: "In many ways, the
period from 1957 to about 1965 was a delightful idyll for almost everyone. . . . Whether or not some of our numbers were wise enough to re-
alize that, like the grasshoppers, we would have to face the horrors of
winter, I do not know." We are no longer asking that the basic researcher be better subsidized but arguing for the necessity of subsidizing him at all (Lustman, 1969).

The issues today are whether institutions or individuals, or both, are
to be funded, and how. The worrisome problem is that government resources are generally made available on the basis of gimmickry, novelty,
or the promise of a solution for a specific disease. The whole process of
sustaining research and education is not regarded as a central issue but
is treated as a by-product. For example, government is authorized to
fund training of psychiatrists because there is a shortage of psychiatric
manpower. The government hardly considers sustaining the environment the country needs to sustain the product. In order to have quality
manpower, a quality research operation is needed. The problem of sustaining this financially has never been directly examined. There are as-
tonishingly few funds available for establishing and maintaining a high
level of quality and expertise in psychiatric research. There are relatively few, if any, endowed centers of excellence for training in psychiatric research. It is hard to raise money specifically for psychiatry; it is
easier to ask for funding for a specific problem area, such as delin-
quency or drug addiction. It becomes clear, then, that whatever the
priorities are, if we want to develop a cadre of research psychiatrists we
will need to give them more than minimum support.

RESEARCH ATMOSPHERES

Departmental and national policies create an emotional ambiance that
influences the investigator's contentment and effectiveness. Our knowledge of appropriate motivational stimuli for cultivating research remains

incomplete, especially for the task of fitting the man to the optimal environment. The researcher cannot be programmed, but the rewards for engaging in research can be increased. The most reasonable thrust for a department in which research is to be valued is to provide an atmosphere in which the goal of improving the field of psychiatry is actively sought.

THE ROLE OF THE CHAIRMAN

There are several articles (Hamburg, 1961; Redlich, 1961; Reiser, 1961; and GAP, 1967b) containing suggestions of how a research atmosphere might be established. Throughout these discussions, the role of the chairman of the department in activating psychiatrists to do research is seen to be of crucial importance. The chairman must balance the different interests, doctrines, and personalities within his department. His own life example is thought to be an indication of the extent to which he will encourage research generativity; this view is adopted especially by chairmen whose research credentials are bona fide. Respect for research, though, cannot be equated with one's personal research achievements. Whether he serves concomitantly as a research model, the chairman of a department can encourage research activities by subtly rewarding residents and graduates "for a searching attitude and creative effort," and letting "them know that good clinical work is all right but not enough" (Redlich, 1961).

The perils of chairmanship are being brought to our attention, if only because of the significant number of vacancies. Redlich (1961), who believes the researcher need not be mature before commencing his professional activities, does believe that the chairman of an academic department should be a man of maturity. The chairman is subject to numerous distractions as are all administrators today. In a portrayal of the half-life and hard times of a medical school dean, Glaser (1969) wrote: "Service on NIH study sections and panels and innumerable conferences, symposia and other gatherings were organized to insure faculty members and deans the air mileage necessary for membership in the 100,000 mile club." (On this subject, see also Grinker, 1967.)

A chairman can and should actively support psychiatric research within his department. He can do this by creating less administrative bureaucracy and by interfering as little as possible with the work itself. The chairman controls the array of visiting lecturers and professors, and whom he selects will again have a decided impact on the direction of his department. It seems to us important to stress at this juncture that not every department need be a center of excellence in training psychiatric researchers. A variability of focus in different departments is healthy for the future of psychiatry as a healing art as well as a science. What is important is that the chairman be receptive to research and encourage it when possible in his students and younger colleagues. A center of excellence in psychiatric research cannot, on the other hand, be divorced from a high value placed on clinical acumen and competence.

Researchers who are not chairmen of departments relish repeating Ostwalt's observations which stem back to the last century. Ostwalt thought it strange that men of talent in American universities were so frequently rewarded by being promoted to positions in which they could not go on doing the work in which they had shown special skill (Gregg, 1941). Search committees looking for the ideal chairman often opt for the young and promising investigator without recognizing the wide range of responsibilities a chairman has. The man's clinical competence and teaching abilities become, at best, secondary considerations. Medical schools, hospitals, and clinics should first carefully assess the kind of department they have, or the one they are willing to underwrite and support in the future, before searching for a chairman, since there is no ideal chairman, only a specific kind of chairman to fill a particular needed position.

TRAINING PROGRAMS AND RESEARCH

There is no time for training in research during a psychiatric residency because within this relatively short period clinical skills must be learned and practiced and service demands must be met: So goes an argument frequently heard. Further, writes Reiser (1961), the therapist in his immediate work with patients "must orient himself to a theoretical framework in a way that does not optimally allow for much agnosticism." The clinical researcher must be able to work adeptly within both the therapeutic and the research attitudes; this is particularly difficult when neither attitude is set. Reiser advocates the conversion of a therapist into a researcher in a postresidency fellowship period. Is there indeed time for training in research procedures and techniques during the psychiatric residency? It is a question of priorities, which involve one's view of the role of research activities within the profession. Ours have already been stated.

Within Reiser's plan, the research fellow can and should undertake a project of his own but—preferably—with the aid of a supervisor who is mature, experienced, intuitive, and intelligent. Whether the researcher is a fellow, resident, or senior investigator, a competent supervisor (he might also be called a consultant) can correct for errors and diminish the anxieties that derive from being a member of a group that is not in the midstream of clinical psychiatry. Ideally the consultant will teach and aid, stimulating rather than stifling initiative and imagination. With skill the experienced advisor can extend the investigator's awareness of the community of research and therapy, both past and present.

Training programs adopt different stances as to when, what, and whether research activities should be a part of residency programs. Several training programs provide a once-weekly research seminar in which systematic instruction in research principles is presented. Hamburg (1961) stated that residents became interested in research after attending such a course; the course he organized was given in the first year of the residency in order to capture the first-year resident's enthusiasm, to develop a "thorough-going research outlook" and to signify to the resident "from the outset that research is highly valued in this institution." Hamburg believed that this kind of course could be a valuable instrument for increasing residents' understanding of and interest in research.

In other institutions—and several years later at the same one Hamburg had described—training programs include a requirement for each resident to undertake a research project during his residency; residents are introduced to research procedures and techniques through their direct participation in an area they have chosen. A research supervisory staff is available to aid the residents. Factors inherent in changing the program from time to time may be as important as the actual program. The staff and students who initiate the change are enthusiastic about their innovation and devote more time and energy to making it work. The requirement itself, to investigate a research problem of relevance to psychiatry, is a good learning experience for some and a necessary evil for others. The resident who was afraid to try his own skills in research may be captivated as he is given that extra push, whereas, the resident with no interest in research is being initiated into research by a slightly different bitter pill than once-a-week sojourns into learning combined with ceiling-staring. In other programs, the arousal and sustaining of research interest is fostered by intimate contact of the trainee in his routine assignments with clinical teachers who are also investigators. Finally, we face a new phenomenon wherein potential researchers—or administrators—are seduced to training programs with little emphasis on clinical competencies or even standards of excellence for clinical prosity and skill.

MODELS, RECOGNITION, AND PUBLISHING

According to Zuckerman's interviews of Nobel prize laureates, a research atmosphere is most likely to be created by the presence of men who have made important scientific contributions in their fields (Merton, 1968). Throughout the literature proclaiming the insufficiency of researchers in psychiatry, major blame is laid on the lack of models of psychiatrists in research. Often the research seminar is led by a psychologist with whom it is difficult for the resident in psychiatry to identify. Within psychiatric departments, the presence of a renowned investigator can add an aura of excitement to the position of a researcher. The illustrious personage inspires his students and colleagues through a seminar, a lecture, a cold sneer, or a warm handclasp, depending on his style. His way of life will act as a model for other psychiatrists who might wish to develop their own research propensities as well as their senior investigator's research projects.

In addition to working with, near, or under a man of accomplishment, recognition tends to reinforce the scientist's efforts to be productive and to maintain high levels of qualitative excellence. On the basis of a study of research scientists, B. Glaser (1964) concludes that scientists who receive lower levels of recognition than the Nobel prize or acceptance into the French Academy of Science can be adequately stimulated to pursue their chosen careers. Acceptance by one's colleagues,[2] a moderate degree of fame in one's own field, space to work, or salaries and positions awarded, can supply the motivation for a departmental star, even though he be not a superstar.

One of the major methods of communication between the psychiatric researcher and his colleagues throughout the field is through his published reports. Publishing also serves as a clarifying measure for the researcher systematizing his own ideas. Thoughts delivered at random do not require the integration or commitment the written word demands. After publication, the researcher's investigation will no longer be his own private affair, but can be courted and violated by others, according, of course, to the purest of scientific traditions. As his apostrophes and hypotheses become available for alien scrutiny, the researcher will be professionally solicited, damned, or, to some the worst of all fates, ignored.

ATTITUDES WITHIN PSYCHIATRY
TOWARD RESEARCH

Despite the need for research in our field and for some psychiatrists to be doing it, psychiatrists are not unanimous in condoning or supporting research activities. C. P. Snow's (1961) town and gown controversy is observed in a split between the clinician-practitioner and the researcher-academician. The service-research dichotomy has always existed. It recently received a strong impetus from the community mental health movement. The latter, supported by the granting agencies (notably the National Institute of Mental Health), has tended to opt for service, thereby reinforcing the researcher's image of himself as an underdog fighting for his rights. From reading papers on the need for research,

[2] Colleagueship, and working as a member of a research team, can cause new problems for the identity of the researcher. Multidisciplinary research in psychiatry is particularly difficult to undertake. (For an extensive discussion of the pros and cons of multidisciplinary research, see Shakow, Chapter 12.)

one gets the uncomfortable feeling that to incite the psychiatrist to do research is equivalent to asking him to swim against a tidal wave of hostility engendered by practitioners of every ilk. Clinical therapists are the primary professional models available for the young psychiatrist and are the friends or acquaintances of the researcher in psychiatry. Who are these "enemies" within our ranks?

ON THE HOSTILE CLINICIAN

Can we label the true clinician as the psychiatrist who really cares? Or is he the man who can treat but has a writing bloc? Perhaps he looks enviously at his colleague whom he thinks writes without pain (though also without much content, he may silently add). Or, says he, I am not a seeker after glory. I do not suffer from the need to become famous which drives the prolific writer. Seeking neither an overabundance of money nor fame, he wants only the humble pleasures of soothing the sick and being the lord of his office or hospital ward. To see a patient improve is to declare that he is a happy man, professionally. But, why the hostility toward the researcher? Or is this, too, a figment of the researcher's pathological conceptualizations?

This topic is a favorite of authors who discourse on the problems of encouraging psychiatrists to engage in research (Redlich and Brody, 1955; Hamburg, 1961; Redlich, 1961; Reiser, 1961; Anthony, 1969; and Stein, 1970). The majority of our teachers of psychiatry are practitioners who regard research as a good choice for the resident who has a low level of clinical ability. Anthony (1969, p. 387) attributes the hostility of clinicians to a stereotyped view of psychiatric research:

There are many clinicians whose experience of the NIMH granting propensities have been such that they are ready to condemn all contemporary research in the field of psychiatry (especially that done by non-psychiatrists) as trivial and given to "methodology." They will warn residents against premature engagement in research before they have acquired adequate clinical background and experience. They will sneer at findings that are no more than elephantine gleams into the obvious. . . . It is also well for the researcher to remember (and, in fact, fatal for him to forget) that for many of them, there is something vaguely threatening about research. They may see it as intrusive, intimidating, dehumanizing and fixated on numbers. Perhaps, its greatest crime in their eyes lies in approaching children in a non-therapeutic way and perceiving them in terms of means and not ends. This would imply an absence of alliances, of working through and of respect for defenses. The patient loses his identity and becomes a figure among other figures and, following the statistical analysis, he is no longer considered to exist. There is no responsibility for him and no concern about his future except in follow-up studies.

Mildly, the attack is against the researcher's methods, purposes, humanity, and integrity. As if this were not enough, researchers are described even in proresearch papers as disagreeable characters. A certain amount of their personal abrasiveness is attributed to their sensitivity to the clinician's hostility as well as to his economic status.

Redlich and Brody (1955) hypothesized that "unconscious guilt and anxiety are more prominent in investigators of human behavior than in other fields." The investigator becomes especially uneasy when working in a group where it becomes necessary for him to discuss interactions between himself and a subject who may have been unaware of the nature of the research project: "clandestine observation takes place in most test situations and it occurs in all of those in which subjects are elected as controls without their knowledge." Redlich and Brody suggest that these anxieties account for various defensive maneuvers observed in investigators of human behavior. We could add a familiar issue: the behavioral scientist in psychiatric (or mental health) research who revenges his mother's disappointment in the son who was not a "real doctor" by overdetermined attacks on psychiatry and overdetermined ignorance of the psychiatric clinician, the clinical tradition, and—worst of all—ignorance of the patient. Others, so motivated, establish liaison with all those forces that consider man's biological heritage as irrelevant or politically and ethically beneath the concern of enlightened humanists.

The research psychiatrist is a doctor. His training is to help people. On the balance, the training toward care of the patient generally is, and certainly should be, in the forefront of all clinical research in psychiatry. It is rare, in fact, that the psychiatric researcher corroborates the practitioner's fears, utilizing a subject for research to the detriment of the individual's health. Healthy interaction between the two might heighten respect for the focus of each and diminish the vision that each has an exclusive entree to truth.

THE NOBLE RESEARCHER SYNDROME

The researcher is altruistically devoting himself to the progress of mankind, to the amelioration of human suffering. Mankind may symbolize to the researcher his mother who had a hysterical backache every morning and/or a vast sea of unknown faces throughout time looking to him for help. Even without providing the definitive answers, he knows that his life is being spent to help others on a larger scale than the role of therapist alone would provide. He has dedicated himself to the improvement of his field, a field that is service oriented.

Of course, the noble researcher syndrome exists also as a defense compensating for a minority group status and a comparative lack of financial rewards. A life with less money is equated to the choice of the nobler path.

Maslow (1966) refers to the "quiet amusement" of the great investigator. Is quiet amusement the other side of frustration? The researcher must continuously reinforce his own beliefs about the particular value of his project. A belief in one's self and one's judgment are necessary when long-term projects lose their original attractiveness as well as their short-term grants. The ability of the researcher to persevere will be tested as he engages in the more arduous aspects of the day-to-day collecting, organizing, and analyzing of data. The clinician-critic can attack the method and in a once-over-lightly presentation disqualify the data. It is harder to evaluate the positive aspects of the data within the context of the methodology and to determine their relevance to a larger framework.

When society denies the researcher's relevance and threatens not to recognize his nobility, it is well to remember that there are two schools of thought. (Alvin Weinberg [1970] recalled these in his "In Defense of Science.") Newton characterized science as an intellectual exercise, a part of high culture. Bacon believed that science is justified because from science "we learn how to make two blades of grass grow where one grew before." As we all will do, we can combine the two statements saying that we are dedicating our lives to our own pleasure and to the benefit of mankind.

Anthony's (1969) answer is for the two roles, that of the researcher and the clinician, to be assumed by the same person. Then we will have true communication between two attitudes that should complement each other.

EMERGENCE OF CLINICAL RESEARCH

In no area of psychiatry is the integration of the clinician's experience and abilities with the researcher's tools and methods more important than in the area of clinical research. A true integration of the two would do much to enhance developments in the field of clinical psychiatry. During the past decade it has become increasingly apparent that in order to tackle some of these clinical problems (for example, schizophrenia), we need to develop a new breed of research psychiatrists. Investigators have all too often shunned participating actively in clinical work, management, and administration. Likewise the clinician-therapist isolated himself from the academic community (within, as well as outside, of the university) and helped develop the wall that separated clinical practice from psychiatric research. Only on rare occasions has the same person functioned equally well in the dual role of the clinician and the researcher. Roy R. Grinker, Sr., is an example of a person who successfully combined both roles.

The inevitable requirements of bureaucracy have tended to mask a cardinal fact of life: that advancing clinical research is a difficult task indeed. It is extraordinarily rare that a successful research program administrator can be a man fundamentally ignorant of research aims, if not methodologies and problems. Many clinical investigations require that the primary investigator, after acquiring his basic skills, tools, and hypotheses, also be the principal administrative implementor of them. Innovation in clinically relevant programs frequently requires both administrative and research responsibilities in the body of one person. This does not make for clearly distinguished functions, but is a pragmatic necessity for much of psychiatric research as we have known it. At different phases of evolving research, a different blend or emphasis of research, clinical, and administrative skills is required.

Learning the techniques of investigators is that part of science that can be taught. Learning how to be a creative scientist is an art. It cannot be taught. A clinician who undertakes to investigate is bound by the same traditions of science used by the more basic sciences. As clinical psychiatry joins the other disciplines in the behavioral and social sciences in expanding its knowledge boundaries, it will need men skilled in utilizing the appropriate methods of designing, undertaking, and concluding research projects.

CONCLUSION

We have described some aspects of the theories about and practices of the psychiatric researcher. Psychiatry should foster the researcher for its own health as a thriving clinical and scientific field, as a community where student, practitioner, patient, and investigator have intrinsic common stakes and bonds. The researcher is identified by his products, his view of himself, and the reviewer's perspectives. The decision as to who is best prepared to engage in psychiatric research is as perplexing as ever, but the issues of training a psychiatric researcher and of financing him are legitimate concerns for psychiatrists and administrators. Neither we nor the literature reviewed have yet come to sufficiently concrete explication of all these issues.

The researcher's qualifications, training, and performance are shaped by the programs and attitudes prevailing in the psychiatric department and the professional community setting in which he works. Positive reinforcements for engaging in research can be structured. These would not only be a boon to the field of psychiatry but would also make it easier for the individual investigator to experience the gratifications inherent in the doing of the work itself. The psychiatric researcher can gain encouragement through an identification with others who investigate the mechanisms of human behavior, with students and receptive clinical colleagues. His position becomes less tenuous as individual emotional reactivity is seen within the context of a more informed understanding of his place within his profession and the cognate areas of inquiry.

Keeping open a network of communication about the life of the researcher serves as a handle for creating and maintaining research heterodoxy. Whatever the researcher's products ultimately are, the valuing of inquiry and activities to that end can in toto take us one step further away from rigidity and dogmatism and one step further in the pursuit of the knowable.

REFERENCES

Anthony, E. J. 1969. "Research as an Academic Function of Child Psychiatry," *Archives of General Psychiatry* 21:385–391.
Bakan, D. 1967. *On Method: Toward a Reconstruction of Psychological Investigation.* San Francisco: Jossey-Bass.
Brosin, H. W., ed. 1959. *Lectures on Experimental Psychiatry.* Pittsburgh: University of Pittsburgh Press.
Chassen, J. B. 1967. *Research Design in Clinical Psychology and Psychiatry.* New York: Appleton-Century-Crofts.
Feigl, H., and Scriven, M., eds. 1956. *The Foundations of Science and the Concepts of Psychology and Psychoanalysis,* Minnesota Studies in the Philosophy of Science, vol. 1. Minneapolis: University of Minnesota Press.
Feigl, H., Scriven, M., and Maxwell, G., eds. 1958. *Concepts, Theories, and the Mind-Body Problem,* Minnesota Studies in the Philosophy of Science, vol. 2. Minneapolis: University of Minnesota Press.
Freedman, D. X. 1968. *Biological Psychiatry and the Community,* Dedication of the Connecticut Mental Health Center, October 1966. New Haven: Yale University Press.
Friedman, N. 1967. *The Social Nature of Psychological Research.* New York: Basic Books.
Group for the Advancement of Psychiatry. 1967a. "Psychiatric Research and the Assessment of Change." In *Clinical Psychiatry: Problems of Treatment Research and Prevention.* New York: Science House.
Group for the Advancement of Psychiatry. 1967b. *The Recruitment and Training of the Research Psychiatrist,* report no. 65.
Glaser, B. G. 1964. "Comparative Failure in Science," *Science* 143:1012–1013.
Glaser, R. J. 1969. "The Medical Deanship: Its Half-Life and Hard Times," *Journal of Medical Education* 44:1115–1126.
Gregg, A. 1941. *The Furtherance of Medical Research.* New Haven: Yale University Press.
Grinker, R. R., Sr. 1953. *Psychosomatic Research.* New York: Norton.
Grinker, R. R., Sr. 1967. "What Are Professors of Psychiatry For?" *Archives of General Psychiatry* 16:261.
Grinker, R. R., Sr. 1970. Personal communication.
Hamburg, D. A. 1961. "Recent Trends in Psychiatric Training," *Archives of General Psychiatry* 4:215–224.
Hamburg, D. A., ed. 1970. *Psychiatry as a Behavioral Science.* Englewood Cliffs, N.J.: Prentice-Hall.
Heseltine, G. F. D., ed. 1969. *Psychiatric Research in Our Changing World.* Amsterdam: Excerpta Medical Foundation.
Hook, S., ed. 1958. *Psychoanalysis, Scientific Method and Philosophy.* New York: New York University Press.

Jahoda, M., Deutsch, M., and Cook, S. W., eds. 1951. *Research Methods in Social Relations*, vol. 2. New York: Dryden Press.

Kaplan, A. 1964. *The Conduct of Inquiry*. San Francisco: Chandler.

Koch, S. 1959–1963. *Psychology: A Study of a Science*, 7 vols. New York: McGraw-Hill.

Kubie, L. S., 1953a. "The Problem of Maturity in Psychiatric Research," *Journal of Medical Education* 28: 11–27.

Kubie, L. S. 1953b. "Some Unsolved Problems of the Scientific Career," *American Scientist* 4:596–613.

Kubie, L. S. 1954. "Some Unsolved Problems of the Scientific Career," *American Scientist* 1:104–112.

Kubie, L. S. 1959. "Research in Psychiatry: Problems in Training, Experience and Strategy." In H. W. Brosin, ed., *Lectures on Experimental Psychiatry*. Pittsburgh: University of Pittsburgh Press. Pp. 213–226.

Levitt, E. E. 1961. *Clinical Research Design and Analysis in the Behavioral Sciences*. Springfield, Ill.: Charles C Thomas.

Lustman, S. L. 1969. "Mental Health Research and the University," *Archives of General Psychiatry* 21: 291–301.

Maslow, A. H. 1966. *The Psychology of Science*. Chicago: Regnery.

Merton, R. K. 1968. "The Matthew Effect in Science," *Science* 159:56–63.

Offer, D. 1969. *The Psychological World of the Teenager*. New York: Basic Books.

Offer, D., and Sabshin, M. 1966. *Normality: Theoretical and Clinical Concepts of Mental Health*. New York: Basic Books.

Pumpian-Mindlin, E., ed. 1952. *Psychoanalysis as Science*. New York: Basic Books.

Rapaport, D. 1959. "The Structure of Psychoanalytic Thought." In S. Koch, ed., *Psychology: A Study of a Science*. New York: McGraw-Hill. Vol. 3, pp. 55–184.

Redlich, F. C. 1961. "Research Atmospheres in Departments of Psychiatry," *Archives of General Psychiatry* 4: 225–236.

Redlich, F. C., and Brody, E. B. 1955. "Emotional Problems of Inter-disciplinary Research in Psychiatry," *Psychiatry* 18:233–239.

Reiser, M. F. 1961: "Research Training for Psychiatric Residents," *Archives of General Psychiatry* 4:237–246.

Rosenthal, R. 1966. *Experimenter Effects in Behavioral Research*. New York: Appleton-Century-Crofts.

Rubinstein, E. A., and Parloff, M. B., eds. 1959. *Research in Psychotherapy*. Vol. 1. Washington, D.C.: American Psychological Association.

Sargent, H. D. 1956. "The Psychotherapy Research Project of the Menninger Foundation II," *Bulletin of the Menninger Clinic* 20:226–233.

Schlien, J. M., ed. 1968. *Research in Psychotherapy*. Vol. 3. Washington, D.C.: American Psychological Association.

Sherwood, M. 1969. *The Logic of Explanation in Psychoanalysis*. New York: Academic Press.

Snow, C. P. 1961. *Science and Government*. Cambridge, Mass.: Harvard University Press.

Stein, M. 1970. "Psychiatrist's Role in Psychiatric Research," *Archives of General Psychiatry* 22(6):481–490.

Strupp, H. H., and Luborsky, L., eds. 1962. *Research in Psychotherapy*. Vol. 2. Washington, D.C.: American Psychological Association.

Webster's New Twentieth Century Dictionary, 2d ed. 1956. Cleveland and New York: World.

Weinberg, A. M. 1970. "In Defense of Science," *Science* 167:141–145.

JUDITH L. OFFER,
DANIEL OFFER,
DANIEL X. FREEDMAN

Appendix 1: Summaries of Research Undertaken by Roy R. Grinker, Sr.

As we groped for the idea of a Festschrift, we initially became curious enough about the man to review his psychiatric research and compulsive enough to attempt a relatively systematic précis and paraphrasing of them. Our selection was skewed to focus on what we thought was his major research in clinical psychiatry, not in neurology or psychoanalysis. A glance at Grinker's curriculum vitae makes it evident that this is a selection; some of his favorite papers and editorials, such as "Psychiatry Rides Madly in All Directions" (1964), are not included. Because our contributors have found these summaries a valuable guide to Grinker, as well as the topical areas of inquiry, they are included in this volume, and any resemblance to the original author—who has not edited these —is hopefully more than coincidental. The value of Grinker's writings and the import of his presence on the thinking and research in psychiatry is not calculable. The summaries may give some idea of the extent to which he has been a pioneer in his field. He is not retired and is busy organizing and participating in an extensive research program on schizophrenia.

UNITARY OR SYSTEMS THEORY
(1956–1970)

Grinker's research has continuously aimed toward encompassing the three major bases for human behavior: the biological, the psychological, and the social. During the early 1950's he brought together distinguished scientists from each of the three fields to attempt to build a "unified theory of human behavior." Their discussions—carried on from 1951 to 1956—led to the book which he edited (Grinker, 1967b). Why was Grinker interested in this particular theoretical point of view? (1) Systems theory offers an eclectic perspective, yet one that basically views psychiatry as part of an overall theoretical biology (von Bertalanffy, 1968). (2) In searching for a theoretical perspective that explains all behavior, systems theory seems to have the potential of encompassing the widest range of relevant data. (3) The frame of reference of the theory is developmental as well as static.

Grinker (1971) has recently summarized his views on systems theory. We have relied most heavily on this article.

DEFINITION OF A SYSTEM

A "system" is considered to be some whole form in structure or operation, concepts or functions, composed of united and integrated parts. As such, it has an extent in time and space and boundaries. A system has a past which is partly represented by its parts, for it develops or assembles from something preceding. It has a present, which is its existence as relatively stable or what might be called its resting form, and it has a future that is its functional potentialities. In space, its form, structure, and dimensions constitute a framework that is relatively stable and timeless, yet only relatively so, for its constituents change during time but considerably slower than the novel or more active functions of the system. To view the change of these functions through time, the frame or background may be artificially considered as stable.

ELABORATION OF THE DEFINITION

Though its parts are in continued activity in relation to one another and the whole, the function of a system in relation to other systems involves mostly its surface or boundaries and its openings. In other words, the function of parts of a system are projected to the surface of the whole. The question remains whether systems are theoretical conceptualizations or real, something outside of us in nature, which do not depend on the process of observation. Grinker discusses whether systems should be defined and demarcated by the focus of the observer; he notes that the observer may shift focus as he pleases, and, therefore, may use any number and sizes of systems. It can be said that the observer defines his system by selecting a definite number of variables. The experimenter can control these factors. He can make, at times, any variable take an arbitrary value. The state of a system is the set of numerical values its variables have at that instant, in the line of behavior specified by a succession of states and the intervals of time between them.

The commonsense view of a system, however, would include those aspects of behavior in which the variables have some naturalness of association. Experiments are possible to ascertain whether empirical data or naturalness of past experience can be verified. This can be done by manipulating certain variables and determining whether changes in them have any effect on the system. If they do then they may be considered significant to certain degrees. If they do not have an effect, they may be neglected.

The dividing line between organism and environment becomes partly conceptual and to that extent arbitrary. That is true of a blood vessel in living tissue as well as the relationship between people, their physical, social, and cultural environment. Though anatomically and physically there may be some distinction between two parts of a system, if we view the system functionally we may ignore the purely anatomical parts as irrelevant and the division into organism and environment becomes vague. A system is the entire complex of the organism and environment. Environment is composed of those variables whose changes affect the organism and which are changed by the organism's behavior. Thus, both the organism and environment are two parts of one system. The system always has an extended environment; extending this, one can always add a variable to the environmental parameter. Within the boundaries of a system, parts of the system can become focal and what are known as system functions become environment to the subsystem under observation.

Sociologists have a tendency to deal with current processes in time and to consider developmental transactions as less important, dealing therefore, with nonhierarchical, nongradient, current action. Parsons stated that there is a distinction between a frame of reference and a class of systems. Action is not concerned with the internal structure of processes of the organism; rather, action is concerned with the organism as a unit in a set of relationships and the other terms of that relationship, which he calls "situation." From this point of view, the system is a system of relationships in action; it is neither a physical organism nor an object of physical perception. But from other points of view, the foci or systems that are identified in a living field must be considered as being derived through evolution, differentiation, and growth from earlier and simpler forms and functions and that within these systems there are capacities for specializations and gradients. Sets of relationships among dimensions constitute a high level of generalization that can be more easily understood if the physical properties of its component parts and their origins and ontogenic properties are known.

Grinker suggested the following three overriding principles of the unified or system theory:

1. Within all systems and in all fields there exists a principle of stability called "homeostasis." The homeostatic principle applies only to stability, and additional theory is required to include change and growth. Homeostatic processes, or the trend toward stability, as the background of change or novelty, is best understood as being accomplished through transactional processes within ever-extending fields. The limits to the fields are based on our positions, our powers of observation, and also our theoretical point of view.

2. The second principle is the transactional. We use the term "transactional" as meaning a reciprocal relationship among all parts of the field and not simply an interaction that is an effect of one system or focus on another. It is a philosophical or theoretical attitude and yet also a system of analysis. If one makes observations on multiple systems as nearly simultaneously as possible, one can see a sequence of change among several systems that involve adjustive processes across boundaries. There is not simply a response to a stimulus, but a process occurring in all parts within the whole field. Any transactional study, of course, can be broken down into the interactional if observations are focused on two systems in order to see the isolated effect of one on the other.

3. The third principle encompasses the processes of communication of information, which vary from signals characteristic of biological systems to symbols characteristic of social systems.

SYSTEMS THEORY AND PSYCHIATRY

The field of psychiatry currently includes diverse areas, such as biodynamics and existentialism. In fact, psychiatry seems to involve a study of all human behavior and is as broad as life. Research in psychiatry has to encompass the behavior of man in his environment, environmental structure, genetics, and neural and physiological systems. Unfortunately, there has been no unified theory, so that controversy still exists between "reductionists," who hope to explain man's behavior exclusively on a physiochemical or cellular basis, and "extensionists" or "humanists," who focus entirely on society and culture. These are polarities leading to futile arguments. The biological or the social sciences alone cannot explain behavior, but psychiatry cannot progress without them.

If we consider the ontogenesis of human personality, health and illness can be viewed within a continuum as polarities indicating a wide range of possible integrations. Some are lethal to the self or others and some are deviant in the creative sense, but diversity is expectable. With an increase in macroscopic parameters there is an increase in individuality—more variables available, which result in a larger system.

The end result, designated as health, illness, creativity, or relative stability and instability, is determined by criteria set up in part by macrocultures and microcultures and societies, and by self-beliefs incorporated from the microculture of the nuclear family. These criteria, which establish judgments concerning the appropriateness of developmental processes, are values varying with cultures and changing over time. The values need to be included in any research on normality as significant variables.

From a practical standpoint the degree and type of integration in the process of development determines the proneness or susceptibility to stress responses in reaction to various conditions, though at present we have low powers of predictability of adult illness from the infantile neuroses. Nevertheless, diagnosis, prognosis, and choice of therapy are related to critical periods of development during which defenses and coping devices are built into specific life styles.

From the research point of view, we require a longitudinal field—the concept of ontogenesis. However, holistic or umbrella-like theories can only be utilized as connectors of several levels of subtheories into general theories. As abstractions they are far removed from processes at developmental levels. Observation, prediction of outcome, and post hoc reconstruction of significant events require different frames of reference.

Society and culture, in general, are invariable factors in that their influences affect every phase of maturation, development, and function. Parsons and Bales (1955) state that the family functions as a small social group in which the child's dependency needs may be temporarily gratified and from which he may be gradually emancipated. The principal stages in the development of personality, particularly on its emotional side, leave a residue that constitutes a stratification of the structure of the personality itself with reference to its own developmental history. As the child grows, he becomes differentiated into systems of action during separate phase patterns of socialization. Every stage of a child's development or socialization requires learning "which is never only a triggering off of hereditary patterns."

The component parts of the personality leading to normality become integrated into viable systems despite defects in genetic constitution, disturbed maturational patterns or developmental stages, or unstable environments, if not too extreme. Diverse activities within a definite pattern can coexist, indicating that psychological structure is dependent on the continuity of integrating processes.

Normality and illness are only polarities of a wide range of integrations; and without any strains, an unlikely hypothetical condition, there is only normality. When strained, the organismic systems respond according to the processes by which the many subvariances have become integrated. There are no new defenses or coping devices; they have already been built into the organism. Thus, the degree of health or illness in the stress responses reveals the quality and quantity of integration.

The stage or phases of development may be viewed from a variety of theoretical positions: psychosexual, epigenetic, behavioral, or learning theories. They all have in common the concept of primary undifferentiation, gradually passing through critical periods of differentiation that are age specific. All include concepts of process not only in differentiation, but in the phenomenon of dedifferentiation. For each phase there are specific scientific disciplines concerned with creating and storing knowledge, specific medical specialties for diagnosis and therapy, and specific psychotherapeutic strategies, as well as corresponding social institutions.

SYSTEMS THEORY AND TOPOGRAPHY
IN PSYCHOANALYSIS

1. The symbolic system has developed from the system of signs by an evolutionary jump-step resulting in preconscious and conscious process as distinctly human phenomena.

2. There is an ontogenetic system of phases of learning from body signs to visual imagery to primitive symbols to creative thinking but the flow of information among these phases persists in all directions throughout life.

3. There are flexible transactional operations among these parts so that all are involved in all forms of thinking.

4. All phases or parts of the symbolic system are in transactional relationship with reality and inner experiences.

5. A disintegration of optimum or effective relations among parts of the symbolic systems may lead to breaking off of transactions (repression) and thereby to distorted thinking and behavior or to temporary acceleration of creativity.

SYSTEMS THEORY AND MEDICAL EDUCATION

Medical education in the past has been based on a dualistic concept. More definitive knowledge about particular organs or systems of functions has increased specialization and isolation of professional groups. Medical education has become additive rather than integrated. A patient becomes a heart, kidney, or lung case, and medicine has continued to be disease oriented. What was missing, and, for that matter, still is, constitutes the psychological realism of emotions and thinking, frustration, disappointment, and hopelessness. Recently the psychological and social have been included in the study of living functions. All this has required multidisciplinary research groups, each component of which focuses on a part of the total. In practice, separate disciplines or departments have entered into cooperation, and group practice has flourished. We speak about this broader approach to problems of man in health and illness as biopsychosocial.

This has introduced complexities that have been denied and even their language spurned. This is still the current error in biomedical thinking and education, understandable because an adequate theoretical umbrella has been ignored and obscured by pressing practical considerations, actually to their detriment. It has taken general medicine a considerable time to begin to use von Bertalanffy's general systems theory. It is still not an accepted theoretical point of view in medicine, though it has become increasingly more relevant to fields such as molecular biology, ethology, physiology, and psychiatry.

Science, like organisms, despite its analytic methods, has a tendency toward the integration of its various disciplines into a whole governed by unitary abstract theory and unifying principles, from cell to society, what von Bertalanffy calls isomorphism. The trend of physical nature is toward increase in disorder and randomness while living systems organize, increase order, and maintain a steady state of negative entropy.

Control and regulation in open systems are achieved by means of communication, transmitting information within feedback arrangements, but involving many complicated reverberatory cycles rather than simple linear chains. The living organism is not only reactive to its environment but also actively searches for changing goals by which it differentiates, grows, learns, and evolves. This proposition contradicts the robot model of man who supposedly searches for needs and relaxation of tension and achieves homeostasis, because there is more to life than maintenance and survival. In fact, the human capacity for symbolic representations functions as a powerful force for change apart from biological evolution. New symbols are transmitted to other generations by means of language. Thus, symbols are the genes of culture.

The essential problems inherent in unitary thinking, include our habitual two-dimensional imagery. What is required is a three-dimensional spatial perception plus a recognition of temporal dimensions. Grinker (1971) has proposed the following transactional model that encompasses process and relationship in order to resolve this problem.

The model would assume the form of a cylinder open at both ends. Height represents levels of complexity. At the bottom extension could occur to the physical world, the top to the infinite cosmos. The depth of the cylinder represents the dimension of differentiation so that the cen-

ter is least differentiated and the surface is the point of maximal differentiation equated with growth and decay or at least change in the life
cycle.

The surface of the cylinder is comprised of three columns: the first
represents somatic variables such as physiological, biochemical, enzymatic, cardiovascular control, nervous systems, drives, and their regulatory and control systems; the second represents the psychological variables such as memory, perception, cognition, motor behavior, and their
controls; the third represents environmental factors that surround the organism stimulating growth and creating stresses and strains. These three
columns also represent the traditional subsystems of the field designated
as somatopsychosocial.

The model does not derive causal relationships between and within
columns, but assumes that a functional relationship does exist among
and within the columns in any direction giving rise to an infinite number of contingent variations depending both on heredity and experience
permitting individual and patterned behaviors with specific meanings. It
is important to stress that behavior is a term not only applicable to total
movement but is also descriptive of somatic and psychological behaviors
called mentation.

The relationships among the parts of subsystems occur at their inter-
faces in a reverberating cyclic manner, each acting on the other. Such
transactions can only be observed between two variables, keeping the
others in mind and holding them reasonably steady, at one particular
point in time. Comparisons cannot be made across temporal zones, that
is, heart rate pattern of today with poor mothering of twenty years ago.

The model permits movement in two directions: from trait to trait or
across systems, as well as across time in one system, permitting a longitudinal study of the changing nature of specific functions over time. It
must be recognized that the designated systems are concepts whose relationship is descriptive and not explanatory. Subsequent explanations
and attribution of cause of purpose in adaptation are permissible teleol-
ogy based, however, on inferences. These are then hypotheses to be
tested by means of appropriate designs.

If we agree, then, that the error of our medical education is the absence of an overarching theoretical system and the lack of recognition of
psychological and environmental factors, we must recognize that all
these defects may at least begin to be rectified by radically changing the
curriculum, not just by permitting electives, but by viewing phases of
the life cycle as they relate to health and illness by means of multidisciplinary approaches. This will ultimately require, even force, unitary
thinking. Obviously, resistance involves vested interest in departmental
or specialty boundaries, but these may be overcome by demonstrable
advantages to both teacher and student.

PSYCHOSOMATIC RESEARCH

In *Psychosomatic Research,* Grinker (1953) presents a theory of psychosomatic organization in man from infancy to maturation. He begins by reviewing historical and current psychosomatic theories. Grinker is critical of most psychosomatic conceptualizations because of their failure to define adequately patterns of relationship among somatic and psychological systems. Investigators frequently conduct inquiries into their own field of specialty while viewing other disciplines and their findings from afar.

Grinker believes the field must not be so severely fragmented. Psychosomatic medicine includes the processes within and between the many systems that effect the total unity. To fill the theoretical vacuum Grinker presents his own formulation of psychosomatic organization. He traces developmental patterns of reactivity within the living organism, defining processes of differentiation and integration. Psychosomatic disturbances are described as signs of failures in complex assemblies of transactions. Finally, Grinker presents an operational model to be utilized for empirical studies of anxiety. An interdisciplinary focus on anxiety and concomitant physiological, biological, and sociological variables would aid in understanding both healthy and disturbed psychosomatic organization and explicate theoretical formulations.

This summary emphasizes Grinker's original thoughts and theoretical position with respect to psychosomatic research and medicine. A summary of Grinker's views on systems theory and his empirical investigations of anxiety appear separately. Following is, in brief, Grinker's statement of the assumptions and hypotheses on which psychosomatic research may be based.

DIFFERENTIATION PROCESSES

The infant is born with an intrinsic visceral behavior pattern that can be measured as a baseline on which subsequent influences impinge. At birth infants may be grossly categorized as to their rapidity of reaction to the environment and the quantity of motor or secretory activity within such response. These differences are due to innate (hereditary) characteristics.

The infant responds to internal and external environmental stress through massive reaction patterns. As an undifferentiated whole, committed only to those structuralized functions that are species bound, the infant has his highest degree of potentiality, but also his highest degree of sensitivity. Stimuli impinging on him early lastingly affect all of him, no matter how much differentiation later occurs. At this stage, the response to stress is primarily global. The lack of system differentiations means that the stress will be handled by the total pattern of action and reaction, out of which elaboration of more complicated functioning will occur (Coghill, 1929).

In Grinker's opinion, psychosomatic problems originate during the period of differentiation from hereditary total response patterns to individual learned patterns and their integration into a new personal system. Any hypothesis concerned with psychosomatic pathology should deal with an intermediate stage of development between the undifferentiated and the fully integrated. It is this period that determines the formation of a healthy, sick, or potentially sick organism.

Stimuli acting on the undifferentiated organism or the organism in the process of differentiation emanate from the external or incorporated environment (mother, food, bacteria, medication, temperature variations, fluids or solids, quantities, timing, and so on) and act not only on the orifices but also on the visceral in-between. The mother-child relationship is incorporated within the organism's subsequent differentiation and integrative processes. The mother's communications are determined by her own personality, her ethnic tradition, her current cultural values, and her own problems in maintaining an integrative capacity in relation to her child.

Somatic differentiation influences the psychological after ego patterns have begun to evolve. Then delay, reality testing, synthesis of conflicting inner forces, and the like, as ego functions, become easier or more difficult depending on the somatic skills and pressures, some of which are constitutional, others stirred by external forces.

THE STRUCTURE AND FUNCTIONS OF THE MOUTH

Grinker discusses the structure and functions of the mouth in order to illustrate the development and functioning of psychosomatic organization. The primary psychosomatic integration that predisposes to health and/or disease is present at the early undifferentiated period when the infant receives its satisfactions, tests the world of reality, and reacts to others largely with its mouth. Furthermore, the mouth as a functioning organ at first enables the infant to obtain an awareness of the nature of the world, of what is inside or outside and how things and sensations are related.

Biological data on the anatomy of the mouth and face are reviewed and evolutionary processes examined. The mouth is seen to have changed from a passive intaking organ to one which is active, biting, and mobile. The development of teeth and facial expressiveness are correlated with environmental variables and emotional expressiveness. For example, early oral satisfactions or frustrations may give rise to soft or bitter expressions around the mouth.

Organs associated with the mouth in supplying substances for growth and the maintenance of activity and the motor or feeding patterns, in which the mouth is greatly concerned, are correlated through brain-stem centers. Their activities are augmented or inhibited by action of the cerebral cortex. Oral functions become intimately associated with all visceral motions, which are rerepresented in integration with emotions through the functions of the rhinencephalon or the central visceral brain.

The primitive oral sense seems to be a perceptory gestalt in the service of testing external objects and obtaining sensual satisfactions. The overt orality seen in experimentation with monkeys whose visceral brain has been extirpated, in the behavior of humans with destruction of the temporal lobe, and in observations of very young infants have supported this view.

PSYCHOLOGICAL IMPLICATIONS OF THE
ORAL FUNCTIONS

What are the psychological implications of the oral functions which are basic to the individual's development?

Mythology and legend have described the mouth not only as the passageway for food but also as the main pathway of the egress of the soul, as the path for impregnation, and as the means of devouring the needed, feared, loved, and hated mother and father. Psychoanalysis postulates little more in the mouth's triad of functions: maintenance of life, sensual satisfaction, and aggressive mastery of reality. However, major advances in modern times are the concepts of mouth functions in determining the structure and stability of the personality. Anthropological, observational, and psychoanalytical methods have been directed toward determining the interrelationship between child and mother, mouth and nipple.

Whereas most modern psychoanalysts have tended to consider the physiological operations of the mouth as literal rudiments of psychic functions, Grinker believes that a transactional learning process better describes the relationship of the mouth to the development of psychic functions. This growth can be observed in the processes involving mother and child. In our discussion only the external maternal energy source and the sensorimotor system of the mouth are considered.

Inner needs of the neonate energize drives which push the organism to search for external objects of supply. This need-searching is necessary for tension reduction. Primary learning theory postulates the complex as need-response-reduction of tension. The sensation acquired from the responding object and its supply, as well as its image or memory, is probably associated with that part of the body through which the aim of the physiological drive is expressed or satisfied. As the needs build up, tension is experienced through gastric hunger contraction, salivation, lip sensations, and so on, and reduced through sensations of nipple between lips, mouth filled with milk, and deglutition. The tension experience and the object reducing tension represent a single psychosomatic complex.

Psychologically, the resulting satiation pleasure or satisfaction constitutes an erotic system in which the response from the object (mother) becomes experienced as love and the body sense (lips) as sensual or erotic. The satisfaction (or disappointment) leads to a preservation of the image within the rudimentary organism and becomes part of the developing self. In this phase of maturation, reality, as far as the organism is concerned, exists only within itself, partly pleasurable, partly painful, depending on the satisfactory or unsatisfactory experiences the primary drives meet. Primary preverbal learning is only briefly limited to responses experienced by the mouth, because tactile, pain, auditory, visual and vibratory perceptual systems very quickly participate in the child-mother relationship.

The next stage of development is described as identification through meaningful attachments to significant others. The child is learning to use and understand cognitive and expressive symbols. He must now be able to distinguish between self and not-self and accept signals from a distant mother, not one who is part of him.

In the development of the ego, ego ideal, and superego, symbols of memory traces, or relationships, are internalized after a process of cognition, appraisal, and cathexis. Identification, thus, does not mean internalization of objects and affects. Internalization is achieved in terms of feelings toward objects and in terms of the value of the relationship.

In maturation, symbolic pictures, words, and abstractions add to, screen, or overlay the visceral components of the body ego and the primitive affects of the psychological ego. The early cognitive pattern expands with ever-increasing organization. When the organization breaks down under severe stress, the arrangement of its parts assumes a different form. The primitive body ego again assumes primacy, and the burden of communication returns to the contact sensibilities; primitive touching and sucking forms of testing are revised. Often the regressive pattern is characteristic of specific mother-child experiences. The mouth-nipple pattern reassumes primacy.

Grinker compares the depressed patient with the newborn and contrasts both with psychosomatic pathology. Both the newborn and depressed are facially unexpressive and presumably both are concerned with fantasies of sucking and biting. It would seem that the crying but nonweeping infant and the depressive are completely given up to self, waiting passively and hopelessly for sources of supply, unable to differentiate angry self from hostile depriver. Psychosomatic disturbances, however, are associated with hypersecretion. The development of free fluid, which is the homologue of weeping, may be shifted by spontaneous psychological means or hypnosis from organ to organ. The psychosomatic patient weeps or secretes fluid, asking for supply from others or giving to himself in fantasy, with appropriate pouring out of fluid as a primitive means of maintaining homeostasis. According to this formulation, his reaction corresponds to a stage when awareness of separation between self and not-self has already developed.

Grinker suggests that an increase in psychosomatic disturbances in our Western culture is related to socioeconomic influences that have changed family structure and function. The major factor in this change is the absence or minimization of the child's relationship with the father. (The links between the sociocultural and the psychosomatic, between the father's function and physiological response, somehow seem yet to be established and spelled out.—Eds.)

IMPLICATIONS FOR THERAPY

On the basis of these theoretical formulations, therapy for psychosomatic disturbances requires not less but more penetration into the deeper levels of the organism's mental structure. Techniques that penetrate far beyond the nuclear oedipal conflict of classical psychoanalysis into the vast area of pregenitality or the age of succorance must be evolved. To reach this, verbal screens must be penetrated. Investigation and treatment of the primary psychosomatic (healthy or disturbed) functions in adults must revive the preverbal visceral undifferentiated total functions in which are contained the rudiments of the organ and psychological system.

INTEGRATION AND FIELD THEORY

Integration is explained as a change in organization of an open system in which there is a continuous exchange of energy with the environment. New organizations of mass are created and energy stored. Between the parts of the whole are constant circular processes of interaction and communication.

LIVING ORGANISMS

The primary source of action of the living unit arises from an internal energy system that is the stimulus for the first interactions and exchanges of the organism with its environment; only later do perceptive systems evolve and function in this process. Thus, constitutionally derived, "wound-up" activity of the organism facilitates its existence before exteroceptive functions develop. As the organism increases in complexity, a greater number of possible responses to a single stimulus are developed. Each organism's selective sensitivity will influence the nature of its response to external stimuli. In addition to selecting the energy

(stimulation) it accepts from the environment, the organism also maintains its own rate of responsiveness. This constitutes a biological and
effective rudimentary defense against disintegration.

Enzymatic biochemistry and hormonal secretions are determined
through their own systems of growth and balance and through interrelationship with the total organism. Grinker proceeds to describe nerve
fiber conduction, autonomic nervous system activities, and the functions
of the 'tween-brain (associated) areas. Each is considered in light of its
responsiveness to danger and its subsequent methods of adjustment
within an energy system that functions to maintain itself. In addition to
the activities of the peripheral receptors and effectors of the nervous system, the cerebral cortex and its functions are outlined.

PSYCHOLOGICAL PATTERNS

In what way and to what degree do the psychological patterns reflect
or rerepresent the biological integrative processes? Do psychological activities utilize and bind energy like any other organic system?

The Freudian psychological model demonstrates considerable congruence with biological concepts. The id represents the constitutionally derived inner organization that first functions without stimuli from the
external environment. The development of more complicated reactivity
with the environment, as in the total organism after its perceptive systems have developed, is a function of the ego that is sensitive to both internal and external stimuli. The ego develops its capacities from a learning self-corrective process in transaction with the living and physical
world about it. The ego then becomes the functioning organization that
perceives inner tension and outer stress, permits or inhibits action, and
synthesizes conflictual tendencies.

INTEGRATION IN THE SERVICE OF EQUILIBRIUM

Our most cogent theoretical formulations indicate that from the intracellular enzymes to the psychological forces, operating through the cortical mantle, processes of integration of part functions tend to maintain
the organism in a steady state. Integration is characterized by the attempts of the organism to maintain its equilibrium within an increasingly complex time and space environment.

These assumptions that the human organism is part of and in equilibrium with its environment, that its psychological processes assist in
maintaining an internal equilibrium, and that the psychological functioning of the organism is sensitive to both internal needs and external
conditions bring us to the realization that investigations of interrelationships between two functions are not enough. Psychosomatic organi-

zation must be understood as a transactional process occurring in a total field. (See the systems theory summary, pp. 236–242.)

CONCLUSION

Behavioral sciences are in our time alive with new vigor through the use of transactional concepts that depend on multiple frames of references and therefore require interdisciplinary research. The evolution of these concepts corresponds somewhat to the evolutionary complexity of the organism. Simpler forms of life processes are not abandoned but become parts of a larger whole. Simpler ways of viewing life as stages of development of science need not be abandoned but become parts of a larger whole. The study of successive fields in time from birth to maturity are needed to furnish a longitudinal dimension to multiple cross-sections. The observer must include his own influence in the circular or cyclic processes of communication among the many systems within the field.

LABORATORY STUDIES OF ANXIETY AND STRESS

These psychosomatic studies of anxiety and stress were undertaken originally because of the need for a methodological model generally applicable to the problems of psychosomatic research. If this model were fruitful in the study of a single emotion, such as anxiety, it might be utilized in studies of other emotions and, perhaps, eventually lead to the formation of more general laws. The choice to continue to investigate anxiety was based on its great importance in the economy of human existence.

"ANXIETY AS A SIGNIFICANT VARIABLE FOR A UNIFIED THEORY OF HUMAN BEHAVIOR" (1959)

To study any one aspect of human behavior, the field must be limited enough to provide meaningful data and yet must include the significant interrelationships between variables. Grinker designates the investigation of anxiety as a model for the investigation of other important behavioral processes.

The affect of anxiety, primarily a psychological concept by its definition, is not to be studied only within the complex of a psychological investigation. Parallel biological and sociological theoretical formulations and observations are needed to understand the nature of anxiety within man. One system may be absorbing the effects of the anxiety while it is hardly observable within another system. Thus, Grinker utilizes anxiety as an example of a process that must be described within the three traditional systems into which we fit human behavior: the biological, the psychological, and the social. In "Anxiety as a Significant Variable for a Unified Theory of Human Behavior" (1959), he presents the theoretical framework derived from his research projects that underlies their very constructions and could influence the design of future investigations.

How does one study a variable within the context of a field approach? First, each one of the systems must be analyzed in terms of their functional processes and part-whole relationships. Then, their transactional intersystem patterns can become the focus of the research.

Grinker categorizes anxiety in qualitative terms. Anxiety responses are described as (1) alertness, (2) apprehension, (3) free anxiety, and (4) panic. The first two categories are generally facilitative; the latter two are destructive or maladaptive.

Psychoanalytic theory has considered anxiety as being of two types. The first, signal anxiety, functions as a warning and tentative preparation for mobilization of defenses against the breakthrough of overpowering, unacceptable, or conflictful impulses. The second, traumatic anxiety, is the result of failure of signal anxiety to stimulate defenses and/or the overwhelming intensity of the drives that cannot be checked.

These definitions lead Grinker to four assumptions on which experimental investigations of anxiety may be based.

1. Anxiety is an ego or boundary function.

2. Its signal threshold is related to ego strengths and sensitivities that are partly constitutional and partly experiential (learned). Cultural identification, through incorporation of value systems, is seen as having a significance (as psychological conditioning does) in establishing thresholds for strain tolerance.

3. Behind traumatic anxiety are a series of antecedent learned experiences: somatic, psychological, and cultural. The genesis of predispositions toward, for example, irrational guilt or shame anxiety is discussed. General somatic predispositions, such as responsivity to change of temperature, physical trauma, and bleeding, are referable to early somatic conditioning. In the social field, an earlier situation inducing a feeling of not belonging could be the cause of later severe reactions to situations perceived as being similar to earlier experiences.

4. Anxiety is triggered in current time by appropriate meaningful stimuli.

Thus, anxiety is seen as a nonspecific response that may be activated by cues suggesting danger to the person and initiating preparatory activities; when these fail, psychological disintegration results. Signal anxiety is a sensitive indicator, an alarm.

Biologically the corresponding process is unconscious and can be recognized by observing the heightened activities of secretory glands (including the endocrines) and of smooth muscles.

Certain investigations have led to the hypothesis that all stress responses, except for local excitatory reflexes directly under the impact of trauma, are mediated by action of the central nervous system. Considerable evidence has connected the increased release of adrenocorticotrophic hormone (ACTH) with strains acting on the central nervous system. According to this hypothesis, a measurement of ACTH or its corticotrophic actions would serve as a somatic concomitant to a psychiatric evaluation of anxiety. Hormonal, as well as central vegetative, functions are seen as sensitive indicators of stress responses and reacting to a wide variety of stimuli. It is suggested that an increase in the intensity or duration of the disruptive stimulation would call into effect a plethora of other stress responders, eventuating in the exhaustion of the participating organs.

Social groups are integrated through verbal and nonverbal communications of information. Rumor is seen as the signal of social tension. Rumor resembles anxiety and corticotropin in being a signal of impending danger as well as, in greater intensity, a symptom of destructive effects.

Biological incorporation of material sources of energy and psychological introjections lead to identifications that, when filtered through ego boundaries, indicate the reservoir of appropriate behaviors in various social roles. Social groups are then formed by idealization of significant others with an acceptance of their values. In the extreme, stress responses would lead to a reversal of the developmental processes of growth and differentiation. This is referred to as physiological, psychological, and social partial regression to what existed before, inhibited but never lost.

There is no sharp dividing line between the responses to stimuli that are strains and the processes that operate to maintain homeostatic equilibrium. When strains reach a variable threshold, disequilibrium thwarts adaptation. More subsystems will then become activated. The timing of onset of strain is difficult to ascertain. In all systems, too, the effect outlasts the stimulus. In addition, there seems to be both a species and individual hierarchical nature of stress responses.

Protective devices against stress responses will be called into action. Grinker describes three categories of psychological defense with suggestions of similar biological and social reactivity. The defense may encompass a denial of the stress, an effort to handle the stress response itself, often with positive or negative feedback necessitating second-order defenses, or an effort to combat the side effects of the stress response.

Grinker asks the teleological question: What is the purpose of the destructive effects of anxiety, rumor, or an increase in corticotropin when external dangers are absent? And he answers it by saying it has no known purpose. (See, Levine. Chap 6, pp. 80–93.)

A field approach theory requires two fundamental operational principles that are difficult to implement. The first is simultaneity of observation of change through time, taking into consideration the need to include single, multiple, and continuous measurements, often necessary for the study of parts of the intact human. The second principle is some kind of uniform system of casting measurements of all variables into numerical symbols for the use of computer systems. Systems for multivariate analysis need to be developed further.

The unified approach demands a coordination of the biological, psychological, and sociological in order to essay formulations of man's nature. Grinker has portrayed how investigations can concentrate on one variable while working within the context of a behavioral total.

THE USE OF ANXIETY-PRODUCING INTERVIEWS

An experiment (Grinker, Sabshin, Hamburg, Board, Basowitz, Korchin, Persky, and Chevalier, 1957) was designed to evoke or augment free-floating anxiety in human subjects, who were anxiety-prone to some degree, in order to determine the level, trend, and change in anxiety and simultaneous changes in several other psychological and somatic variables.

The subjects were nineteen anxious or anxiety-prone inpatients at the Psychosomatic and Psychiatric Institute of Michael Reese Hospital. Each subject was interviewed for four successive days. The first preexperimental day was utilized to acclimatize the patient to the laboratory. It was planned to produce graded increments in the level of anxiety by the use of appropriate verbal and attitudinal stimuli during the successive daily psychiatric interviews. For the four days, two psychiatrists observed from behind a one-way mirror; one of the two would conduct evaluative interviews before and after the laboratory sessions. Each of the observers took extensive notes and rated anxiety, anger, and depression evidenced in the subjects. A nineteen-point scale devised for the experiment was utilized for the affect ratings. Ratings were to be based on both verbal and nonverbal communications. The patient's therapist was involved in direct observations and/or through reports of the patient's response to the experimental procedure.

Each subject was given an area judgment test on the preexperimental day and then in the before- and after-stress periods. This psychological procedure was designed to measure perception of external stimuli and decision-making accuracy and speed regarding size and distance of an experimental object.

Physiological measures were taken continuously in order to determine changes in heart-rate, blood pressure, respiratory rate, general bodily movements, and various other indices of physiological disturbances. Blood samples were taken during the preexperimental day and then before, during, and after the stress interviews. Urine samples were also evaluated. Plasma hydrocortisone, plasma-bound iodine, and urinary hydrocorticoids were measured.

SIGNIFICANCE OF PREEXPERIMENTAL STUDIES

On the preexperimental day the subjects were first introduced to the laboratory and laboratory procedure (Grinker, Sabshin, Hamburg, Persky, Basowitz, Korchin, and Chevalier, 1957). It became clear early in the research that measurements from this day could not be regarded as basal measurements. At different times the viewing apparatus, mirror, blood-drawing, decision-making, immobilization, or interaction with the investigators served to evoke anxiety, depending on the significance of these cues for the unique conflict areas of the individual subject. On the other hand, certain patients responded with hopefulness or pleasurable affect, viewing the experiment as a relief from boredom or in a therapeutic context. Overall, however, anxiety was relatively high for the subjects on the preexperimental day in the laboratory as compared with the subsequent three experimental days. In addition, plasma hydrocortisone showed a striking reversal of diurnal pattern on the first day in the laboratory. Control anxiety-prone patients not exposed to the laboratory setting showed the expected diurnal fall. In the subject population, urinary hydrocorticoids were elevated above values that were normal for a control population, but were not significantly higher than the experimental day values.

Both the inherent novelty and ambiguity of the experiment and laboratory situation were seen as constituting a stress situation. The results support the conclusion that a preexperimental day cannot be considered as a psychologically neutral condition for the assessment of a basal or resting state, at least in the type of experimental population employed in this study. This becomes particularly important in investigations where subtle changes are to be measured.

THE STRESS INTERVIEW

The interviews (Grinker, Persky, Hamburg, Basowitz, Sabshin, Korchin, Herz, Board, and Heath, 1958) were transactional in nature, with the psychiatrist, generally Grinker, taking his cues from the subject's past history, present conflicts, and degree of arousal obtained in his line of questioning or prodding. The techniques utilized were to probe the subject's unconscious problems and disturbances, to attack his defenses, and to distort or block communications with him.

In practice, graded increments in the level of anxiety could not be manipulated. The stimulus interview became a rapidly changing field in which the interviewer constantly utilized feedback from the patient to determine whether he should become more reassuring or more stressing. Subjects received the psychiatrist's attempt to provoke them in different manners. Mitigating their anxiety was, for some, the belief that they were receiving special attention from the director of the Institute. Certain subjects perceived the stressor's frankness as therapeutic. Also counteracting the effectiveness of the stressor in producing anxiety was his traditional role as therapist and helper. He felt a responsibility for the patients themselves as well as a desire to maintain their emotional stability to a sufficient degree so that continued participation in the project would not be harmful for them. In other words, the stressor's desire to produce anxiety in the subject was countered by a desire to maintain the subject's anxiety within controllable bounds.

Emotional turmoil was induced but so, too, were defenses against further disturbances strengthened. Thus, a basic stability of neurotic organization was observed. Predicted stimuli did not necessarily work in the expected direction and shifts in interviewer techniques were prevalent. Overall, controlling amounts of anxiety produced was seen as an extremely complex and difficult undertaking. Psychological and physiological change needed to be correlated with actual responses of the subject rather than the intended graded strength of the stress stimuli.

CLASSIFICATION AND RATING OF EMOTIONAL EXPERIENCES

Reliability of Observation

Central problems in psychosomatic research have been poor observation of behavior, lack of checks under similar conditions or by means of a second observer, and lack of quantitative estimates for psychological variables. This project (Grinker, Hamburg, Sabshin, Board, Korchin, Basowitz, Heath, and Persky, 1958) was designed to minimize these faults by (1) utilizing two nonparticipant psychiatric observers, (2) specifying the criteria for each affect as clearly as the phenomena would permit, and (3) developing rating scales for anxiety, anger, and depression. The rating scales were constructed to facilitate comparisons of the affects with each other, as well as correlations with nonaffective variables, and were designed to make explicit the sort of quantification that is common and useful in clinical work. Several analyses of the affect ratings were made in order to determine the extent to which two observers can agree independently when exposed to the same data.

Affect ratings were made to compare the levels of anxiety, anger, and depression for one subject from day to day, as well as intersubject comparisons. Ratings on the scales were relative, change in direction was seen as meaningful. Stringent criteria were utilized in evaluating interrater agreement. Consensus of the two raters was compared with each of the observer's individual ratings.

The study clearly indicates that independent observers are able to make quantitative judgments of the level of anxiety, anger, and depression with a high degree of interrater reliability. The highest reliability was obtained through independent observers when change in affect level from period to period was involved. The two observers could agree more reliably on direction of affective change than on absolute level.

The original seven-point rating scale was expanded to a nineteen-point scale. Suitable subjects tended to be rated within the lower range of the scale. A higher rating precluded participation because this degree of disturbance always involved motor activity that made cooperation impossible. It was thus important to indicate changes of emotions from period to period more exactly by expanding the original scale to include plusses and minuses around each ordinal.

The consensus rating compared more favorably with measurements of plasma hydrocortisone response to the stress situation than did the individual rating scores. Thus, from the point of view of agreement with a hormone variable, a consensus rating was seen to be superior.

Affective Responses as Rated

On the whole, the ratings on all affects were rather low. Anxiety seemed to be fairly generally raised on the last two days, while anger was most evoked on the second experimental day. In general, the three variables form a cluster, and the occasions of testing produce parallel changes on emotional variables in the group. On the other hand, the range in level of one affect for an individual was only slightly associated with corresponding changes in other affects.

Physiological Results

The only significant physiological differentiations seen were on measurements of heart rate. The relatively less disturbed subjects psychologically showed more focal cardiac response to the stress situations. The highly anxious patient was characterized by generally diffuse reaction to the experiment as a whole, with cardiac responses starting high and gradually decreasing.

Perception and Decision Accuracy and Speed

There was little evidence of change in the psychological test performance, particularly on the decision-time scores, in response to the stress interview. Though variations were minor, the greatest drop in accuracy did appear on the days most affect was aroused, and the least errors were made on the days little affect was aroused.

Endocrinological Findings

Degrees of increase in anxiety, anger, depression, and a combined affect rating were found to be significantly and linearly related to the change in the plasma hydrocortisone level. The plasma level of hydrocortisone appeared to be increased by any type of emotional arousal. However, data were presented that indicated a particularly striking effect when anxiety of a disintegrative nature was developed in the course of the stress interview. Stimulus-hormone relationships were also found to occur, but these relationships were generally less striking than response-hormone relations. A significant stimulus-hormone relationship occurred when the stimulus produced a relatively intense emotional response in the subject. These findings are of special interest because this experiment included only a moderate range of emotional responses. The limited data available on more extreme responses suggest that an even greater degree of adrenocortical activation occurs.

PERCEPTUAL DISTORTION AS A SOURCE OF ANXIETY

A technique was developed to alter the anxiety levels of patients and normal subjects by exposing them to a stimulation in which the adequacy of their perceptual judgments about quite simple external events was challenged (Grinker, Korchin, Basowitz, Hamburg, Sabshin, Heath, and Board, 1958). Briefly, visual materials were presented for judgment, and again, for more detailed examination, to test the correctness of the initial judgment. Unknown to the subjects, the material might be altered between the first and second viewing so that it would seem as though their earlier perception had been inaccurate, and, at times, projective.

Subjects were eleven psychiatric in-patients who had been diagnosed as anxiety-prone and five normal young men. The latter were U.S. army enlisted men. A larger group had been screened and those thought to be most clinically normal were included in the study.

The first day was a preexperimental one without the perception distortion procedures. On the following three days the subjects were given a length estimation task. Errors were built into the experiment in order to give the subject the impression that his perceptional ability was decreasing each day. In this task he first indicated which of two lines viewed tachistoscopically was the longer. Distinctions could be easily made. Then, on the second viewing, the subject would measure the lines with a ruler to check the accuracy of his choice.

After completing the length estimation task on the third day, the subject was told that he had done poorly so he would be given another test that would bear more directly on his ability to make accurate judgments within life situations of social importance. The picture distortion procedure was then administered. An original threatening version of a picture would be seen tachistoscopically. The subject would be asked to describe it. Then, unknown to the subject, a nonthreatening version of the same picture would be supplied for him to view directly and check his original description. Thus, for example, a picture of a man holding a gun to his head would be supplanted by a picture of the same man in a similar position but this time holding a pipe near his head. Eight pairs of pictures were used.

In conjunction with the perceptual distortion task, affective, psychological, and endocrinological measurements were taken as in the preceding experiment. Here, too, subjects were observed from behind a one-way mirror by two psychiatrists who kept continuous behavioral records. Subjects were interviewed during the pre- and poststress periods.

The experimental stress was found to be convincing and to lead to discernible emotional response both in the patients and in the normals. Typically, testing on the second and third day was in a grimmer atmosphere, with the subjects determined to do better and more visibly perturbed by each "error." Decision-time directly following a perceived error was often a little longer for the next presentation. In the picture description performance, one defensive maneuver adopted was to report less in the original account, thus limiting the amount of potential error. Anxiety, though, did increase and generally lasted beyond the time of the final interview later in the day. Few actual errors were made at any time.

Changes in emotional ratings, perceptual behavior on an enlarged version of the area judgment task, and hormonal activity were similar to changes produced in the stress-interview experiments. Response differences to these two different types of anxiety-producing experiences were too small to substantiate specificity theories.

In contrasting the two subject groups, the normal subjects' responses were more specific and appropriate to the focal stresses, and in general more adaptive, as revealed in the pattern of before-and-after measures. The patients were more responsive to the situation in general and less to the specific stress effects. The fact that more anxiety was not observed is discussed in terms of (1) the needs of the subjects, and the relevance of the stress to their frustration, (2) the subject's ego strength and defenses, and (3) the psychological context of the experiment.

RELATION OF PHYSIOLOGICAL RESPONSE
TO AFFECT EXPRESSION

Are physiological changes that occur during emotional arousal heightened when there is failure to express feelings outwardly? Or, on the contrary, when we see intense anxiety as a response to stress, can we assume a concomitant high involvement of the muscular and autonomic systems in the reaction to the stress? A central notion in psychosomatic theory is that an anxiety unexpressed psychologically is expressed physiologically. An investigation (Grinker, Oken, Heath, Herz, Korchin, Sabshin, and Schwartz, 1962) was designed to discover whether individuals who express high degrees of affective response would have less physiological response than those individuals who fail to express feelings outwardly.

Thirty-nine male college students were rated on their affectual responses during a psychiatric interview. Out of the thirty-nine students interviewed, eighteen subjects were chosen; nine had been rated as highs in affective response; nine were rated as lows. Ratings were done according to scales developed previously for estimating degrees of affect.

Three interviews scheduled over a three-week period were given to each of the subjects. The first was designed to emotionally distress the subject. The setting was one of maximum ambiguity and lack of support. During this session, the subject was instructed to swallow a capsule. Afterwards, he was given the erroneous information that this was a new experimental drug designed to produce profound psychological effects including visual hallucinations. The implication was of dire consequences to follow. In addition to a variety of further harassments, the subject was shown a surgical film that involved enucleation of an eye and insertion of a metal prosthesis.

After the stress period, a psychiatric interviewer first elicited the subject's reactions without giving further information, and then supportively explained the hoodwinking procedures the subject had undergone. Continual reassurance was given, and the two remaining interviews were described.

In the second interview, the laboratory was heated to 95°F to provide a physical stress.

The final interview was a control one given in air-conditioned comfort.

Measurements for nine physiological variables and ratings of six psychiatric response categories were made during the experimental sessions. Physiological variables measured were heart rate, respiratory rate, skin resistance, GSR, finger blood flow, muscle (calf) blood flow, and stomach motility. A Valsalva procedure was administered. Blood samples were taken on two occasions within the first interview. Mean levels of biochemical variables for the three interviews for the high and low groups were also measured. Psychological variables were rated by the two psychiatrists who had continuously observed from behind a one-way mirror; one of them also would conduct the evaluative interviews. Psychological variables measured were anxiety, anger, situational cathexis, defense intensity-primitivity, defense against awareness, and active coping.

Stress responses occurred generally as anticipated for both types of stimuli. The psychological stress produced the greatest affective and defensive stimulation, and there was little arousal on the control day. Physiological responses differed on the two days in a manner related to the nature of the stimuli.

The central hypothesis that the low group (less propensity for development of affect states) would have higher physiological responses was not supported. The high group showed slightly but insignificantly greater physiological stress, except for peripheral vascular resistance. Three separate measures indicated that peripheral vascular resistance does seem to manifest a trend suggesting a link to affective inhibition.

Autonomic response stereotypy and specificity were checked. Statistical evaluations verified the tendency for individuals to maintain relatively stable patterns of physiological levels under varied conditions. However, the specific variable exhibiting maximum levels showed but slight constancy. Further, examination of the magnitude of stress reactions failed to reveal hierarchical constancy. Though response specificity and stereotypy appear to be established phenomena, the stability of the levels characteristic of an individual may be the prime factor in their operation.

Data was examined on symptom specificity, the degree to which subjective complaints correspond with objective physiological disturbance. No correspondence was found. To some extent, negative results are understandable. Subjects are more sensitive to certain physiological reactions (for example, heart rate) than to others (for example, perspiration). Thresholds vary for awareness of physiological effects.

From the data, the role of specific affects could not be delineated. The group as a whole exhibited general affective constraint. It was not possible to single out one or two affectual responses as most characteristic of particular individuals.

For this investigation, the primary result for most variables measured denied a correlation between a tendency toward, or against, affective response and high or low physiological responsivity.

CLINICAL JUDGMENTS AND SELF-RATINGS

Within the context of the above experiment, a subsidiary study (Heath and Korchin, 1963) was conducted. The study describes the reactions of the subjects as reported on paper and pencil checklists. The purpose was to compare trait and state self-ratings of subjects who had been judged emotionally labile or stable on the basis of a psychiatric interview. Trait ratings refer to ratings done on an adjective checklist long-form in which the subject was asked to evaluate himself according to his usual or characteristic mood; the implication was that he should use the hypothetical judgments of others as a basis for making judgments. State ratings refer to the results of an adjective checklist shortform in which the subject was asked to rate his affective state during a particular interval. In these situations he would be more apt to respond in terms of his own range of emotional experiences. State instructions were employed under both stress and control conditions.

Adjectives that proved to differentiate the highs and lows on their self-rating of traits yielded the following picture: The lows emerged as more differentiated, mature, autonomous persons, in general exemplifying the personal and social competence described for the population as a whole (Grinker, 1962). By contrast, the highs suggested a pattern of neurotic incompetence, marked by stronger negative affects in a general context of inability and inadequacy.

Under stress conditions the state ratings of the lows reflected a pattern of concentration and some anxiety, appropriate to the nature of the stimuli. The highs, who were initially more activated and troubled, showed paradoxically more varied, pleasant, and active affects, seemingly quite inappropriate to the psychological atmosphere. On all three state ratings, regardless of the imposed condition or the nature of the mood described by the specific adjective, the highs claimed more of it. Higher ratings suggest greater fluctuation between feelings of well-being and discontent and, in general, a wider and more intense range of feelings.

In contrasting self-ratings, this study has further contributed in differentiating by another technique between responses of more and less anxious persons under stressful conditions. The patterns of adjective ratings suggest that the lows react more specifically to the particular qualities of each of the sessions, while the highs appeared more diffusely aroused.

PHYSIOLOGICAL RESPONSES IN ANXIOUS
WOMEN PATIENTS

One project (Goldstein, 1964) examined the question of whether stress or rest is a better discriminator of physiological responsiveness among anxious and nonanxious women. Forty-two female subjects were selected. Twenty-one were patients who rated extremely high on anxiety symptoms as measured by two manifest anxiety scales. Twenty-one were normal women (hospital personnel) who had low manifest anxiety scores on the two scales. The two groups differed from one another primarily in degree of manifest anxiety.

Recordings of skin resistance, respiration rate, blood pressure, and EMGs from seven skeletal muscles were taken at rest and during one minute of white noise. Noise significantly increased the activity of all variables, except frontalis muscle tension and diastolic blood pressure.

For six out of the seven muscles, the patients had a greater reaction to the noise than did the controls. During the resting period, however, there was little difference in the skeletal muscle tension of the two groups. Patients were able to relax skeletal muscles almost as well as the normals.

Autonomic reactivity measures do differentiate patients from normals both during resting periods and during stimulation. Systolic and diastolic blood pressure, respiration rate, and heart rate all tended to be higher in the anxious patients both at rest and during reactivity. Skin conductance was actually lower in patients under the two conditions.

The question of whether stress or rest is a better discriminator of physiological responsiveness among anxious and nonanxious groups seems to depend upon which system one is measuring.

STUDY OF PSYCHOPHYSIOLOGY OF MUSCLE TENSION

Response Specificity

The purpose in the next study (Grinker, Shipman, Oken, Goldstein, and Heath, 1964) was to examine response specificity and stereotypy within the skeletal muscles and across both skeletal muscular and autonomic variables combined. Does an individual tend to respond to various stimuli with his maximum level of response within the same muscle? Does an individual maintain a consistent hierarchy or pattern of muscle tension? When the skeletal muscular measures are combined with autonomic measures, will response specificity and / or stereotypy occur? Previous investigators have reported response specificity and stereotypy in the autonomic nervous system but very little is known about its applicability to the skeletal muscular system.

Subjects were fifteen psychiatric patients with depression as a major symptom. Physiological rates were measured during four sessions. The first session was designed to stimulate affect. The interviewer who was familiar with the patient's psychiatric history manipulated the conversation to areas which would cause the most discomfort to each subject.

The second session was designed to induce self-control. When the subject would express affect, the interviewer would respond with such comments as "Get hold of yourself" or "Control yourself." The third session consisted of neutral discussion. In the final session, the subject remained alone in a quiet, resting state. Psychiatric ratings showed that the three interview sessions produced significantly different psychological responses.

During each session, recordings were made of heart rate, systolic blood pressure, body movement, and muscle action potentials in seven widely separated muscles.

In measuring levels of response, specificity was observed within the skeletal muscles and within skeletal muscular and autonomic variables combined. Response stereotypy was likewise effective within the muscular system and across systems. In measuring change or reactivity to various stimuli, there also was evidence for response specificity and stereotypy. However, change was less prevalent among autonomic than muscular responses. Possibly this is due to the more uniformly low basal level of the muscular system when at rest. Autonomic variables exhibit greater differences at rest.

One can conclude that response specificity is a phenomenon that has wide applicability among physiological systems. More important, however, are the implications of this finding. A tremendous range of response modes was revealed. Some individuals responded primarily by means of the autonomic nervous system, some by muscle tension, and others with overt activity (movement). Results show the need for utilizing multiple simultaneous physiological recordings of both voluntary and involuntary systems.

Personality Factors

A question that arose in conjunction with these investigations (Grinker, Shipman, Oken, Goldstein, and Heath, 1964) was whether certain psychological variables might be related to certain physiological responses. When and to what degree is the neuromotor system involved in the nonvoluntary expression of emotions and attitudes? Is elevated muscle tension more related to impulse expression or control in a group of subjects under stress? Here subjects were not divided by their likelihood of responding with high or low degrees of affect, but state responses were evaluated in a group of undifferentiated subjects. How is each aspect of emotion, alone or in combination, related to muscle tension? On the basis of a review of pertinent literature, the investigators hypothesized a connection between high measures of change and hysteria.

An examination was made of the data on the fifteen patients who had been chosen on the basis of an acute illness in which intense depression was a major component. In addition to the four sessions (see preceding explanation), a series of personality trait assessments were made. Twenty-five trait scales were compiled. This group was characterized by being very depressed and anxious, moderately hysterical, mildly underactive, and poorly emotionally controlled.

Though the expected correlation between anxiety and increased heart rate was seen, the hypothesized link between anxiety states and muscle tension elevation was not found. Group averages for muscle tension on the affect arousal and self-control interviews were not different from those recorded during the neutral conversation session. There was an exceptionally large standard deviation of the muscle values within each of the sessions. Within any one session, when ratings of emotional states were correlated with muscle tension, only a few significant relationships appeared.

When the personality trait scores were compared with values of muscle tension, many significant findings emerged. These tended to be between four muscles (frontalis, trapezius, biceps, and quadriceps) and four trait scales (emotional stability [E], manifest anxiety [M], Rorschach movement, and Rorschach barrier concern [B]). On any day, high muscle tension tended to appear in the emotionally stable, least anxious (trait) people with an active fantasy (that helps them delay and plan) and who have a clear sense of personal limits. The hysterically inclined patients had the lowest muscular and cardiovascular values at rest. For these subjects, the control function related to muscle tension seemed to be a characterological inner strength rather than a temporary attitude.

THE RELATIONSHIP OF MUSCLE TENSION AND
AUTONOMIC ACTIVITY TO
PSYCHIATRIC DISORDERS

An investigation of the physiologic responses of several groups of psychologically disturbed subjects was conducted in order to learn how their responses differ from normal subjects (Goldstein, 1965). Would the patients be unresponsive, hyperactive, or very similar to a comparable group of individuals assumed to be in good mental health? The focus was on skeletal muscular responses.

Subjects were selected for three experimental groupings. The first sixty subjects were those diagnosed as psychotics, neurotics, character disorders, or normals. Fifteen subjects represented each subgrouping. Anxiety was an important factor in differentiating these subjects. The second grouping was composed of sixty-three subjects diagnosed as depressives, nondepressives, or normals. The nondepressives and depressives were selected with similar anxiety levels. Finally, nineteen females classified as hysterics, nonhysterics, or normals were selected. Hysterics and nonhysterics were matched for anxiety. Some subjects participated in more than one aspect of the experiment. Patient subjects were applicants to the outpatient clinic and had not yet begun psychotherapy. Many were later hospitalized. Normals were selected from hospital personnel.

Physiological recordings of palmar resistance, respiration rate, heart rate, blood pressure, and muscle action potentials in seven different skeletal muscles were recorded during a resting state and in response to a white noise.

Noise produced an increase for all the physiological responses except diastolic blood pressure. Comparative results show that the psychologically disturbed individuals were at least as physiologically responsive as the normals. None of them was less responsive; psychotic and depressive disorders were characterized by exaggerated responses. This was particularly true for the skeletal muscles in response to the auditory stimulus. During rest the maladjusted and normal subjects responded at similar physiological levels.

SPECIFICITY OF RESPONSE TO STRESSFUL STIMULI

A study (Oken, Heath, Shipman, Goldstein, Grinker, and Fisch, 1966) was designed to investigate the psychophysiological theory that postulates specificities of response patterns arising from the differentiated impact (meaning) of the stimulus. Will subjects respond similarly to all stressors or does the meaning of the stress initiate its own specific response? Analysis here is directed toward whether simple psychological stimuli are appropriate indicators of stress response and to pursue the hypothesis of Lacey regarding differential patterning of response to noxious stimuli vs. stimuli requiring external cognitive attention.

Thirty-three subjects were selected from applicants for psychiatric outpatient treatment. During a single session and with resting periods interspersed, the subjects were exposed to a one-minute white noise followed later by a fifty-minute situation of simulated danger. The subject was led to believe, through mild electrical shocks, visual experiences, and the voiced fears of the experimentor, that he was in danger of being exposed to high electric voltage from an apparatus attached to his leg. Two observers watched from behind a one-way mirror. One was the person who conducted the poststress interview which served as a primary basis for making psychiatric ratings.

Subjects were rated on self-control, active coping, and credulity. The reported affective response to the white noise was a transient initial feeling of surprise followed usually by a more alert attitude. Only a few described mild irritation. During the contrived situation, subjects sighed, moaned, and groaned, offered desperate advice about corrective measures, pleaded in quavering voices for the study to be ended, sought reassurance, and, in general, looked frightened. Thus, the stimuli were clearly divergent. The first was primarily a sensory stimulus, while the latter was an unequivocal psychological stressor inducing observable anxiety.

The white noise produced a rise in all physiological variables except diastolic blood pressure, which fell. The anxiety stress was associated with rises in all the physiological measures to an extent that exceeded the response to the white noise. With the exception of the earlier fall of diastolic blood pressure, the differences of response were of magnitude only. It is suggested that the diastolic blood pressure level is also changing according to magnitude of anxiety stimulus. Thus, the diastolic blood pressure may fall with a little anxiety, while rising with a greater degree.

Though the researchers still are inclined toward some theory of stimulus-induced specificity, specificity of patterning is clearly not confirmed by their data. Lacey's theories of situational specificity were not corroborated. Both stimuli utilized can be viewed as producing the same physiological stress response, the anxiety situation being only more potent than the noise stimulus.

"MUSCLE TENSION AND PERSONALITY: A SERIOUS SECOND LOOK"

Results from the earlier studies on fifteen depressed patients indicated that a correlation existed between muscle tension and six criterion psychological variables. The six criteria were (1) psychiatric rating of emotional lability vs. control; (2) scores on the Thurstone emotional stability scale; (3) Rorschach barrier; (4) Holtzman movement; (5) Holtzman barrier; and (6) Minnesota multiphasic personality inventory affectivity scale. Subjects were chosen to represent highs and lows of these criteria. High muscle tension was believed to be related to inner control traits; therefore, the hypotheses were that muscle tension would be greater for the highs than for the lows on the first five criteria, and that the opposite would obtain for the A scale, since a high score on this scale indicates emotionality (Heath, Oken, and Shipman, 1967).

A total of forty-four subjects participated. These were outpatient clinic applicants. Included within the forty-four were ten high and ten low subjects for the variables involved. The scores of the others were slightly less extreme in the direction of high or low.

The same laboratory procedure as described in the earlier study was followed. Subjects were given strong indications of danger from high voltage electricity in an apparatus attached to their legs. Muscle tension scores and psychiatric ratings were obtained.

All results failed to confirm the hypotheses derived from the earlier study. Attempts were made to explain these negative findings on the basis of possible artifacts either in the earlier study or in the present study, and/or on the basis of a different population. The chief conclusion was that findings concerning depressives do not necessarily apply to other patient groups. The essential fact is that the electrical activity of each muscle is regulated by a multitude of influences. Thus, the previous finding, which was later reviewed to confirm its statistical validity, could not be generalized to a nondepressed population as had been anticipated.

CONCLUSION AND COMMENTARY

The investigators have tried to determine the component parts of anxiety by measuring a number of variables, not only the biological, psychological, and behavioral, but also the degree of anxiety itself. Several techniques for evoking anxiety were devised. A tool was designed for the rating of affect with interrater reliability. Difficulties in obtaining basal measurements were discussed. A correlation between plasma hydrocortisone response and anxiety responses was observed. Emphasis was shifted from the importance of the stimulus to the importance of the response or stress effects.

Possible relationships between personality traits and states and the evocation of anxiety were examined. An inner strength as represented by six personality traits was correlated with certain physiological responses in a depressed population. These correlations could not be duplicated in a more varied population. Further experiments revealed that patient populations were no less physiologically responsive than were normal subjects. Muscle tension became a focus of research attention; muscular specificity and stereotypy within individuals was observed. The muscular system proved no more easily amenable to predictable one-to-one relationships between the arousal or repression of a particular affect and the reaction of a particular muscle than had studies of the autonomic system. Likewise, stimuli-specific responsivity was not seen. These results continued to give credence to the hypothesis that it is individual patterning or conditioning of responses in an organism at an early period of life that is the critical factor in psychosomatic disorders rather than a specific emotional constellation.

ANXIETY AND STRESS

Basowitz, Persky, Korchin, and Grinker (1955) undertook an interdisciplinary study of anxiety and stress as seen in a life situation. Previous studies of men in war and of patients with neurotic anxiety provided the incentive to investigate further the psychological and physiological nature of anxiety. The study of anxiety promised increasing knowledge of the problems of neuroses and a contribution to general personality and psychosomatic theory.

For the purposes of this research project, anxiety was defined as the conscious and reportable experience of intense dread and foreboding, conceptualized as internally directed and unrelated to an external threat. Its functional role was seen as (1) a signal or precursor of defensive and adjustive processes and (2) a symptom or consequence of a breakdown of defensive and adaptive processes. "Stress" was defined as the stimulus condition likely to produce anxiety.

METHOD

The life situation selected for this research project was that of paratroop training for physically healthy young American men. A series of experiments were designed to study this group as they underwent the stresses of paratroop training, to assess changes in psychological and biochemical functioning, independently and as they were related, and to evaluate the particular changes as a function of initial personality. The locale was the Airborne Department of the Infantry School, U.S. army, Fort Benning, Georgia. The training program consisted of three weeks. Included in the requirements for completion of the course were jumps first from training towers and later from aircraft in flight. Added to the physical threat were the threats of failure to receive one's graduation wings or to lose face either by refusing to jump or otherwise performing poorly.

The research was executed in two phases, corresponding to two separate four-week field trips to Fort Benning in July–August 1951 and March–April 1952. The second group of experiments (phase 2) were designed to obtain more data and to examine specific aspects of earlier results. Certain unproductive procedures were modified and new techniques were utilized. In addition, control data was obtained from sixteen airborne housekeeping cadremen, soldiers who had already completed the training course (experiments 5 and 6) and from a group of twenty infantry soldiers undergoing a ground battlefield stress (experiment 7). A total of eighty-five trainees were selected as subjects out of a population of 1,300 training candidates.

Clinical psychiatric interviews were utilized for evaluating levels of free anxiety, before the onset of training, and previous coping behavior. On the basis of the interview data, free anxiety, nature of defense, and adequacy of defense were measured. Interviews during phase 2 concentrated on distinguishing between anxiety related to an apprehension of failure and anxiety related to an apprehension of physical danger. The distinction was seen as a distinction between shame and guilt anxiety.

In addition to the evaluations of the interviewers, each subject was asked to rate himself on a seven-point scale in response to a series of questions concerning his level of conscious fear experienced during the tasks of the training program. Subjects were given a stress tolerance test which had been devised for an earlier study of anxiety; the men were asked to make up a story when shown certain pictures. During phase 2, Rorschach testing was abandoned and the stress tolerance test was redesigned so as to consist of pictures specifically related to the stresses of paratrooper training. Failure and harm were suggested by the pictures. Serial subtraction, memory for digits, tachistoscopic closure, and tachistoscopic Bender-Gestalt tests were administered at specific points during several of the experiments. For one experiment (experiment 2) a group of soldiers who had failed previously but were now being given a second opportunity were observed. For all subjects, performance records and personal data from army records and a biographical data sheet were recorded. In addition, behavioral observations were filed daily by noncommissioned officers living with the men and by the research team observing the candidates in the field.

The researchers believed that one biochemical process, if consistent in its response and position in the total homeostatic regulation and in the response to stress measurements, could be utilized as an index of the total process. They chose the measurement of hippuric acid as a valid biochemical index of processes associated with anxiety. In earlier studies of military patients with severe and prolonged free anxiety, hippuric acid and its immediate precursors were found to be important indicators of the biochemical functions related to anxiety. During the course of the research experiments with the paratrooper trainees, hippuric acid tolerance tests were administered. Plasma glycine, plasma amino acid, blood reduced glutathione, and blood eosinophil levels were determined. The biochemical tests were arranged to compare stress and nonstress measures as well as weekly variations. In phase 2 an experiment (experiment 3) was designed which utilized hippuric acid excretion rather than anxiety as the independent variable. Two extreme groups, high and low in hippuric acid synthesis, were studied throughout the course of airborne training. In experiment 4, the biochemical tests were designed primarily to determine whether blood glutathione, a compound that is a potential donor of glycine needed for the formation of hippuric acid, would vary as a function of the stress of airborne training.

RESULTS

Results were evaluated in terms of group and individual variances for both the psychological and biochemical procedures. Intraindividual variability was also assessed.

Psychiatric Measurements
Overall, paratrooper training did not create a high level of anxiety. The findings were very different from those revealed by the earlier studies of patients with combat anxiety. On comparison, what was seen in the paratroop candidates was a lower level of anxiety, less intense, and more temporary. Even in the anxiety self-ratings, the men judged them-

selves to be at the lower end of the available scale, agreeing with the evaluations of the clinical observers.

Not only was little anxiety expressed but the locus of anxiety was found to be allied with anticipations of failure rather than of death and destruction. This conceptualization of anxiety as either shame anxiety or harm anxiety was an important finding during phase 1. Phase 2 revealed that the subjects experienced anxiety as a unitary state of emotional distress without discrimination among specific threats. It is postulated, therefore, that whatever the anxiety may express psychodynamically, the more anxious individual generalizes his distress to all aspects of the situation. A further hypothesis that harm anxiety may be more biologically disruptive than shame anxiety is suggested.

Biochemical Reactivity

Despite the relatively low level of anxiety, stress did affect both biochemical and psychological functioning. Marked eosinopenia occurred in a number of training situations. There was also considerable heightened group variability in eosinopenia measures during the stress periods.

Glutathione levels dropped characteristically during the focal stresses of training. More intensive studies revealed that the maximum drop occurred some ten hours after a focal stress. The stress response of glutathione was, of all the biochemical measures, second only to the eosinophils.

Contrary to original expectations, hippuric acid excretion did not change between stress and nonstress periods of measurement.

Though hippuric acid did not change in response to stresses in training, when groups were preselected for high and low hippuric acid, the groups differed in a variety of psychological variables. Heightened hippuric acid individuals were more anxiety-prone. Thus, heightened hippuric acid values might be indicative of a characteristic mode of reaction rather than a temporary state of disturbance at the time of measurement.

Psychological Measures

Trainees under stress performed less adequately on digit memory and perceptual discrimination than did control subjects who were not involved in the stress situation. The trainees would improve after the day's jumps when retested. The psychological measures, particularly memory for digits and serial subtraction, were distinctly related. Intensive individual case studies suggest biochemical and psychological relationships in meaningful sequences for particular subjects.

Many of the psychological procedures utilized failed to reveal changes in stress during training; constant improvement followed in general what appeared to be a learning curve. Similarly, for biochemical variables, the base period could not be seen as a stressless period. Only for the recognition of conscious anxiety itself were ratings consistently lower during the base period.

End Phenomenon

There was a distinct increase in anxiety in the period following graduation. This affective response was referred to as the end phenomenon. The researchers have related the end phenomenon to the reduced control needed once the stress itself has been removed.

Quality of Performance

Though the differences were small, group and individual variability revealed that individuals who were unable to complete the course successfully usually showed greater disturbances, both psychologically and somatically. The airborne trainees who subsequently failed the training course performed less adequately from the outset of training. Biochemical variables proved to somewhat more clearly distinguish passing from failing subjects than did psychological measures. Measures of blood reduced glutathione consistently related well with performance. Distinctions between those who failed because of refusal to jump and those who failed for other reasons were suggested. The evidence, though, was not sufficient to confirm the hypotheses.

Stress responses can be measured by comparing the ordering of individual scores within a group distribution. With this scoring, most groups revealed a relatively low level of stress. Though individual test scores were not conclusive, the stress-resistant individual should be one who is better able to maintain stable functioning.

Statistical Methodology

The statistical methodology integral to this research model called for the development of rating scales to reduce the large quantity of primary psychiatric and psychological data. This involved complex statistical evaluations as well as testing for precision to demonstrate the reliability of the rating scales themselves. Original techniques of data evaluation were contributed in the studies of consistency and variability of groups and individuals. Differences in stability of the various measures observed were noted. Furthermore, the degree of variability of at least some of the measures was significantly changed under conditions that were not stressful enough to alter the absolute magnitude of the measures.

The demonstration of variability between measurements brought into question the concept of threshold levels as absolute measures. The criterion of absolute response of a system should not be used as the only sign of the system's activation or participation in a stress response. Attention should be paid more to changes in variability as first, preparatory responses that may in themselves constitute intermediate links in the detonation of other systems. Continuous transactions among multiple systems create an extremely complex pattern of total response.

CONCLUSION AND COMMENTARY

In general, paratrooper training was not a universal stress in itself. Since a strictly scheduled exposure to danger and possible failure evoked such a wide variety of responses, it became apparent that stress does not conform to an a priori value judgment of what should happen, but can only be determined by observations of what does happen. Often subtle personality characteristics define the manner and degree to which men respond psychologically, and hence somatically, to stressful situations. However, the researchers believe that it is reasonable to postulate a continuum of events ranging from situations that evoke anxiety in everyone to those that are meaningfully unique for the individual. The effects likewise form a continuum; in the psychological field—universal apprehension to occasional free anxiety—and in the somatic field—universal eosinopenia to occasional elevation in hippuric acid production. Psychological and somatic changes need not occur concomitantly.

Specificity of physiological or hormonal responses to psychological experience was virtually impossible to document. Psychological responses could be predicted only with far greater knowledge of the special situation and the subject than was here possible. The nature of affect may be integrally tied to the social structure. The dynamics of the small group and its meaning to the individual will greatly influence stress reactivity. For example, this study indicates that personalities have predispositions to various types of anxiety based on past experience, but that the structure of social groups usually emphasizes success and hence stirs up shame anxiety. The results of the paratroop studies have led to a reevaluation and improvement of the original research model. More complex hypotheses and more controlled research conditions were shown to be necessary in order to take the next step in psychosomatic sophistication. The newer conceptual models must allow for operational manipulation of the basic cyclic, two-way circular relationships among variables or between any one of them and the central nervous system. The advantage of naturalness in the study of a life situation is diluted by the defect of less control and by the problem of the extent of the field. Laboratory studies are needed for the study of anxiety (or any other consciously experienced and measurable affective state). Observations of different levels of anxiety in the same individual or experimentally inducing or increasing anxiety in an anxious or anxiety-prone subject should be evaluated in relation to the degree and direction of responses of physiological and psychological measures. Conversely, the effects of a disturb-

ance of a somatic or psychological function on manifest anxiety and
other variables related to it should be tested under laboratory conditions. Highly complex models are needed in order to obtain a complete
picture of the whole elastic, rebound activity of any one system.

MEN UNDER STRESS

In *Men under Stress,* Grinker and Spiegel (1945) examine the psychological functioning of Air Force personnel during World War II. From the
study of men under the stress of war, lessons can be learned regarding
the methods by which men adapt themselves to all forms of stress,
either in war or in peace. Each man has his own limit of endurance.
Though for each the Achilles's heel may differ. A failure of adaptation
will be evidenced by neurotic symptoms.

Grinker and Spiegel's work was with air force combat forces overseas
and with veteran returnees seen at an air force hospital in the United
States. These young American soldiers had been reared within a culture
that traditionally emphasized the importance of individualism and self-
assertion. Then they were placed in a battlefield where a group orientation was necessary to their emotional and physical survival. Each man
was assigned to a combat unit. The combat unit became a family substitute with members feeling toward one another as brothers and looking
toward their leader as a father figure. Hostilities aroused by members of
one's own unit were repressed or projected outward onto the enemy.

Although fear and anxiety frequently related to aircraft failures or
human inefficiency, the greatest source of fear stemmed from enemy activity, specifically from the enemy fighter planes and flak attacks. Once
attacked by flak, the men experienced a sense of virtual helplessness.
The sight of friends being killed resulted in further distress. The soldier
had to bear both a sense of bereavement for the friend and the keener
realization that what happened to his friend could have happened to
him.

Severity of combat missions could not be directly correlated with inci-
dences of psychological breakdown. It was the nature of the stress combined with the individual's own background that was significant. Also of
interest today are Grinker and Spiegel's commentaries on World War II:
"This is no war of idealism with lofty slogans and stirring patriotic
songs." The combat soldier fights for his group and "so I can go home."

A select group of case histories are presented. The first signs in combat that indicated that the soldier was suffering from disabling tension
or anxiety often were psychosomatic in nature. The serviceman would
consult the flight surgeon for relief from, for example, diarrhea, head-
aches, dizziness, fatigue coupled with anorexia or insomnia with recur-

rent nightmares. The flight surgeon evaluates the flyer's complaint. The risk to the individual, the effect on group morale and the needs of the air force should guide the flight surgeon in his decision as to whether the flyer should continue in his present assignment.

Grinker and Spiegel provide a simple psychiatric plan for the flight surgeon to utilize in maintaining the flyer in combat while reducing his debilitating symptoms. The chief points of the plan are (1) to increase insight, (2) to neutralize ineffective or crippling methods of dealing with anxiety, (3) to permit abreaction of anxiety and hostility, (4) to modify the superego, (5) to support the dependent needs, and (6) to estimate the limit of tolerance for anxiety. Thus, the treatment should be supportive, helping the patient to cope with the ongoing situation of intense danger. The soldier is helped to see his symptoms as manifestations of anxiety; he is allowed to express his anxiety; and he is reassured that his anxiety is a natural consequence of external events.

The flight surgeon may decide the soldier needs at least temporary removal from combat and the more skilled treatment of a combat psychiatrist. The psychiatrist's first task is diagnostic. He must determine whether the flight surgeon has correctly evaluated the situation and the flyer is actually unable to tolerate any additional combat stress. On the basis of an evaluation of the strength of the patient's ego, the psychiatrist decides on further treatment and assignment needs. The psychiatrist functions by the same principles as the flight surgeon. He is aided, however, by the skill which comes from experience and by the wider range of possibilities for dispositions open to him. As stated above, therapy should aim toward abreaction in the context of a supportive relationship of the therapist with the patient. Sodium pentothal was often given to inform the therapist and to relieve the intense emotional repression burdening the young flyers. Narcosynthesis could be achieved through the use of sodium pentothal.

What could be accomplished by pentothal narcosynthesis and/or brief psychotherapy depended on the interplay of three main factors: the initial weakness and dependence of the ego, the amount of increased strength gained from treatment, and the degree of stress that would be encountered in any future assignment. The psychiatrist must juggle these three factors and extract from the mixture a disposition that is a reasonable compromise of the need of the flyer and the need of the military situation.

In the hospital in the United States, Grinker and Spiegel saw soldiers who had broken down during combat and those whose emotional disabilities arose on their return home. There was rarely a single psychological trend manifested in any large group of patients, but rather there were mixtures of several trends. All the disturbances were regressive in a psychological sense, in that the individual no longer had a mature and adult capacity to discriminate reality and adapt to his environment; he used infantile reactions or lower level visceral techniques, which brought him into new conflicts causing anxiety or producing crippling physical symptoms. For the purpose of clarification, the patients are divided into five groups as to their predominant psychodynamic description: (1) passive-dependent states, (2) psychosomatic states, (3) guilt and depression, (4) aggressive and hostile reactions, and (5) psychotic-like states.

The role of past conflicts and patterns of adaptations in psychiatric casualty differed in the context of the battlefield on the one hand and the stateside hospital on the other. Overseas the stress of war is so near and so intense that former experiences rarely come to consciousness. After his return to the United States, the pattern of behavior resulting from stress becomes definitely linked to the soldier's precombat personality and his previous habitual manner of solving conflicts. At home, the treatment of war neuroses was described as an uncovering technique in contradistinction to the therapy most successful close to the front lines, which was called the cover-up method. Based on their observations and treatment programs developed, Grinker and Spiegel concluded that the mode of functioning in war can be interpreted in light of past psychological mechanisms established by the soldier in civilian life. There was a clear-cut relationship between previous personality trends and the types of war neuroses. Every war neurosis was seen as a psychoneurosis since the old unsolved conflicts of the past are stimulated by stress to assist in the production of a neurotic reaction and in its persistence.

NORMALITY

In 1958 Grinker studied a group of mentally healthy young men to whom he gave the name "homoclites." "Homoclite" is meant to refer to a person who follows the common rule. The purpose of Grinker's study (1962) was to describe and explain the mental state of this group of young adult males. Grinker encountered these students within the context of a psychosomatic study. They intrigued him by their "health" and seeming adaptation to their environment. His curiosity was aroused, and he developed a research program in order to further explore their psychological development.

Psychiatrists had rarely studied nonpatients. Psychiatric tools, methods, concepts, and theories dealt with the normal man only from a distance. In studying the mentally healthy young male, Grinker was searching for operational hypotheses. Thus, the research was designed to suggest tenable hypotheses as well as testing established theories.

The subjects were male students at a small YMCA college in Chicago which trained YMCA leaders. In 1958, Grinker interviewed 31 preselected subjects, and in 1959 gave a 700-item questionnaire to 80 subjects, of whom 34 volunteers were later interviewed. Thus, a total of sixty-five students were interviewed by a psychiatrist in a semistructured interview. Questionnaire and interview responses were subsequently statistically analyzed.

On the basis of the questionnaire data, differences between those interviewed and those not interviewed were found to be minimal. Three statistically significant subgroups were delineated from the group as a whole: very well adjusted, fairly well adjusted, and marginally adjusted groups. Comparisons among these groups resulted in significant differences, which could be correlated with variations in behavior, performance, and feelings. Yet this population as a whole was relatively homogeneous. Despite the existence of subgroupings, the members of this population were more alike than different.

In general, the subjects were action oriented. Their world was conceived of as being external and calling for action; they were not introspective youths. One does something about problems. The response to affect was to "kick it off." When the students did feel depressed, they coped with the affect by (1) muscular action, (2) denial ("It doesn't matter."), and (3) isolation through withdrawal, sleep, fantasy, or concentration on music. Anxiety referred primarily to fear of failure in school or athletics. Likewise, school work and athletics were primary foci of interest.

These students were not in conflict with their families. They had had good affectionate relationships with their parents. Discipline in their homes had been firm and consistent. The boys felt closer emotionally to their mothers, while identifying strongly with their fathers and, subsequently, father figures. With few exceptions, passage through puberty and adolescence had been smooth and devoid of turbulence.

The subjects' present interpersonal relationships revealed good capacities for adequate human relationships. Heterosexual experimentation was minimal.

The lives of these students would be similar to those of their parents, though not identical. Their life at school was in many ways a continuation of their lives at home, at work, or in the YMCA summer camps. The boys had incorporated the family ideals of contentment and, to a lesser degree, responsibility. Further, they felt a strict religious commitment. They wanted "to do good, to do well and to be liked."

The homoclites exhibited mild dependency. The capacity for leadership was obtained by virtue of the strength they received from identifying with a cause greater than themselves. They had developed strong feelings of self-worth which would not easily be thwarted because there was little discrepancy between goal-setting and action. The behavior of these students revealed goal-seeking rather than goal-changing ambitions.

Grinker suggests that intense commitment to change in itself might be one of the elements in neurosis building. His healthy population moved ahead slowly at a pace that did not overly strain them. The nonhuman environment to which his subjects had gravitated and in which they felt comfortable was patterned after the environment of their childhood. Perhaps, these students would have displayed emotional crises had they been placed in a more complex environment that could threaten some of their earlier conceptualizations. Though his subjects display one kind of mental health, a reasonable adaptation to their environment, Grinker indicates that a criterion for mental health might be the fit into several or changing environments.

SUMMARY AND CONCLUSION

Grinker (1962) has observed, studied, and described a population of relatively "healthy" young male adults. He concludes:

Future research among many kinds of populations will be necessary to delineate other kinds of "health" or "normality." Combining the results of many such studies may enable us to develop suitable abstractions and theories. . . . Limited theory, whether organic, psychological, psychoanalytic, or sociological, does not encompass all the significant variables or their many possible transactions, or designate the operational procedure to be used in such research. Ultimately, of course, a scientific study of values involved in concepts of "health" and "illness" is necessary.

TRANSACTIONAL APPROACH
IN PSYCHOTHERAPY

In "A Transactional Model for Psychotherapy" (Grinker, 1959b) and *Psychiatric Social Work*, Grinker (Grinker, MacGregor, Selan, Klein, and Kohrman, 1961) presents his thinking on how the psychotherapist and the patient might best understand and learn from each other. Grinker states that psychoanalytic theory is the essence of the best modern psychodynamic theory. However, he continues, the theory has a minor place in the operational procedures of psychotherapy. When translated into psychotherapeutic procedures, the guidelines of psychoanalytic theory are weak, confused, and even disruptive to a short-term treatment program. He proposes a transactional model for psychotherapy derived from empirical operations involved in psychotherapy. Communications between patient and psychotherapist provide the tools for treatment and the givens in the setting. The setting, the persons involved, and what goes on between them are observable variables from which the transactional model is built. The relationships among field theory, role theory, and communications theory are epitomized as transactional in its broadest sense, involving nondimensional space bounded by people enacting a variety of roles and traversed by verbal, nonverbal, and paralingual communications.

The utilization of field theory emphasizes the extent of influences surrounding the two-person system of therapist and patient. Though this total field of interaction has considerable extension beyond the interviewing room, what one observes and hears depends on one's position or frame of reference at the time. Role theory is useful in illuminating that position and in understanding the individual's efforts to adapt within his field of interaction. The psychotherapist's office provides a new field of interaction where new and old patterns of behavior will be elicited. From the first encounter, explicit and implicit roles already have been adopted by the psychotherapist and the patient; each has expectations of the other and of himself in relation to the other. Complementarity of roles when established within the office setting represents stability and harmony and is conducive to communication of information. When one's expectations from another are disappointed, tension, anxiety, and self-consciousness develop. Disequilibrium because of noncomplementarity of roles results in attempts at reequilibrium, representing the disruption of an old repetitive process and, eventually, the establishment of a new system. According to Spiegel (1945), when the reequilibrium is attained by mutually achieved modification, a learning process has occurred.

Though activities within a transactional process are related to current reality and start with well-defined explicit roles, they expose the repetitive nature of the patient's unadaptive behavior and stimulate his recall of past experiences. Some of these are preconscious and some are unconscious; but the orientation of the therapist remains in the present transaction within the field in which both members of the transaction find themselves. The transactional approach evokes implicit expressive or emotional roles, inciting repetition of old patterns and illuminating the genetic source of the current behavior. However, it does not require focusing on and interpreting a vast uncharted area of unconscious. On the contrary, it enables the therapist to orient himself in a special situation with a specific person in a transactional relationship which can be understood by commonsense evaluation of ordinary modes of communication.

Communications theory emphasizes the wealth of messages contained within the expressions of an individual. The psychiatrist as a participant in the communication process, as well as an observer dealing with verbal and nonverbal communication, needs to be an expert in understanding language as it is related to thinking and behavior. Transactional theory refers the foci of the communication to the individuals as they are in relatedness with each other within a specified environment. A feedback system is in operation. The process is reciprocal, cyclical, and ever-changing. Communications within the transaction express forms of role performance having explicit and implicit meanings indicating past learning and identifications as well as current relearnings, which are termed therapy. Thus, the therapeutic field consists of mutual understanding of transactions in which role processes, their antecedents, and patterned identifications are communicated and changed.

Grinker's formulations resulted from six years devoted to the study and supervision of the communications between psychotherapists and patients at the outpatient clinic of the Division of Psychiatry at the Michael Reese Hospital and Medical Center. In order to demonstrate the relevance of the transactional model for psychotherapy, Grinker, MacGregor, Selan, Klein, and Kohrman (1961) investigated in detail one case that was presented by the psychotherapist in a continuous case conference. Each session of the therapy was discussed, and a critical appraisal was made of the therapist's handling of the case. The uncondensed case reportage and seminar responses serve as an excellent teaching device. Transactional roles can be observed as they are adopted by the activities, verbal and nonverbal, of the psychotherapist and the patient. The actions of the psychotherapist are evaluated and at times criticized when she has failed to recognize the import of the transactional communication.

The psychotherapist, particularly the inexperienced one who is familiar with the model of the silent analyst, is tempted to sit back and await free associations. The lead to primary conflicts and to subsequent patient regression may then become too much for the therapist to handle, eventually proving to be untherapeutic and causing the patient to flee from further treatment. The psychotherapist should set the original focus and continue to be active in focusing the patient on particular problems of interpersonal relationships. Material from the unconscious should not be encouraged; whereas relevant material from the patient's present life outside of the once or twice weekly sessions should be discussed within the sessions. In the transactional system there is a two-way communication that can be expressed in behavior by each person enacting an explicitly assumed role more or less concealing and revealing a variety of implicitly assumed roles. At the same time, the partner assigns to the other an explicitly assigned role, but again more or less implicitly assigning other roles concurrently. The more information received and elicited regarding the life field of the patient, the easier his implicit roles are to decipher. The implicit and shifting role functions are made explicit in the reverberating transaction, enter into awareness, and then are modified and controlled. It is the task of the psychotherapist to be in control of his role structure, the processes of communication, and the degree of permitted regression.

The attempted adoption of previously learned theoretical stances is seen particularly as the psychotherapist slips into inducing and interpreting a transference neurosis. Grinker believes that the encouragement of a transference neurosis with its regressive pulls for the patient rather than the intercommunication which is taking place is not appropriate for the psychotherapy case.

SUMMARY AND CONCLUSION

Grinker was searching for an operational approach for the teaching and learning of psychotherapy. The transactional model is defined theoretically and operationally in order to meet the needs of teaching staffs in supervising psychotherapy and psychotherapists in gaining a technique suited to help their patients with their current problems. In essence, the transactional approach requires an understanding of the tactics of skilled relationships. Its underlying basic theories involve field, role, and communication theories. The transactional system indicates the reciprocal, reverberating cyclic processes going on between patient and therapist and several possible resolutions. It restricts the use of psychodynamic theory to the understanding of underlying motivations, conflicts, and defenses (the dynamics of transactions) without the confusing use of modified psychoanalytic techniques for psychotherapy cases. The more we understand theoretical psychodynamics and the less we are influenced by it operationally, the better we may understand our patients and ourselves.

THE PHENOMENA OF DEPRESSION

The Phenomena of Depression (Grinker, Miller, Sabshin, Nunn, and Nunnally, 1961) consists of a description and analysis of the authors' pilot study on depressed patients and their subsequent investigations of a larger number of patients. The depression study was instigated in order to describe accurately the clinical processes of depressions and to evolve logical and useful classifications of subgroups or categories. The research team constructed and tested instruments for obtaining reliable data on what patients felt and how they behaved. Interviews of depressed patients, ward observations, psychodynamic evaluations, individual case histories, reliability checks, emotional and behavioral traits, in isolation and in their interrelationships, all form part of the resource matrix for this investigation. Trial and error would reveal the relevance of the project designs to a comprehension of patients classified as depressed. The authors believe that empirical research is necessary in order to avoid psychodynamic stereotypes. Their efforts were aimed toward breaking down the global syndrome of depression into discrete, empirically derived, subcategories. What is provided within *The Phenomena of Depression* are leads for a better understanding of the syndrome of depression.

PILOT STUDY

The pilot study explores what psychiatrists call depression and how the syndrome of depression can be most effectively investigated. Seven psychiatrists, who then comprised the research group, interviewed twenty-one patients hospitalized at the Psychosomatic and Psychiatric Institute of Michael Reese Hospital. The patients were selected at random from those who at the time of their admission to the hospital had received a diagnosis of "Depression" or "Depressive Reaction" as the predominant problem by *both* the attending psychiatrist and the resident who admitted the patient. The assumption was that the diagnosis of depression made by two psychiatrists on admission represented the way in which contemporary psychiatrists perceived this syndrome.

One of the research psychiatrists interviewed the patients while another of the psychiatrists observed from behind a one-way mirror. The interview was semistructured and continued until it became apparent that no further information for the study could be obtained. Subsequent interviews were planned to concentrate upon areas which had been insufficiently explored. Whenever possible, relatives were interviewed to obtain additional information. A complete descriptive report was presented by the interviewer to the entire research group.

It soon became obvious from the quantity of data collected that statistical procedures would be necessary for accurate evaluations. The investigators then developed rating scales for the clinical material. A Q-sort technique was adopted to rate 111 traits dealing with "feelings and concerns" and 87 traits describing characteristics entitled "current behavior." These trait lists were regarded as two separate Q-sorts. In addition, trait check lists were formulated for 59 symptoms, 40 dream types, 65 precipitating events, and 18 meanings of precipitating events. Raters, thus, were recording both behavior which actually was observed during the interviews and inferences or interpretations which needed to be drawn from the patients' verbalizations or nonverbal behavior. Four psychiatrists rated each patient. Two of the psychiatrists had had direct experience with the patient. The other two were members of the research group who had participated in the discussions of each case. The written report of the interviewer was available to all of the raters.

RESULTS OF THE PILOT STUDY

The reliability of the data depended upon the agreement of the ratings recorded by the four different raters. The psychiatrists' ratings on the feelings and concerns list were highly reliable and could be used in factor-analytic studies. This was not true for the current behavior trait list where the degree of agreement between raters was not sufficiently high to warrant factor analysis of the data. Psychiatrists were seen to be less adept at observing behavior than at inferring and interpreting feelings and concerns.

Factor analysis of the feelings and concerns trait list revealed three factors which made good clinical sense. The existence of these sub-groupings as well as the reliability of the ratings in this area was a significant finding.

To evaluate whether the ratings were dictated by stereotypes already held by the researchers, each psychiatrist rated his concept of the "typical" depressed patient on the trait lists. The feelings and concerns ratings depicted a commonly held stereotype which was similar to one of the three factors determined by the factor analysis on the patient data. The existence of the other two factors in the patient data showed that the ratings had not been totally dictated by stereotypes but that to some extent the psychiatrists had been able to ignore their stereotypes in order to make ratings of the characteristics of particular patients.

There was much less agreement on a stereotype of behavior expected of a depressed patient. This lack of agreement on behavioral stereotypes probably intensified the discrepancies between behavioral ratings recorded for the depressed patients who were participating in the pilot study. If data has to be filled in by stereotypes and these are varied, then the resulting ratings will show great variance.

Analyses of check lists concerned with dreams, premorbid personality, psychodynamics, and precipitating causes were not fruitful due to lack of information for the first two and absence of consensus for the others. The authors believe that precipitating causes and psychodynamics discussed in teaching case conferences or reported in the psychiatric literature often reflect interpretations seriously lacking in rigor and reliability.

METHODOLOGY OF FULL-SCALE STUDY

The pilot study gave valuable information around which techniques for the full-scale study could be designed. In itself the larger number of patients to be evaluated required a modification of the original detailed design. The pilot study had indicated how the project procedure could be condensed without the loss of reliable data or important variables.

Reliability studies had shown that the feelings and concerns trait list ratings could be completed by one psychiatrist. In the full-scale study, the psychiatric resident assigned to the patient was asked to do this as part of the routine hospital diagnostic study. The form began with a few blanks to be filled in with personal data about the patient. Then followed the trimmed-down feelings and concerns check list now consisting of forty-seven traits to be rated on a four-point scale. Each fourth patient was also interviewed and rated by one of five of the original research group in order to check reliability within the expanded group.

The current behavior trait list and symptom lists were joined and restated in a clearer form which could be answered by "yes" or "no." The resident assigned to the patient and the head nurse on the patient's unit *jointly* completed the behavioral check list five days after the patient's admission. The pilot study had indicated that the interview situation alone did not reveal sufficient data for adequately characterizing the depressed patient's behavior.

During the regular psychiatric work-up, the psychiatric resident asked specific questions about precipitating events and made simple declarative statements on the patient's chart about no more than three precipitating events.

Stereotype ratings were again used as control devices. Each of the twenty-one psychiatrists (sixteen residents and five members of the research group) who would be filling out the feelings and concerns trait lists was asked to rate his concept of the "typical" depression. Head nurses and residents recorded on the current behavior check lists their concepts of how the "average" depressed patient behaves. These stereotype checks were made at the onset of the project procedure.

A total of 120 in-patients were included in the research project. Ninety-six were patients who were admitted and discharged with the diagnosis of depression. Data from the other patient groups was utilized for comparative purposes. Eight patients were admitted with a diagnosis of depression but discharged with another diagnosis. Six were admitted with a diagnosis other than depression and discharged with a diagnosis of depression. Ten were not seen as depressed patients either upon admission or discharge.

RESULTS OF THE FULL-SCALE STUDY

The three factors found in the pilot study as well as two other factors were revealed in the factor analysis of the feelings and concerns trait list. Essentially, Factor I describes the dismal, hopeless, self-castigating person; II, the person who attributes his depression to external events or persons; III, the person with powerful guilt feelings over aggressions who is attempting restitution; IV, the almost pure culture of the anxiety-laden person; and V, the clinging, demanding, and angry person.

Factor I was the strongest statistically. Factor I may be a quantitative index of the degree of depression and the other factors indicators of types differentiated by varying attempts at defense and resolution of depression.

Differentiation of the patient groups showed that psychiatrists tend to diagnose depression when high anxiety (Factor IV) is present, whereas depression may exist with low anxiety and high anxiety need not necessarily imply depression. Patients consistently diagnosed as other than depressed were low in all factors except for clinging dependency (Factor V) which, it is suggested, might reflect the dependency of almost any newly admitted patient.

Reliability studies on the feelings and concerns check list revealed that the ratings performed by the psychiatric residents showed a higher degree of reliability than did the research psychiatrists' ratings. Thus, the residents' ratings were the ones utilized in the factor analysis. The reliability of the ratings done by the members of the research group was too low to be acceptable for research purposes. Stereotype checks showed high consistency in all concepts of the typical depression but with the psychiatrists from the research group maintaining stereotypes of even greater homogeneity. Stereotypes were almost identical to Factor I from the patient data but would not have explained the existence of the other factors.

Differences in rating reliability are discussed. The possibility of the older psychiatrists utilizing a more fixed stereotype into which they tended to fit the individual patient is considered. More importantly, the familiarity of the resident with the patient's case is seen to over-shadow the advantage of the greater experience of the research psychiatrist. Subsequently, hospital records in hospitals with a psychodynamic orientation were examined and found to be very poor. Reports are kept primarily in the minds of the psychiatrists with a sparsity of detailed data communicated in a written form. Thus, a psychiatrist reading the patient's chart cannot have an adequate representation of the live patient. The authors believe that for the ratings of depressed patients the residents had the necessary information for high reliability; the research psychiatrists had much less information, and the hospital records were too vague or bare for the reconciliation of differences. Protocols thoroughly describing each patient had been a part of the pilot study.

From the current behavior trait list with its 139 items, 10 factors were elicited which correspond well with clinical descriptions. These factors are composed of groups of traits which characterize subgroupings. For the purposes of subgrouping, traits which occur least and most frequently were not included. Briefly, the ten behavioral factors can be epitomized as (1) isolated and withdrawn, (2) slowed and retarded in speech and thought, (3) general retardation in behavior and gait, (4) angrily provocative and demanding, (5) hypochondriacal, (6) cognitively disturbed (questionable organic brain changes), (7) agitated, (8) rigid, immobile, (9) somatic disturbances such as dry skin and hair plus other minor physical abnormalities, and (10) clinging, ingratiatingly pleading for love and attention.

Reliability of the current behavior factors was high, although not as high as for the feelings and concerns trait list. Similarly, behavioral stereotypes of the depressed patient were not as well-formed. Again, the stereotypes did not reveal the same ten factors as the patient data did. The one strong factor of the four found within the stereotypes of behavioral traits was a mixture of the ten factors. This factor from the stereotype ratings was more a list of symptoms in rank order of frequency than a description of a behavioral syndrome. There were no significant differences between stereotypes of current behavior held by the sixteen psychiatric residents and the nine graduate nurses.

Behavioral differentiation between the diagnosed-as-depressed upon admission and discharge and the control groups was much less than had been seen in the factor analysis of the feelings and concerns. Only two significant findings could be obtained. Both groups discharged-as-depressed were high in apathy, and those never considered depressed were lowest in agitation.

Correlations between feelings and concerns and current behavior factors with personal data were weak. There was some tendency toward greater correlation between behavior and life situation than between feelings and concerns and the personal data. Age and sex did suggest certain behavioral characterizations. The interpretation is made here as elsewhere that external manifestations may be determined by circumstances in the life of the patient which have little to do with the basic pathology. This is a crude hypothesis for which more study, more data, and analyses by subgroups, which might have obscured possible correlations in this investigation, are needed.

Correlations between factors derived from *within* each check list revealed certain obvious relationships. These correlations supported the hypothesis that Factor I, characterized primarily by dismal hopelessness, is the central or basic theme of depression as manifested in feelings and concerns traits.

The two sets of factors proved to be independent, with no inter-correlations possible. A complicated mathematical analysis was then undertaken in order to determine whether typing or clustering in depressed patients was obscuring correlations between the feelings and concerns factors and the current behavior factors. In this way, the researchers hoped to determine which patients "go together" in the same sense as the previous analyses had determined which traits go together. The sum of the factors from each list was utilized in this computation.

This combination of the fifteen factors yielded four factor *patterns*. They are:

A	Feelings	dismal, hopeless, loss of self-esteem, slight guilt feelings
	Behavior	isolated, withdrawn, apathetic, slow thought and speech, some cognitive disturbances
B	Feelings	hopeless with low self-esteem, high guilt feelings, high anxiety
	Behavior	agitation, clinging demands for attention
C	Feelings	abandonment, loss of love
	Behavior	agitated, demanding, hypochondriacal

D	Feelings	gloom, hopelessness, anxiety
	Behavior	demanding, angry, provocative

Most patients were mixtures of the four patterns in varying degrees, although for Factor Pattern A there were a considerable number of relatively "pure" types.

The clinical profiles from which these four factor patterns were elicited are then described. The authors suggest premorbid personality characteristics, precipitating events, extent of pathology, nature of defenses, therepeutic potential, and personal statistics which might be associated with each of the four factor patterns. These latter are presented as theoretical possibilities which might prove to be empirically valid. The four factor patterns can serve as the subgroupings within the depressive syndrome to which other characteristics may be associated. The factor patterns can be utilized for building hypotheses which can be tested.

SUMMARY

Grinker et al's investigation provides clinical definitions and descriptions of depression based on careful analyses of behavior and feelings and concerns of in-patients diagnosed as suffering from a state of depression. Statistical analyses of data resulted in a viable division of the syndrome into subcategories. Additional collected data related to experiences precipitating hospitalizations, the meaning of these experiences to the individual, genetic, and developmental origins of the manifest disequilibrium, meaningful others involved in precipitating the illness, the premorbid personality, sociocultural factors, and demographic indices. These variables, as they are better understood and associated with symptomotology, can further our comprehension of depression. Methodological problems encountered throughout the project reflect the difficulties of observing, recording, and conceptualizing behavior patterns in the mentally ill and the elusiveness of clear definitive premorbid personalities, precipitating factors, and unstereotyped psychodynamics. The results of this study argue for improving and extending clinical psychiatric research, without which meaningful sociological, psychodynamic and biological correlations are impossible.

THE BORDERLINE SYNDROME

What is the borderline syndrome? A research program (Grinker, Werble, and Drye, 1968) was designed to answer this question. The primary goal of this research was to retrieve the term "borderline" from its use as a depository for clinical uncertainty and, in the process, determine if the retrieval was indeed worthwhile. Does a borderline syndrome exist? If so, what are its attributes? Secondarily, it was hoped to ascertain if sub-categories could be delineated and to define them, if possible.

No systematic study of the borderline had ever been attempted. The available professional literature did suggest that the borderline was a specific syndrome with a considerable degree of internal consistency and stability. The prominent characteristic of the borderline syndrome, as described within the existing literature, was not regression but rather an arrested development of ego functions. Therefore, the utilization of a framework of ego psychology based on psychoanalytic theory promised a better understanding of the syndrome and its subcategories. A study was developed to investigate the ego functions of borderline patients as they would be exposed by ongoing behaviors.

A STUDY IN EGO FUNCTIONS

The subjects were fifty-one patients whose diagnoses had been clini-cally unclear on admission to the Michael Reese floor at the Illinois State Psychiatric Institute. Instead of using the data of dyadic introspec-tion from interviews, or various forms of psychotherapy, or tapping his-torical or anamnestic data, Grinker and associates described and coded observable behaviors. They redefined ego functions into observable be-havioral variables. Ward personnel observed and described behaviors extensively and then rated traits. The period of observation lasted for two and one-half years and involved eighty-four staff members who were collecting data on ninety-three ego function variables from the fifty-one patients.

Prominent in the picture of the borderline syndrome as it emerged through the study was the presence of anger as the main or only affect in object relationships, the absence of indications of self-identity, and an abundance of depressive loneliness.

Complex statistical analyses designed for the data gathered in this in-vestigation revealed four distinct subgroups. When translated into clini-cal syndromes, the subgroups coincided with clinical experience. Group 1 was closest to the psychotic border. Members of this group gave up attempts at relationship but at the same time overtly, in behavior and affect, displayed anger toward the world. Their ego integrations were endangered by this strong affect.

Clinically, group 2 was seen as representing the core process of the borderline. Persons in group 2 were inconsistent. They would move to-ward others for relations, and then with acted-out repulsion initiate a re-turn to a lonely and depressed state of isolation. Thus, shifts from anger to depression would accompany their continual back and forth move-ments.

Group 3 was the most adaptive, compliant, and lacking in identity. These patients seemed to have given up their search for identity. Their world was an empty place. They did not have the angry reactions characteristic of group 1. Instead they passively awaited cues from others and behaved in complementarity—"as if."

Subjects in group 4 were the closest to the neurotic border. They were searching for a lost and unattainable symbiotic relation with a mother figure. They revealed what might be called an anaclitic depression, weeping and feeling neglected and sorry for themselves. These patients had sometimes been mistakenly categorized as suffering from depression.

DYNAMIC IN-DEPTH CORRELATIONS

The next step was to check the characteristic behavioral traits of the borderline category and its subgroups and factors against total case reports. For this purpose, sixteen patients were hospitalized at the Michael Reese Psychosomatic and Psychiatric Institute. These patients displayed varying types of clinical disorders that had been diagnosed under the borderline rubric. The patients were examined intensively using both psychiatric and psychological techniques. The full resources of a general psychiatric hospital were utilized: including social services, medical consultation, electroencephalograph, psychological tests, individual and group psychotherapeutic experiences, somatotherapy, and adjunctive therapy as indicated.

The resulting sixteen total case reports as well as the protocols of the original fifty-one patients, an experimental group previously studied, one patient in treatment, and published case histories were then categorized. By utilizing the four statistically isolated behavioral categories, the researchers were able to place all the patients in appropriate groups. Thus, the statistical differentiation of the whole syndrome and of the four groups made logical clinical sense.

FAMILY STUDIES

In addition to the study of the ego functions of the patients, several parameters of the families of a sampling of patients were examined. Because the study of the family had not been included in the original design, routine social service data were the only material available for this aspect of the study. Specific criteria were selected for the independent rating of this data.

The families of the borderline showed the usual range of concern about the patients' illnesses. No specific type of family was correlated with any of the borderline groups. Nevertheless, a by-product of this part of the study was a technique for family analysis that would discriminate family types. As for the borderline patient in the nuclear family, he married infrequently and, when married, was an inadequate spouse and a poor parent.

FOLLOW-UP STUDY

A follow-up study provided a perspective on the borderline over time. A majority of the residual patient sample (86 per cent) and about 40 per cent of the dropouts were interviewed for follow-up information. Essentially, in the time span of one to three and one-half years after hospitalization, the borderline had not become schizophrenic, except for two patients in group 1. Despite therapy in the hospital oriented toward improving the social aptitudes of the patients, the borderline subjects remained for the most part socially isolated. Yet most of them returned to school or employment, and with some psychiatric contacts, maintained their instrumental roles successfully. Thirteen patients had to be rehospitalized, five of whom had had a previous history of hospitalization prior to the time of our project. Of the dropouts, only one could be classified as a borderline.

CONCLUSION AND COMMENTARY

The method of study proved successful. A workable concept of the borderline syndrome was achieved. The borderline patient can be described, and the diagnosis need not be discarded. The diagnoses of "borderline" can now be made with more confidence, and treatment and further research can proceed from more accurate diagnoses.

REFERENCES

Basowitz, H., Persky, H., Korchin, S. J., and Grinker, R. R., Sr. 1955. *Anxiety and Stress.* Philadelphia: Blakiston.
von Bertalanffy, L. 1968. *General Systems Theory.* New York: Braziller.
Coghill, G. E. 1929. *Anatomy and the Problem of Behavior.* London: Cambridge University.
Goldstein, I. B. 1964. "Physiological Responses in Anxious Women Patients." *Archives of General Psychiatry* 10:382.
Goldstein, I. B. 1965. "The Relation of Muscle Tension to Autonomic Activity in Disorders." *Psychosomatic Medicine* 27:39.
Grinker, R. R., Sr. 1953. *Psychosomatic Research.* New York: Norton.
Grinker, R. R., Sr. 1959a. "Anxiety as a Significant Variable for a Unified Theory of Human Behavior." *Archives of General Psychiatry* 1:537–546.
Grinker, R. R., Sr. 1959b. "A Transactional Model for Psychotherapy." *Archives of General Psychiatry* 1:132–148.
Grinker, R. R., Sr. 1962. " 'Mentally Healthy' Young Males (Homoclites)." *Archives of General Psychiatry* 6:405–453.
Grinker, R. R., Sr. 1964. "Psychiatry Rides Madly in All Directions." *Archives of General Psychiatry* 10:228–237.

Grinker, R. R., Sr. 1967a. "Normality Viewed as a System." *Archives of General Psychiatry* 17:320–324.

Grinker, R. R., Sr. 1967b. *Toward a Unified Theory of Human Behavior* (1956). New York: Basic Books.

Grinker, R. R., Sr. 1968. "Psychiatry: The Field." In *International Encyclopedia of Social Sciences*. Vol. 12. Pp. 607–613.

Grinker, R. R., Sr. 1969. "Symbolism and General Systems Theory." In W. Gray, L. Duhl, and C. Rizzo, eds., *General Systems Theory and Psychiatry*. Boston: Little Brown.

Grinker, R. R., Sr. 1971. "Biomedical Education as a System." *Archives of General Psychiatry* 24:291–298.

Grinker, R. R., Sr., Goldstein, I. B., Heath, H. A., Oken, D., and Shipman, W. 1964. "Study in Psychopathology of Muscle Tension: I. Response Specificity." *Archives of General Psychiatry* 11:322.

Grinker, R. R., Sr., Hamburg, D. A., Sabshin, M., Board, F. A., Korchin, S. J., Basowitz, H., Heath, H. A., and Persky, H. 1958. "Classification and Rating of Emotional Experiences." *Archives of Neurology and Psychiatry* 79:415.

Grinker, R. R. Sr., Korchin, S. J., Basowitz, H., Hamburg, D. A., Sabshin, M., Heath, H. A., and Board, F. A. 1958. "Experience of Perception Distortion as a Source of Anxiety. "*Archives of Neurology and Psychiatry* 80:98.

Grinker, R. R., Sr., MacGregor, H., Selan, K., Klein, A. and Kohrman, J. 1961. *Psychiatric Social Work*. New York: Basic Books.

Grinker, R. R., Sr., Miller, J., Sabshin, M., Nunn, R., and Nunnally, J. C. 1961. *The Phenomena of Depression*. New York: Hoeber.

Grinker, R. R., Sr., Oken, D., Heath, H. A., Herz, M., Korchin, S. J., Sabshin, M., Schwartz, N. 1962. "The Relation of Physiological Response to Affect Expression." *Archives of General Psychiatry* 6:336–351.

Grinker, R. R., Sr., Persky, H., Hamburg, D. A., Basowitz, H., Sabshin, M., Korchin, S. J., Herz, M., Board, F. A., and Heath, H. A. 1958. "Relation of Emotional Responses and Changes in Plasma Hydrocortisone Level after Stressful Interviews." *Archives of Neurology and Psychiatry* 78:207.

Grinker, R. R., Sr., Sabshin, M., Hamburg, D. A., Board, F. A., Basowitz, H., Korchin, S. J., Persky, H., and Chevalier, J. H. 1957. "The Use of an Anxiety-Producing Interview and Its Meaning to the Subject." *Archives of Neurology and Psychiatry* 78:101.

Grinker, R. R., Sr., Sabshin, M., Hamburg, D. A., Persky, H., Basowitz, H., Korchin, S. J., Chevalier, J. H. 1957. "Significance of Pre-experimental Studies in the Psychosomatic Laboratory." *Archives of Neurology and Psychiatry* 78:207.

Grinker, R. R., Sr., Shipman, W., Oken, D., Goldstein, I. B., and Heath, H. A. 1964. "Study in Psychopathology of Muscle Tension: II. Personality Factors." *Archives of General Psychiatry* 11:330.

Grinker, R. R., Sr., and Spiegel, J. P. 1945. *Men under Stress*. Philadelphia: Blakiston.

Grinker, R. R., Sr., Werble, B., and Drye, R. C. 1968. *The Borderline Syndrome*. New York: Basic Books.

Heath, H. A., and Korchin, S. J. 1963. "Clinical Judgments and Self-Ratings of Traits and States." *Archives of General Psychiatry* 9:390.

Heath, H. A., Oken, D., and Shipman, W. 1967. "Muscle Tension and Personality: A Serious Second Look." *Archives of General Psychiatry* 16:720–727.

Oken, D., Heath, H. A., Shipman, W. G., Goldstein, I. B., Grinker, R. R., Sr., and Fisch, J. 1966. "The Specificity of Response to Stressful Stimuli. *Archives of General Psychiatry* 15:624.

Parsons, T., and Bales, R. F. 1955. *The Family* Glencoe, Ill.: The Free Press.

Spiegel, J. P. 1954. "The Social Roles of Doctor and Patient in Psychoanalysis and Psychotherapy." *Psychiatry* 17:369–394.

Appendix 2: Roy R. Grinker, Sr.'s Curriculum Vitae and Bibliography

Curriculum Vitae

Name:	Roy R. Grinker, Sr.
Title:	Director, Institute for Psychosomatic and Psychiatric Research and Training and Chairman of the Department of Psychiatry of Michael Reese Hospital.
Born:	August 2, 1900

College Education:	1919	B.S., University of Chicago
Graduate Education:	1921	Rush Medical College
Post-Doctorate Education:	1921	Internship, Chicago Psychopathic Hospital
	1921–1922	Internship, Chicago Wesley Memorial Hospital
	1922–1924	Internship, Cook County Hospital
	1924–1925	Universities of Vienna, Zurich, Hamburg, London
	1933–1935	Rockefeller Fellowship, Vienna, London
Licensure:	1923	State of Illinois
Board Certification:	1941	American Board of Neurology and Psychiatry
	1950	American Board of Mental Hospital Administrators

Honor Societies:	Sigma Xi
Professional Societies:	American Medical Association
	Central Neuropsychiatric Association (Vice-President, 1931)
	American Psychiatric Association
	American Neurological Association
	American Psychoanalytic Association
	American Academy of Psychoanalysis (President, 1961–1962)
	American Association of Neuropathologists
	American Psychosomatic Society (President, 1952)
	American Association for Research in Nervous and Mental Diseases
	National Association for Mental Health
	Chicago Neurological Society (President, 1941)
	Illinois Psychiatric Society (President, 1955)
	Life Member, Chicago Psychoanalytic Society
	Chicago Institute of Medicine
	Chicago Psychoanalytic Society (President, 1950–1951)
	Society of United States Medical Consultants in World War II
	Fellow, American Association for the Advancement of Science
	Fellow, New York Academy of Sciences
	Fellow, American College of Neuropsychopharmacology
	American Psychopathological Association

Academic Appointments:	1924–1927	Instructor in Nervous and Mental Diseases, Northwestern University Medical School
	1929–1931	Assistant Professor of Neurology, University of Chicago
	1931–1935	Associate Professor of Neurology, University of Chicago
	1935–1936	Head of Division and Associate Professor of Psychiatry, University of Chicago
	1936–1950	Lecturer in Psychiatry, University of

		Chicago, Social Service Administration
	1945–	Training Analyst and Supervisory Analyst, Chicago Institute for Psychoanalysis
	1951–1956	Clinical Professor of Psychiatry, University of Illinois, Medical School
	1961–1966	
	1969–	Professor of Psychiatry, Pritzker School of Medicine, University of Chicago

Professional
Experience:

1926–1928	Attending Neurologist, Cook County Hospital
1927–1936	Attending Physician, Billings Memorial Hospital
1946–1948	Attending Psychiatrist, Chicago Psychopathic Hospital
1926–1929	Attending Physician, Michael Reese
1937–	Hospital

Special Committee
Appointments:

1947–1950	Member of original Research Study Group NIMH and various other study sections
1950–	Psychiatric Council, State of Illinois, Department Mental Health
1957–1967	State of Illinois Psychiatric Training and Research Authority
1961–1964	Psychopharmacology Study Section, National Institutes of Health (Chairman)
1956–1969	Chief Editor, *Archives of General Psychiatry*
1962–1969	Member, Editorial Board, *Journal of Psychosomatic Medicine*
1962–1969	Member, Editorial Board, *Family Process*
1950–1956	Committee for the Development of Unified Theory of Human Behavior
1948–1952	Group for the Advancement of Psychiatry
1955–1961	Consultant, Surgeon, Fifth Army
1930–1961	Editor, neuropsychiatric section, *Tice's Practice of Medicine*
1968–1971	Board of Directors, Foundation Fund for Research in Psychiatry and Chairman 1969

Military Service:

1918–1920	Student Army Training Corps
1942–1945	Commissioned Major, M.D. September 8, 1942, and assigned to AAF.

Langley Field, Va.: Assigned as Consulting Psychiatrist for Twelfth Air Force, AAF, and Army Ground Forces during Allied North African Expedition beginning November 1942; Assigned AAF School of Aviation Medicine, Randolph Field, Texas, August 31, 1943. Assigned as Chief of Professional Services and Psychiatry, Don Ce-Sar Convalescent Center, Florida, February 14, 1944. Later Commanding Officer. Promoted to Lt. Colonel, March 24, 1943. Designated Flight Surgeon, January 7, 1944. Separated from Service with rank of Colonel, September 1945.

Awards:

Legion of Merit (United States of America) 1944

Major R. F. Longacre Award for Scientific Contribution To Aviation Medicine, 1955

Professional Achievement Award of the University of Chicago Alumni Association, 1969

Gold Medal Award—Society of Biological Psychiatry, 1970

Salmon Medal, New York Academy of Medicine, 1970

1947

Louis Gross Memorial Lecture, "Psychiatric Objectives of Our Time." Jewish Hospital, Montreal, Canada, October.

1949

Visiting Psychiatrist, The Mount Sinai Hospital, New York, New York, week of January 24.

1952

Sigma Xi Lecture: "Scientific Model on Operations in Psychiatry." Illinois Institute of Technology, Chicago, February 20.

Loevenhart Memorial Lecture: "Irritability, Tension and Anxiety." Madison, Wisconsin, April 29.

1961

Franz G. Alexander's 70th Birthday Celebration: "Psychoanalytic Theory and Psychosomatic Research." Los Angeles, November 10.

1963

Keynote Address, Western Division Meeting of the American Psychiatric Association: "Psychiatry Rides Madly in All Directions." San Francisco, May.

1966

Karen Horney Lecture: "Open System Psychiatry." The Association for the Advancement of Psychoanalysis, New York, New York, March.

1969

Visiting Professor, Upstate Medical Center, State University of New York, Syracuse, July 17, 18.

Visiting Scholar, Center for Studies of Schizophrenia, National Institute of Mental Health, Bethesda, November 24, 25.

1970

Visiting Professor, Clarence P. Oberndorf Visiting Psychiatrist, Mount Sinai Hospital, New York, Week of April 6.

Bibliography
of
Roy R. Grinker, Sr., M.D.

1925

1. "Über einen Fall von Leuchtgasvergiftung mit doppelseitiger Pallidumerveichung und schwerer Degeneration des tieferen Grosshirnmarklagers." *Zeitschrift für die gesamte Neurologie und Psychiatrie* 98 (1925):433.

1926

2. "Chronic Arachno-Perineuritis with the Syndrome of Froin." *Journal of Nervous and Mental Disease* 54 (1926):615.

3. "Parkinsonism Following Carbon Monoxide Poisoning." *Journal of Nervous and Mental Disease* 64 (1926):17.

4. "Pernicious Anemia, Achylia Gastrica and Combined Cord Degeneration and Their Relationship." *Archives of Internal Medicine* 38 (1926):291.

1927

5. "The Microscopic Anatomy of Infantile Amaurotic Idiocy with Special Reference to the Early Cell Changes and the Intracellular Lipoids." *Archives of Pathology and Laboratory Medicine* 3 (1927):768.

6. "The Pathology of Amyotonia Congenita with a Discussion of Its Relation to Infantile Progressive Muscular Atrophy." *Archives of Neurology and Psychiatry* 18 (1927):982.

7. "Sprain of Cervical Spine Causing Thrombosis of Anterior Spinal Artery." *Journal of the American Medical Association* 88 (1927):1140. (With W. B. Guy.)

1928

8. "Acute Toxic Encephalitis in Childhood. *Archives of Neurology and Psychiatry* 20 (1928):244. (With T. Stone.)

9. "Amaurotic Idiocy." *Archives of Neurology and Psychiatry* 19 (1928): 185.

10. "Neurosis of the Bladder." In *Eisendrath-Rolnick Textbook of Urology* (1928), 2d ed. Philadelphia: Lippincott, 1930.

11. "Pathologic Changes in the Pineal Gland." *Archives of Neurology and Psychiatry* 19 (1928):743.

12. "The Pathology of Post Infectious Acute Toxic Encephalitis in Children." *Transactions of the Chicago Pathological Society* 13 (1928):15.

13. "Tuberculomas of the Brain." *Archives of Neurology and Psychiatry* 20 (1928):415. (With C. Liefendahl.)

1929

14. "Mucoid Degeneration of the Oligodendroglia and the Formation of Free Mucin in the Brain." *Archives of Pathology and Laboratory Medicine* 8 (1929):717. (With E. Stevens.)

15. "The Nervous Regulation of Sugar Metabolism II." *Archives of Neurology and Psychiatry* 22 (1929):919. (With F. Hiller.)

16. "Paroxysmal Cerebellar Dysfunction." *Archives of Neurology and Psychiatry* 22 (1929):616.

17. "The Proper Use of Phenobarbital in the Treatment of the Epilepsies." *Journal of the American Medical Association* 93 (1929):1218.

1930

18. "Human Rabies and Rabies Vaccine Encephalomyelitis." *Archives of Neurology and Psychiatry* 23 (1930):1138. (With P. Bassoe.)

19. "The Pathology of Functional Vascular Disturbances of the Brain." *Archives of Neurology and Psychiatry* 23 (1930):634. (With F. Hiller.)

20. "The Technique of Neurologic Examination." In *Blumer's Practitioner Series*, vol. 2. New York: Appleton Century, 1930. Chap. 41.

21. "Tumors of the Optic Nerve." *Archives of Ophthalmology* 4 (1930):497.

22. "Tumors of the Retina." In *Penfield's Cystology of the Nervous System*. New York: Hoeber, 1930. Sect. 22.

1931

23. "Disseminated Encephalomyelitis and Its Relation to Other Infections of the Central Nervous System." *Archives of Neurology and Psychiatry* 25 (1931):723. (With P. Bassoe.)

24. "Gliomas of the Retina." *Archives of Ophthalmology* 5 (1931):920.

25. "Electrische Reizung der Basalganglien bei einem Falle von Anenkephalie." *Zeitschrift für die gesamte Neurologie und Psychiatrie* 135 (1931):5736.

26. "Manual of Neurological Technique." Chicago: University of Chicago Mimeographic Office, 1931.

27. "The Negative Results from Transfer of Material from Human Acute Multiple Sclerosis to Mecacus Rhesus under Optimum Conditions." *Archives of Pathology and Laboratory Medicine* 16 (1931):373. (With N. P. Hudson.)

28. "The Pathology of Experimental Vaccinial and Rabies Encephalitis." *Proceedings of the Society for Experimental Biology and Medicine* 28 (1931):349.

29. "Regenerative Possibilities of the Central Nervous System." *Archives of Neurology and Psychiatry* 26 (1931):489. (With R. W. Gerard.)

30. "Some Experiences with 'Jake' Paralysis." *Archives of Neurology and Psychiatry* 25 (1931):649.

1933
31. "The Pathology of Spasmodic Torticollis with a Note on Respiratory Failure from Anesthesia in Chronic Encephalitis." *Journal of Nervous and Mental Disease* 78 (1933):630. (With E. Walker.)
32. "Vitamins, A, B, and C Complex Deficiency." *Archives of Neurology and Psychiatry* 30 (1933):1287. (With E. Kandel.)

1934
33. *Neurology.* Springfield, Ill.: Charles C Thomas, 1934.
34. "Pernicious Anemia: Results of Treatment of the Neurological Complications." *Archives of Internal Medicine* 54 (1934):851. (With E. Kandel.)
35. "Role of the Hypothalamus in Regulation of Blood Pressure: Experimental Studies with Observations on Respiration." *Archives of Neurology and Psychiatry* 31 (1934):54–86. (With L. Leiter.)

1935
36. "Syphilis of the Nervous System." In *Tice's Practice of Medicine.* Hagerstown, Md.: Prior, 1935.

1937
37. Editor, neuropsychiatric section, *Tice's Practice of Medicine.* Hagerstown, Md.: Prior, 1937.
38. "Infections of the Nervous System." In *Tice's Practice of Medicine,* Hagerstown, Md.: Prior, 1937.
39. *Neurology* (1934), 2d ed. Springfield, Ill.: Charles C Thomas, 1937.
40. "Psychiatry." In *Encyclopaedia Britannica Year Book.* 1937.
41. "Technique of Neurological Examination." In *Tice's Practice of Medicine.* Hagerstown, Md.: Prior, 1937.
42. "A Method for Studying and Influencing Cortico-Hypothalamic Relations." *Science* 87 (1938):73.
43. "Myoclonic Epilepsy." *Archives of Neurology and Psychiatry* 40 (1938):968.
44. "Psychiatry." In *Encyclopaedia Britannica Year Book.* 1938.
45. "Studies on Corticohypothalamic Relations, in the Cat and Man." *Journal of Neurophysiology* 1 (1938):579–589. (With H. M. Serota.)

1939
46. "A Comparison of Psychological 'Repression' and Neurological 'Inhibition'." *Journal of Nervous and Mental Disease* 39 (1939):765.
47. "Cranial Nerve Disturbances Due to Arteriosclerosis of the Intracranial Arteries." *Illinois Medical Journal* 75 (1939):453–457. (With J. Reich.)
48. "Hypothalamic Functions in Psychosomatic Interrelations." *Psychosomatic Medicine* 1 (1939):19.
49. "Nervous System." In *Encyclopaedia Britannica Year Book.* 1939.
50. "Newer Knowledge of the Central Vegetative Nervous System." *Journal of the Kansas Medical Society* 40 (1939):321.
51. "Newer Methods of Neuropsychiatric Diagnosis and Treatment." *Journal of the Michigan Medical Society* 38 (1939):385–389.

52. "The Psychoses." In *Tice's Practice of Medicine*. Hagerstown, Md.: Prior, 1939.

53. "Various Manifestations of Cerebral-Arteriosclerosis." *Journal of the Kansas Medical Society* 40 (1939):453.

1940

54. "Acute Syphilitic Infections of the Nervous System." *Urology and Cutaneous Review* 44 (1940):485. (With N. A. Levy.)

55. "The Course of a Depression Treated by Psychotherapy and Metrazol." *Psychosomatic Medicine* 2 (1940):119. (With H. V. McLean.)

56. "Mental Hygiene in Old Age." *Mental Health Bulletin* 1940.

57. "Sigmund Freud: A Few Reminiscences of a Personal Contact." *Journal of Orthopsychiatry* 10 (1940):850.

58. "Tumors of Retina, Optic Nerve and Choroid." In *Modern Trends in Ophthalmology*. London: Butterworth, 1940.

1941

59. "Cisterna Magna Lead for Electroencephalography." *Confinia Neurologica* 3 (1941):257.

60. "Electroencephalographic Studies of Corticohypothalamic Relations in Schizophrenia." *American Journal of Psychiatry* 98, no. 3 (1941):385.

61. "Interrelations of Psychiatry, Psychoanalysis and Neurology." *Journal of the American Medical Association* 116 (1941): 2236.

62. "Nervous System." In *Encyclopaedia Britannica Year Book*. 1941.

63. "Neurogenic Disturbances of the Gastro-Intestinal Tract. In S. Portis, ed., *Diseases of the Digestive and Gastro-Intestinal Tract*. Philadelphia: Lea and Febiger, 1941.

64. "Neurosurgical Treatment of Certain Abnormal Mental States," panel discussion. *Journal of the American Medical Association* 117 (1941):517–527.

65. "The Present Status of Electroencephalography in Clinical Diagnosis." *Diseases of the Nervous System* 2 (1941):276. (With H. M. Serota.)

1942

66. "Disturbances in Brain Following Convulsive Shock Therapy." *Archives of Neurology and Psychiatry* 47 (1942):1009–1027.

67. "Graduate Education." *Illinois Medical Journal* 82, no. 4(1942).

1943

68. "Psychological Observations in Affective Psychosis." *Journal of Nervous and Mental Disease* 97 (1943):623–637. (With N. A. Levy.)

69. "War Neurosis in North Africa: The Tunisian Campaign, January–May 1943." Prepared by the Josiah Macy, Jr. Foundation, New York, September 1943. (With J. P. Spiegel.)

70. *Neurology* (1934), 3d ed. Springfield, Ill.: Charles C Thomas, 1943.

1944

71. "Brief Psychotherapy in War Neurosis." *Psychosomatic Medicine* 6 (1944):123–131. (With J. P. Spiegel.)

72. "The Management of Neuropsychiatric Casualties in the Zone of Com-

bat." In *Saunder's Manual of Military Neuropsychiatry*. Philadelphia: Saunders, 1944. (With J. P. Spiegel.)

73. "Narcosynthesis: A Psychotherapeutic Method for Acute War Neurosis." *The Air Surgeon's Bulletin* 1, no. 1 (1944). (With J. P. Spiegel.)

74. "Treatment of War Neurosis." *Journal of the American Medical Association* 126 (1944):142.

1945

75. "War Neurosis in Flying Personnel Overseas and after Return to the U.S.A." *American Journal of Psychiatry* 101 (1945):619. (With J.P.Spiegel.)

76. "A Dynamic Study of War Neurosis in Flyers Returned to U.S." *Proceedings of the Association for Research in Nervous and Mental Disease,* 1945.

77. "Ergotamine Tartrate in the Treatment of War Neurosis." *Journal of the American Medical Association* 127 (1945):158. (With R. J. Spivy.)

78. "Foreword." In S. J. Beck, *Rorschach Method*, vol. 2. New York: Grune & Stratton, 1945.

79. "The Mental Health of Our Veterans." *Northwestern University Reviewing Stand* 6, no. 3 (1945).

80. *Men under Stress*. Philadelphia: Blakiston, 1945. (With J. P. Spiegel.)

81. "Men under Stress." *Proceedings of the Conference at Santa Ana Air Base*, May 1945.

82. "Psychiatric Disorders in Combat Crews Overseas and in Returnees." *Medical Clinics of North America*. Philadelphia: Saunders, 1945.

83. "The Psychiatric, Medical and Social Aspects of War Neurosis." *Cincinnati Journal of Medicine*, 26 (1945):241.

84. "Rehabilitation of Flyers with Operational Fatigue." *Air Surgeon's Bulletin* 2, no. 1 (1945).

85. *War Neurosis*. Philadelphia: Blakiston, 1945. (With J. P. Spiegel.)

86. "War Neurosis or Battle Fatigue?" *Journal of Nervous and Mental Disease* 101 (1945):442.

1946

87. "Breath-holding Time in Anxiety States." *Federation Proceedings* 5, no. 1 (1946).

88. "Indications for Specific Psychiatric Therapy." *Bulletin of the Chicago Medical Society*. July 1946.

89. "Narcosynthesis." In *Psychoanalytic Therapy*. New York: Ronald Press, 1946.

90. "A Note on the Derivation of Speech Patterns." *Lancet* 66 (1946):370.

91. "Peace of Mind," *Northwestern University Reviewing Stand*. October 1946.

92. "The Returning Soldier—Dissent." *Hollywood Quarterly* 1 (1946):321. (With J. P. Spiegel.)

93. "Sedation as a Technique in Psychotherapy." *N.Y. Academy of Medicine Bulletin* 22 (1946):185.

94. "The Stress Tolerance Test with a New Projective Technique Utilizing Both Meaningful and Meaningless Stimuli." *Psychosomatic Medicine* 8 (1946):3–15. (With M. R. Harrower.)

95. "Studies of Psychological Predisposition: I. Officers. II. Enlisted Men." *American Journal of Orthopsychiatry* 16 (1946):191. (With B. Willerman, A. Fastofsky, and A. D. Bradley.)

96. "Wartime Lessons For Peacetime Psychiatry." *University of Chicago Round Table,* no. 445, September 1946. (With H. W. Brosin and W. C. Menninger.)

97. "War Neurosis." In *Encyclopaedia Britannica Year Book* 1946.

98. "War Neurosis." In E. A. Spiegel, ed., *Progress in Neuropsychiatry.* New York: Grune & Stratton, 1946. (With J. P. Spiegel.)

1947

99. "Brief Psychotherapy in Psychosomatic Problems." *Psychosomatic Medicine* 9 (1947):98.

100. "Every Doctor a Psychiatrist." *Institute of Medicine of Chicago* 15, no. 16 (1947):438.

101. "Psychiatric Objectives of Our Time," Louis Gross Memorial Lecture. *Canadian Medical Association Journal* 56 (1947):153.

102. "Psychoses." In *Tice's Practice of Medicine.* Hagerstown, Md.: Prior, 1947. (With H. M. Serota.)

103. "So Now You're Middleaged." *Survey Graphic* 36 (1947):12.

104. "Teaching of Psychiatry to Physicians," panel discussion. *American Journal of Orthopsychiatry* 17 (1947):617.

1948

105. "Psychosomatic Medicine." *Mental Health Bulletin* 26 (1948).

106. "Semantics in Psychiatry." *ETC,* 6, no. 1 (1948):39–54.

107. "Studies on Hypothalamus." In F. Alexander and T. French, eds., *Studies in Psychosomatic Medicine.* New York: Ronald Press, 1948.

1949

108. "Alcoholism—Report of Committee." *Institute of Medicine of Chicago* 17, no. 10 (1949).

109. "Headache and Muscular Pain." *Psychosomatic Medicine* 11 (1949): 45. (With L. Gottschalk.)

110. *Neurology,* (1934) 4th ed., Springfield, Ill.: Charles C Thomas, 1949. (With P. C. Bucy.)

111. "Neurosis Following Head and Brain Injuries." In S. Brock, *Injuries of the Brain and Spinal Cord,* 3d ed., Baltimore: Williams & Wilkins, 1949. (With J. Weinberg.)

112. "The Soldier—by Science, Not by Flags." *The Survey,* August 1949.

1950

113. "The Excretion of Hippuric Acid in Subjects with Free Anxiety." *Journal of Clinical Investigation,* 29 (1950):110. (With H. Persky and M. I. Mirsky.)

114. "Some Psychodynamics in Multiple Sclerosis." *Proceedings of the Association for Research in Nervous and Mental Disease* 28 (1950):456–460. (With G. C. Ham and L. L. Robbins.)

115. "Life Situations, Emotions and the Excretion of Hippuric Acid in Anxi-

ety States." *Proceedings of the Association for Research in Nervous and Mental Disease* 29 (1950):297–306. (With H. Persky, I. A. Mirsky, and S. R. Gamm.)

116. "Syphilis." In *Tice's Practice of Medicine.* Hagerstown, Md.: Prior, 1950. (With I. Sherman.)

1951

117. "Problems of Consciousness." In H. A. Abramson, ed., *Transactions of the Second Conference,* Josiah Macy, Jr. Foundation, March 19–20, 1951.

118. "The Psychiatrist's Contribution to the Concept of Health and Disease." *University of Chicago Round Table,* no. 688, June 3, 1951.

1952

119. "Correlation Between Fluctuation of Free Anxiety and Quantity of Hippuric Acid Excretion." *Psychosomatic Medicine* 14 (1952):1. (With H. Persky and S. R. Gamm.)

120. "The Diagnosis and Treatment of Spontaneous Cerebrovascular Accidents." *Medical Clinics of North America,* 36, no. 1 (1952). (With I. Sherman.)

121. "Psychotherapy in Medical and Surgical Hospitals." *Diseases of the Nervous System* 13 (1952):9.

1953

122. Editor, *Mid-Century Psychiatry.* Springfield, Ill.: Charles C Thomas, 1953.

123. "The Effect of Infantile Disease on Ego Patterns." *American Journal of Psychiatry* 110 (1953):290.

124. *Psychosomatic Research.* New York: Norton, 1953.

125. "Response to a Life Stress: The Experience of Anxiety." *American Psychologist* 8 (1953):318. (With H. Basowitz and S. J. Korchin.)

126. "Some Current Trends and Hypotheses of Psychosomatic Research." In F. Deutsch, ed., *The Psychosomatic Concept in Psychoanalysis,* Monograph Series of the Boston Psychoanalytic Society and Institute, no. 1. New York: International Universities Press, 1953.

127. In F. Alexander and H. Ross, *Twenty Years of Psychoanalysis.* New York: Norton, 1953.

1954

128. "Anxiety in a Life Stress." *Journal of Psychology* 38 (1954):503. (With S. J. Korchin and H. Basowitz.)

129. "Anxiety and Other Factors Modifying Electroshock Seizure Latency in Man." *American Journal of Physiology* 179 (1954):647. (With R. F. Jeans and J. E. Toman.)

130. "Problems of Consciousness: A Review, an Analysis and a Proposition. In *Macy Foundation Conference on Consciousness,* 4 (1954).

131. *Psychosomatic Case Book.* New York: Blakiston, 1954. (With F. P. Robbins.)

132. In S. J. Beck, ed., *The Six Schizophrenias,* research monograph, no. 6, American Orthopsychiatric Association, New York, 1954.

1955

133. "Anxiety in 'Consciousness.'" *Transactions of the Fifth Conference.* Josiah Macy, Jr. Foundation, 1955.

134. *Anxiety and Stress.* Philadelphia: Blakiston, 1955. (With H. Basowitz, H. Persky, and S. J. Korchin.)

135. "Growth, Inertia and Shame." *International Journal of Psychoanalysis* 36 (1955):1.

136. "Neurohumoral Factors in Emotions." *Archives of Neurology and Psychiatry* 73 (1955):123. (Symposium with G. W. Harris, W. R. Hess, K. Akert, P. D. MacLean, and I. A. Mirsky.)

1956

137. "Adrenal Cortical Function in Anxious Human Subjects." *Archives of Neurology and Psychiatry* 76 (1956):549. (With H. Persky, D. A. Hamburg, M. Sabshin, S. J. Korchin, H. Basowitz, and J. A. Chevalier.)

138. "Freud and Medicine." *Bulletin of the New York Academy of Medicine* 32 (1956):878. (Also in I. Galdston, ed., *Freud and Contemporary Culture.* New York: International Universities Press, 1957.)

139. "Inflammatory and Infectious Diseases of the Nervous System." *Tice's Practice of Medicine.* Hagerstown, Md.: Prior, 1956. (With L. Boshes.)

140. "The Institute for Psychosomatic and Psychiatric Research and Training, Michael Reese Hospital, Chicago." *Mental Hospitals* 7 (1956):27.

141. "Psychosomatic Approach to Anxiety." *American Journal of Psychiatry* 113, no. 5 (1956):443–447. (Also in *Journal of the American Society of Psychosomatic Dentistry* 4 [1957]:13.)

142. "A Theoretical and Experimental Approach to Problems of Anxiety." *Archives of Neurology and Psychiatry* 76 (1956):420–431. (With S. J. Korchin, H. Basowitz, D. A. Hamburg, M. Sabshin, H. Persky, J. A. Chevalier, and F. A. Board.) (Also in J. H. Masserman and J. L. Moreno, eds., *Progress in Psychotherapy.* Vol. 2. *Anxiety and Therapy.* New York: Grune & Stratton, 1957.)

143. *Toward a Unified Theory of Human Behavior.* New York: Basic Books, 1956.

1957

144. "On Identification." *International Journal of Psychoanalysis* 38 (1957):379.

145. "Significance of Pre-experimental Studies in the Psychosomatic Laboratory." *Archives of Neurology and Psychiatry* 78 (1957):207. (With M. Sabshin, D. A. Hamburg, H. Persky, H. Basowitz, S. J. Korchin, and J. A. Chevalier.)

146. "Temporal Heart Rate Patterns in Anxious Patients." *Archives of Neurology and Psychiatry* 78 (1957):101. (With M. Glickstein, J. A. Chevalier, H. Basowitz, S. J. Korchin, M. Sabshin, D. A. Hamburg.)

147. "The Use of an Anxiety-Producing Interview and Its Meaning to the Subject." *Archives of Neurology and Psychiatry* 77 (1957):406–419. (With M. Sabshin, D. A. Hamburg, F. A. Board, H. Basowitz, S. J. Korchin, H. Persky, and J. A. Chevalier.)

148. "Use of Drugs in the Treatment of Neurosis and in the Office Manage-

ment of Psychosis." *Modern Medicine* 25 (1957):190. (With F. J. Braceland and L. J. Meduna.)

149. "Visual Discrimination and the Decision Process in Anxiety." *Archives of Neurology and Psychiatry* 78 (1957):425. (With S. J. Korchin, H. Basowitz, J. A. Chevalier, D. A. Hamburg, M. Sabshin, and H. Persky.)

150. "Discussion." In H. D. Kruse, ed., *Integrating the Approaches to Mental Disease.* New York: Hoeber, 1957.

1958

151. "Classification and Rating of Emotional Experiences." *Archives of Neurology and Psychiatry* 79 (1958):415. (With D. A. Hamburg, M. Sabshin, F. A. Board, S. J. Korchin, H. Basowitz, H. A. Heath, and H. Persky.)

152. Experience of Perceptual Distortion as a Source of Anxiety." *Archives of Neurology and Psychiatry* 80 (1958):98. (With S. J. Korchin, H. Basowitz, D. A. Hamburg, M. Sabshin, H. A. Heath, and F. A. Board.)

153. *Interdisciplinary Team Research.* Edited by M. B. Luszki. New York: National Training Laboratories, 1958.

154. "Discussion." In I. Galdston, ed., *Panic and Morale.* New York: International Universities Press, 1958. (Also member of editorial committee.)

155. "A Philosophical Appraisal of Psychoanalysis." In J. H. Masserman, ed., *Science and Psychoanalysis.* New York: Grune & Stratton, 1958.

156. "Relation of Emotional Responses and Changes in Plasma Hydrocortisone Level after Stressful Interviews." *Archives of Neurology and Psychiatry* 79 (1958):434–447. (With H. Persky, D. A. Hamburg, H. Basowitz, M. Sabshin, S. J. Korchin, M. Herz, F. A. Board, and H. A. Heath.)

1959

157. "Anxiety as a Significant Variable for a Unified Theory of Human Behavior." *Archives of General Psychiatry* 1 (1959):537–546.

158. *Cliniques Psychosomatiques.* Paris: Presses Universitaires de France, 1959. (With F. Robbins.)

159. "Editorial." *Archives of General Psychiatry* 1 (1959):1.

160. "Effect of Two Psychological Stresses on Adrenocortical Function." *Archives of Neurology and Psychiatry* 81 (1959):219. (With H. Persky, S. J. Korchin, H. Basowitz, F. A. Board, M. Sabshin, and D. A. Hamburg.)

161. *Neurology* (1934), 5th ed. Springfield, Ill.: Charles C Thomas, 1959. (With P. C. Bucy and A. L. Sahs.)

162. "A Transactional Model for Psychotherapy." *Archives of General Psychiatry* 1 (1959):132–148.

1960

163. "Stress Response in a Group of Chronic Psychiatric Patients." *Archives of General Psychiatry* 3 (1960):451. (With D. Oken, H. A. Heath, M. A. Sabshin, and N. Schwartz.)

1961

164. "A Demonstration of the Transactional Model." In M. I. Stein, ed., *Contemporary Psychotherapies.* New York: The Free Press of Glencoe, 1961.

165. "The Early Years of Psychiatric Social Work." *Social Service Review* 35 (1961):111. (With H. MacGregor, K. Selan, A. Klein, and J. Kohrman.)

166. *The Phenomena of Depression.* New York: Hoeber, 1961. (With J. Miller, M. Sabshin, R. Nunn, and J. C. Nunnally.)

167. "Die Physiologie der Affekte." *Psyche* 14 (1961):38.

168. "The Physiology of Emotions." In A. Simon, C. C. Herbert, and R. Straus, eds., *The Physiology of Emotions.* Springfield, Ill.: Charles C Thomas, 1961.

169. *Psychiatric Social Work.* New York: Basic Books, 1961. (With H. Mac-Gregor, K. Selan, A. Klein, and J. Kohrman.)

170. "A Transactional Model for Psychotherapy." In M. I. Stein, ed., *Contemporary Psychotherapies.* New York: Free Press of Glencoe, 1961.

171. *Psychosomatic Research,* rev. ed. New York: Grove Press, 1961.

1962

172. "Emotional Stress." In T. T. Tourlentes, G. H. Pollock, W. Himwich, eds., *Research Approaches to Psychiatric Problems: A Symposium.* New York: Grune & Stratton, 1962.

173. "Homosexuality." *Counseling* 20, no. 2 (1962).

174. "Introduction to Symposium on Psychoanalytic Education." In J. H. Masserman, ed., *Science and Psychoanalysis,* vol. 5. New York: Grune & Stratton, 1962.

175. "'Mentally Healthy' Young Males (Homoclites)." *Archives of General Psychiatry* 6 (1962):405–453.

176. "The Phenomena of Depressions." In *Proceedings of the Third World Congress of Psychiatry,* Montreal, Canada, June 1961. Toronto: University of Toronto Press, 1962. Vol. 1, p. 161.

177. "The Private Mental Hospital: A Setting for Clinical Research." *Mental Hospitals* 13 (1962):483. (With D. Oken.)

178. "The Relation of Physiological Response to Affect Expression." *Archives of General Psychiatry* 6 (1962):336–351. (With D. Oken, H. A. Heath, M. Herz, S. J. Korchin, M. Sabshin, and N. Schwartz.)

1963

179. "A Dynamic Story of the 'Homoclite'." In J. H. Masserman, ed., *Science and Psychoanalysis,* vol. 6. New York: Grune & Stratton, 1963. Pp. 115–134.

180. "Editorial." *Archives of General Psychiatry* 8 (1963):527. (With S. Rado and F. Alexander.)

181. "A Psychoanalytic Historical Island in Chicago (1911–1912)." *Archives of General Psychiatry* 8 (1963):392.

182. "The Simon Wexler Psychiatric Research and Clinic Pavilion." *Mental Hospitals* 14 (1963):473. (With E. Gordon.)

1964

183. "Bootlegged Ecstasy." *Journal of the American Medical Association* 187 (1964):192.

184. "Psychiatry Rides Madly in All Directions." *Archives of General Psychiatry* 10 (1964):228.

185. "Psychoanalytic Theory and Psychosomatic Research." In J. Marmoston and E. Stainbrook, eds., *Psychoanalysis and the Human Situation*. New York: Vantage Press, 1964.

186. "Recent Medical Books: Psychiatry." *Journal of the American Medical Association* 188 (1964):266.

187. "Reception of Communications by Patients in Depressive States." *Archives of General Psychiatry* 10 (1964):576.

188. "A Struggle for Eclecticism." *American Journal of Psychiatry* 121 (1964):451–457.

189. "Study in Psychophysiology of Muscle Tension: I. Response Specificity." *Archives of General Psychiatry* 11 (1964):322. (With I. B. Goldstein, H. A. Heath, D. Oken, and W. Shipman.)

190. "Study in Psychophysiology of Muscle Tension: II. Personality Factors." *Archives of General Psychiatry* 11 (1964):330. (With W. Shipman, D. Oken, I. B. Goldstein, and H. A. Heath.)

191. "The Testability of Psychoanalytic Hypotheses." In A. Abrams, H. H. Garner, and J. P. Toman, eds., *Unfinished Tasks in the Behavioral Sciences*. Baltimore: Williams & Wilkins, 1964.

192. "Timberline Conference on Psychophysiologic Aspects of Cardiovascular Disease" (contributor). *Psychosomatic Medicine* 26 (1964):405.

1965

193. "Identity or Regression in Psychoanalysis?" *Archives of General Psychiatry* 12 (1965):113.

194. "Research Potentials of Departments of Psychiatry in General Hospitals." In M. R. Kaufman, ed., *The Psychiatric Unit in a General Hospital*. New York: International Universities Press, 1965.

195. "The Sciences of Psychiatry: Fields, Fences and Riders." *The American Journal of Psychiatry* 122 (1965):367.

196. Foreword in Bailey, P. *Sigmund the Unserene*. Springfield, Illinois: Charles C Thomas, 1965.

1966

197. "Psychosomatic Aspects of the Cancer Problems." In C. B. Bahnson and D. M. Kissen, eds., "Psychophysiological Aspects of Cancer." *Annals of New York Academy of Medicine* 125 (1966):876–883.

198. "Continuities in Culture Evolution." *Current Anthropology* 7 (1966): 67.

199. Foreword. In D. Offer and M. Sabshin. *Normality*. New York: Basic Books, 1966.

200. "The Psychosomatic Aspects of Anxiety." In C. D. Spielberger, ed., *Anxiety and Behavior*. New York: Academic Press, 1966.

201. "What Do We Do with It?" *Archives of General Psychiatry* 15 (1966):449.

202. "'Open-System' Psychiatry." *The American Journal of Psychoanalysis* 26: (1966):115.

203. "The Specificity of Response to Stress Stimuli." *Archives of Gen-*

eral Psychiatry 15 (1966):624. (With D. Oken, H. A. Heath, W. G. Shipman, I. B. Goldstein, and J. Fisch.)

204. *Neurology,* 6th Ed., Springfield: Thomas, 1966. (With A. L. Sahs.)

1967

205. "Normality Viewed as a System." *Archives of General Psychiatry* 17 (1967):320–324.

206. "The Psychodynamics of Suicide and Attempted Suicide." In L. Yochelson, ed., *Symposium on Suicide.* Washington, D.C.: George Washington University, 1967.

207. *Toward a Unified Theory of Human Behavior* (1956), 2d ed., New York: Basic Books, 1967.

1968

208. *The Borderline Syndrome.* New York: Basic Books, 1968. (With B. Werble and R. C. Drye.)

209. "The Phenomena of Depressions." In M. M. Katz, J. O. Cole, and W. E. Barton, eds., *The Role and Methodology of Classification in Psychiatry and Psychopathology.* Chevy Chase, Md.: U.S. Department of Health, Education and Welfare, 1968.

210. "Psychiatry: The Field." In *International Encyclopedia of Social Sciences.* New York: Macmillan, 1968.

1969

211. "Conceptual Progress in Psychoanalysis." In J. Marmon, ed., *Modern Psychoanalysis.* New York: Basic Books, 1968.

212. Ed., *Psychiatric Diagnosis: Therapy and Research on the Psychotic Deaf.* Special pamphlet, Washington. D.C.: Social Rehabilitation Service of HEW, 1969.

213. "An Editor's Farewell." *Archives of General Psychiatry* 21 (1969):641–646.

214. "Emerging Concepts of Mental Illness and Models of Treatment: The Medical Point of View." *American Journal of Psychiatry* 125 (1969):865–869.

215. "An Essay on Schizophrenia and Science." *Archives of General Psychiatry* 20 (1969):1–24.

216. "Psychiatric Research in Our Changing World." *Excerpta Medica: International Symposium* 187 (1969):13–26.

217. "Psychiatry." In *Britannica Yearbook of Science and the Future.* 1970. Pp. 281–284.

218. "Psychiatry in Our Dangerous World." In G. F. D. Heseltine, ed., *Psychiatric Research in Our Changing World.* Amsterdam: Excerpta Medica Foundation, 1969.

219. "Psychoanalysis and the Study of Autonomic Behavior." In A. A. Rogow, ed., *Politics, Personality and Social Science in the Twentieth Century.* Chicago: University of Chicago Press, 1969.

220. "Symbolism and General Systems Theory." In W. Gray, L. Duhl, and C. Rizzo, eds., *General Systems Theory and Psychiatry.* Boston: Little, Brown, 1969.

1970

221. "Diagnosis and Schizophrenia." In R. Cancro, ed., *The Schizophrenic Reactions*. New York: Brunner, Mazel.

222. "Discussion." In *Psychiatry as a Behavioral Science*. D. Hamburg, ed., Englewood Cliffs, N.J.: Prentice-Hall.